Death by Corporation

Killing Humankind in the Age of Monster Corporations

Dr. Brian Moench

Dear John,

Thank you for your generosity and friendship for all these years. And thank you for being one of the best people I know.

Brian

ISBN-13: 9781792171956

DEDICATION

This book is dedicated to my wife Shauna and our four children and granddaughter, Monica, Dylan, Duncan, Creighton, and Saffron, who, betrayed by my generation, will be inheriting a much more difficult world than the one I grew up in.

Little known, relevant facts:

- The American Revolution was as much a rebellion against British corporations as it was against the British monarchy.
- The tobacco industry is still responsible for the deaths of 1,200 Americans every day. The progress that was once made in reducing the plague of smoking is now being undone by the surge in teen vaping, spurred on by Big Tobacco.
- The pharmaceutical industry has infiltrated the medical community and thoroughly corrupted the education of physicians and the practice of medicine to create the most drug-intensive healthcare in the world, to the detriment of patients.
- The drug dealers that President Trump should be railing against are not illegal immigrants but pharmaceutical reps in Brooks Brothers suits.
- An extra 500,000 people died from cancer between 2008 and 2010 as a result of the 2008 recession.
- The fossil fuel industry's gift to humanity has evolved from one of creating modern civilization to one of ending it. Most accurate climate models are those that predict "worst case scenarios" for the climate crisis. The planet will likely warm a catastrophic 7° F on average, by 2100.
- Air pollution from fossil fuel combustion is responsible for one in six deaths worldwide, 9 million people annually.
- Pollution particles can lodge in virtually any organ, including the brain and the placenta. Studies show that the typical urban dweller's brain has millions of pollution particles embedded per gram of brain tissue.
- Virtually every person in the developed world now excretes Monsanto's toxic herbicide Roundup in their urine, along with over 2,000 other dangerous chemicals—pesticides, flame retardants, PCBs, dioxins, heavy metals, plasticizers, and endocrine disruptors.
- The average newborn has over 200 toxic chemicals and heavy metals in their blood at the time of birth.
- Eighty to ninety percent of cancer is environmentally caused.

- Human sperm counts are dropping precipitously worldwide. Experts are genuinely concerned that the long-term trend threatens global human reproductivity.
- Lead is not just a potent neurotoxin damaging the brains of our children. The legacy lead contamination of our environment is still triggering cardiovascular disease, causing the death of more than 400,000 Americans per year according to a new study.
- The concept of the public needing guns for self-defense is pure mythology. A gun in the home dramatically increases the risk of death for family members.
- It was cost cutting at the Union Carbide chemical plant in Bhopal, India, that directly caused the disaster that resulted in the death of 30,000 people and destroyed the health of another 120,000, the deadliest industrial accident in human history.
- Domestic companies that are prohibited from manufacturing or selling dangerous products in the U.S. routinely turn around and sell those same products in foreign countries. Many of those products still end up back in the United States, exposing American consumers.
- Big Ag meat producers and ranchers are feeding their cattle cement dust, plastic, oil, sewage, and radioactive waste to get their weight up and to cut costs.
- Beef production is responsible for 80% of the deforestation of the Amazon. Every day 80,000 acres of tropical, virgin rainforest is destroyed.
- Asbestos is so toxic it kills not just exposed workers but some of their family members as well. It is still not banned by the EPA.
- The Volkswagen "Dieselgate" scandal involved several major automakers and has been responsible for up to 380,000 deaths worldwide every year.
- Psychopathy is 10-20 times more common among Wall Street business leaders than the general population--as common as the rate among prison inmates.
- Once placed in a position of power, people actually develop psychopathic-like emotions, revealed by neurostimulation techniques.

TABLE OF CONTENTS

Prologue

There's nothing that makes me angrier than waking up in the morning and having nothing to be angry about. My wife assures me that's never happened. She says that she will make sure these words are carved on my tomb stone: "Here lies my husband, who succeeded in his constant struggle against happiness." If you're like me, then this book will help you succeed as well.

A lot of people are able to put others to sleep. Most people who stand up to speak in my church for example. I had a neighbor who could do that every time she opened her mouth. I can say with complete confidence, regardless of the place, the time of day, or the day of the week, somewhere right now she is boring some poor soul to death. A lot of people can also pass gas. My grandfather was superb at it. But in order to have all these skills—passing gas, putting people to sleep, and flirting with bringing them seriously close to death—you pretty much have to be an anesthesiologist. But despite a relative lack of glamour and glory as medical professions go, hardly anyone enjoys the thought of being awake during surgery, and most people rather enjoy the thought of being alive when surgery is over, so selling the merits of the profession to the public is usually a fairly easy task.

By day I am an anesthesiologist. I have a nice, satisfying career putting people to sleep by administering heavy-duty, mind-altering, blood-pressure-and-heart-rate manipulating drugs. The standard operating room joke is that anesthesiologists try to keep patients alive while the surgeon tries to kill them. Anesthesiologists are control freaks, just like airline pilots. I'm just guessing about airline pilots because I'm not one, but I hope they are control freaks, at least the pilots of the planes that I'm a passenger on. In the operating room, everything has to be just right before "take off"--the patients history, lab data, vital signs, all the dials, alarms, vaporizers, monitors, cables, IV tubing, height of the table, the position of the nurse, etc.

The analogy to airline pilots goes further. Much of anesthesia is routine monitoring of vital signs and machines, much like computers and gauges in an airplane cockpit. Actually looking at the patient can also be very informative, much like noting clouds passing by a cockpit window or a mountain that the airplane is about to hit. It can be monotonous when everything goes well. But in those uncommon circumstances when it doesn't go well, like when a surgeon pokes a hole in the iliac artery, it becomes life threatening and terrifying, sometimes within seconds. It can be fairly characterized as the art and science of bringing people as close to death as they will ever come, and still keeping them alive. Good outcomes in the operating room depend as much on the skill of the anesthesiologist as of the surgeon. And patients always expect good outcomes, just like they always expect their pilot not to run into more than one mountain per trip. In those uncommon instances where the patients never arrive in the recovery room, explaining to family members what happened can be a soul-crushing, guilt-ridden nightmare.

But the practice of anesthesia didn't fulfill all my professional and creative urges. I dabbled in a few medical device inventions, only to be beaten to patents by other colleagues. I loved comedy, and thought about stand up, but in Salt Lake City where I live, The Mormon Tabernacle is hardly "Comedy Central" for stand ups, and Mormons, while having a lot of fine qualities, are rather deficient in their ability to see the humor in themselves. One day I had two open-heart operations cancel, and so I came home in the mid-morning with nothing planned and no opportunity for skiing on our beautiful mountains, it being the middle of July. Being incompetent at golf and a compulsive non-recreator, I spent the afternoon returning to one of my teenage hobbies--cartooning.

Out of that afternoon, eventually grew a decent sized start-up, small business--a humorous greeting card company called, "In Your Face Cards." I drew the cartoons and collaborated with my wife and kids on the jokes. We eventually ended up selling over 35 million cards, including those that we licensed to larger card companies.

I began to receive some national notoriety for my cartoons. I placed some in national medical journals. I sold my cards, T-shirts and calendars at national medical conventions and a thousand different stores. Other doctors came to visit me from out of state. I signed a contract with a syndicate to do a humorous daily cartoon strip with a hospital setting, called "Gang Green," like a hospital-based Dilbert. I was going to write a sitcom about quirky, dysfunctional doctors only to find out *Scrubs* was already being rolled out for the next fall. CNN came to my home to do a human interest story (that's when CNN used to have interesting humans) on my unusual dual career--a former faculty member of the Harvard Medical School by day, and an off-the-wall cartoonist by night.

At one point, I was faced with a decision about whether quitting medicine and devoting myself full time to the comedy/card business made sense. I was making more money selling greeting cards than I was as a physician. Oddly, I was enjoying the operating room more than ever because I was always on the look-out for new material to unleash the creative juices. Then one day my life abruptly changed forever and the search for humor became a distant memory.

The hospital phone rang in the operating room and the nurse told me my wife needed to talk to me. My wife, Shauna, never called me during surgery unless something really bad had happened. And by bad, I don't mean keys locked in the car, or the basement was flooding. I mean the world was coming to an end type "bad." A "parade of horribles" flashed in fast forward before my eyes. As I walked to the phone I thought to myself, "Shauna's drained all my bank accounts and run off to Mexico with the pool boy." No, we didn't have a pool, and the bank account wasn't worth draining. "My mother-in-law was moving in with us." No, she had passed away a few years ago, may God rest her soul. I was in the middle of trying to come up with a third possibility when Shauna said, "Are you sitting down? You need to sit down." Up to that point, I had made it my life's work to never do what I was told. I am alive today only because I never had to serve in the army where, as I understand it, you always do everything you are told. For 50 years, I had accepted being bossed around with all the grace and dignity of a two year old. So naturally, I did exactly as I was told. After a minute of silence, Shauna's words were forever seared into my brain as if from a branding iron, "Our daughter has cancer." Monica, our beautiful, prime of her life, budding Hollywood actress, law school student, 27-year-old daughter, had breast cancer.

Every organ in my body seized up with the news. In a zombie-like trance I sat motionless for what seemed like an hour. As blood flow returned to my brain I felt smothered by denial and despair. For the first time in my life, this control freak had completely lost control of life and destiny. I felt like I was watching an old, slow-motion video of a nuclear explosion. There is peace and tranquility, then all of a sudden, a terrifying, unspeakable evil rises to obliterate all other considerations.

As a parent, I thought I had failed in my greatest responsibility. When it comes to disease, I had thought that a doctor's family has some kind of inherent home court advantage. My father was a physician and when I was a 10-year-old Little League football player, I remember thinking that players on the other team couldn't really injure me because my dad was a doctor, and he wouldn't allow it. Until the time I graduated from medical school myself, I had continued to trust that my dad had somehow provided his eight children with an invisible cloak of protection that warded off everything inconvenient: trauma, tragedy, disappointment, disillusionment, and most certainly death. Furthermore, I felt like it was my inherent right to expect I had that same cloak of protection to wrap around my own children. But in the blink of an eye, I had just become an utter failure as a doctor, a protector, and as a parent.

I don't remember anything about driving home that day. Hours later when I was first able to talk to Monica, I don't think I heard anything she said. I was overcome with images of surgery, chemo, radiation, her head bald, her arms riddled with IVs, tubes in her throat and a ghost-like face of hopelessness. Thankfully, within a few hours Monica helped me come to grips with the reality. She said, "Dad, we're going to fight this, you're going to help me, and together we'll find out what to do about it."

My daughter is worse than me in not taking orders from anyone and assuming that no one else knew what they were talking about. In her first consultation with the surgeon, she was given the opinion that her treatment could be handled with just a lumpectomy. But at her insistence, she and I pored over the medical literature. Nothing that we learned was very comforting. At the time, the most current research suggested 80 percent of women who were diagnosed with breast cancer would eventually die of it. That's one thing for a 70-year-old woman, but it's quite another if you are only 27. It seemed like every study we read talked about five year survival rates, as if that was the gold standard for success. Again, five year survival might sound hopeful if you're 70, but if you're 27, that's hardly what you want to hear. After weeks of research, she and I concluded that if saving her life was the top priority, and we both agreed that it was, then a lumpectomy would not be enough. When we met with the surgeon and showed him the evidence, he changed his mind and agreed with us.

Prologue

I held my daughter's hand as she "celebrated" her 28th birthday being wheeled down the surgery hall for a double mastectomy. Monica's last words to me before going under were, "Dad, don't worry, you'll be OK." She didn't say she would be OK, only that I would be. But she was wrong about that. Monica is still alive, and so far there is no evidence of recurrence. But I've never been OK since that day, and I never will be, because I remain determined that what had happened needlessly to Monica should not happen to anyone else.

American physicians are as well trained as any in the world. The United States is home of the most well respected medical schools and medical research. I graduated from the University of Utah Medical School, did an internship in internal medicine, did a residency in anesthesia, a fellowship in intensive care medicine at Massachusetts General Hospital, and spent time on the faculty at the American medical mecca, Harvard Medical School. During none of that training was I given any understanding of the relationship between environmental degradation and public health. Poring over research papers with Monica is where I learned that 80-90 percent of cancer is environmentally--as opposed to genetically--caused.

After her surgery, Monica had the BRCA gene test, a DNA analysis that identifies genetic mutations known to be associated with breast cancer susceptibility. Women with these genes have a much higher risk of developing breast and ovarian cancer. Inherited BRCA gene mutations are responsible for about 5 percent of breast cancers and about 10 to 15 percent of ovarian cancers. We needed to know if Monica would be at higher risk for ovarian cancer in the future. Her BRCA test was negative, meaning that her cancer was almost undoubtedly caused by an environmental exposure.

The most likely environmental exposure for Monica was birth control pills, more specifically the estrogen in them. Convincing the American public, especially women, that artificially consumed estrogen was not only safe, but beneficial, could be considered as much of a criminal conspiracy as the decades-long deceit engineered by the tobacco companies. I find that I cannot look at my daughter's face and not think of the travesty of what has happened to her. I'm sure whenever she looks at her mastectomy scars she feels the same way.

When I began investigating how estrogen, a known carcinogen, was mass marketed to American women for profit, I stumbled into the skeleton closet of an industry whose ethical construct is no better than a psychopath. Behavior like this in an individual would have provoked a life-long jail sentence.

Several years later as I became more deeply engaged in the battle of the public health consequences of environmental degradation, I saw time and time again the same willingness to pursue profits with stunning and malevolent disregard for human life, spanning virtually all major industries. The willingness of corporations to destroy us all for the sake of their balance sheet is as cold, robotic, and as void of humanity as the cancer cells that set up shop to take my daughter's life. In common conversation, modern day corporations are often referred to as "those greedy bastards." But it's worse than that. They are in fact Frankenstein monsters that are killing us, and they know it. The following chapters are just a few of many examples, with a final chapter on what we can do to save ourselves.

My daughter Monica

Introduction

"This massive ascendancy of corporate power over democratic process is probably the most ominous development since the end of World War II, and for the most part 'the free world' seems to be regarding it as merely normal."

—Wendell Berry,
Bringing it to the Table:
On Farming and Food

On Dec. 3, 1984, 40 tons of a deadly gas began leaking from the Union Carbide plant in Bhopal, India. Within 72 hours, 8,000 nearby poor residents of Bhopal were dead, and within the next several months, at least 30,000 had died. The number of people disfigured, disabled and made chronically ill and eventually dying prematurely was likely another 200,000. The cause of the gas leak was directly related to corporate cost cutting by Union Carbide.

On April 24, 2013, at least 1,134 people died and 2,500 more were injured in a collapsed garment factory in Dhaka, Bangladesh, the deadliest accidental structural collapse in world history. As of this writing, the largest American clothing corporations, Gap, Walmart, and Target, who sell the products made in these death trap factories, are still unwilling to commit to any meaningful safety improvements. In fact, Walmart has a history of actually obstructing efforts to improve safety at thousands of Bangladeshi clothing factories.[1]

Over the course of four months, 64-year-old retired accountant Stephen Paddock, meticulously planned what turned out to be the worst gun massacre in American history, renting hotel rooms above outdoor concert venues, researching SWAT tactics, and stockpiling weapons. On Oct. 1, 2017, using 23 guns, many of them assault rifles, and thousands of rounds of ammunition, Paddock slaughtered 59 people and wounded almost 800 more. He had purchased his military-grade arsenal legally. Nonetheless, despite a recent national movement by high school students who survived a subsequent mass shooting in Parkland, Florida, no meaningful gun control legislation has been passed, and none appears on the horizon.

All of these horrific headlines have a common denominator. Corporate profits are allowed primacy over all other considerations. The death certificate of all these victims--at Bhopal, Dhaka, and Las Vegas— should read, "Death by Corporation."

The word "corporation", derived from the Latin *corparae,* means to physically embody. In his *History of the Corporation*, Bruce Brown notes how in the first thousand years after the fall of the Roman Empire, "the world's most powerful corporations were all trying to embody the Christian God." In 1534, St. Thomas More spoke of Jesus Christ as the ultimate corporation. "He [Jesus] doth...incorporate all christen folke and hys owne bodye together in one corporacyon mistical."

Needless to say, in the 21st century, corporations as creations of civilization make no pretense of embodying the Christian God. In fact today, corporations come much closer to embodying Mary Shelley's Frankenstein than Jesus Christ. Ironically, created by and managed by humans, corporations have become almost robotic monsters, perpetrating, even feeding off human misery, threatening every aspect of human life--the air we breathe, the water we drink, the food we eat, the climate and the earth that makes life possible, and even the future of humankind itself. Under the traditional "property model" of corporate law, "a solvent corporation is viewed as a vehicle with the sole purpose of maximizing the wealth of its owners, the shareholders,"[2] and nothing more. Noam Chomsky describes corporations as "private tyrannies." How did we arrive at the point where corporations wield such enormous power over governments, society, world affairs, human rights, and life and death itself, without any accountability for the common good?

In 15th century Europe, corporations emerged as tools of the British monarchy, chartered for the express purpose of exploiting the vast resource wealth of the "New World" and exporting that wealth back to Europe. Europe's financial elite formed joint-stock corporations to spread the risk of colonizing the Americas and other foreign lands. Given their long distance from the rule of their European monarchies, the managers and owners of these corporations were granted enormous power at the pleasure of the kings. They passed laws, invoked tax authority, and had literal armies to secure, control, and defend property, resources and trade. These early corporations had names familiar to just about any high school graduate—Hudson Bay Company, Royal African Company, and the Massachusetts Bay Colony. The East India Company became perhaps the most notorious, conquering India, establishing government-like control of the territory and a complete trade monopoly. At its peak, The East India Company ruled over one fifth of the world's population using a private army of 250,000 men.[3]

Understandably, these corporations made few friends among either the colonists who had left Europe for the express purpose of escaping royal oppression, or the indigenous people who they exploited. Indeed, American revolutionaries' resentment of British imperialism was very much a resentment of British corporate charters that mandated shipping of raw materials to Britain, and forcing the colonies to purchase the finished goods. The iconic Boston Tea Party of 1773 wasn't just about "taxation without representation." It was a protest against the anti-competitive advantage given by the Crown to the East India Company over colonial tea businesses. The American Revolution was as much the toppling of rule by British corporations as it was the toppling of rule by the British Crown. Adam Smith in his *Wealth of Nations* published in 1776, decried large businesses because they limited competition, and "The pretense that corporations are necessary to the better government of the trade is without foundation."[4]

Brian Murphy, a history professor at Baruch College, says that the founding fathers, "saw corporations as corrupting influences on both the economy at large and on government,"[5] and they had no interest in perpetuating corporate power in the new America. Early corporate charters in the new country were "democratized" and given to schools, cities, and charitable organizations for the purpose of building the common good, like establishing infrastructure, hospitals, and universities.

Decades later, the Industrial Revolution transformed a nation of farmers into a nation of employees, and with that came the attendant anxiety of "unemployment," something easily and fully exploited by corporate employers.

In the mid 19th century American courts steadily drifted towards corporate ascendency, and in 1886 a landmark decision by the Supreme

Court, *Santa Clara County v. Southern Pacific Railroad*, became precedent for the legal concept of corporate "personhood," used repeatedly afterwards to shield corporations from social responsibility and stymie regulations.

Capitalism and corporations were "unchained," with robber barons, railroad and oil tycoons, monopolies and cartels converting the nation's natural resources into personal fortunes. Tens of thousands were forced to live in "company towns," and were paid starvation wages for dangerous and debilitating work. Labor organizers were blacklisted, and industrialists raised private armies to keep their workers from protesting or striking. Big companies bought local newspapers that delivered company spin and propaganda. Corporate officials bribed elected and appointed officials alike to get their share of government spending, and paid "borers" to burrow their way into Congress and state houses to enact favorable legislation protecting corporations from liability and citizen oversight and to extend the duration of their charters.

In 1941, a Congressional committee wrote this critique of the state of corporations, "The principal instrument of the concentration of economic power and wealth has been the corporate charter with unlimited power."[6] The Great Depression of the 1930s, the labor movement, and WWII had turned the social and legal tide against corporate supremacy for the next 40 years.

But meanwhile, US corporations still controlled land, militaries and infrastructure in poorer countries, blatantly ignoring their national sovereignty. United Fruit pioneered the business model of the "Banana Republic," where it controlled the government and populations of poor countries through bribery, blackmail, and mercenary militias willing to commit murderers and massacres of protesting workers. United Fruit provoked and supported a right-wing coup in Guatemala in 1953 and other Central American countries. This "successful" business model became a template for other US industries, especially the oil industry, to brutally exploit the resources of third-world countries, including toppling other non-cooperative governments in countries like Iran.

Antipathy towards United Fruit played a big role in the success of leftist revolutionaries such as Fidel Castro and Che Guevara. United Fruit was a primary conspirator in the failed the Bay of Pigs invasion of Cuba to topple Castro. The corporation was eventually dismantled for its anti-trust activities, but it paved the way for corporate exploitation of the third world. The arrival of the Reagan administration in 1980 signaled the beginning of the end for meaningful corporate regulatory restraint and the dawn of the epidemic of Washington corporate lobbying. In 2017, a record $3.36 billion was spent by corporations lobbying Congress and the White House (not counting campaign contributions) by an army of over 11,500 lobbyists: over 20 lobbyists for every member of Congress.[7]

A nakedly partisan Supreme Court then released the "corporate Kraken" in 2010 with their notorious "Citizen United" ruling, which essentially set fire to a 100-year-old legal precedent restricting independent political spending by entities like corporations and unions. The five conservative judges went far beyond what the victorious Citizens United attorneys even asked for and willfully blinded themselves to something a kindergartner could grasp—the flow of money would hollow out to its core the integrity of American democracy, and an already perverse level of corporate power would increase exponentially. To the same extent that wealth became concentrated, democracy would dwindle to extinction.

The late Supreme Court justice Antonin Scalia was perhaps best known for being a fierce proponent of "originalist theory," which holds that judges should interpret the Constitution to align with the public meaning it had when originally written. Yet if Scalia and other judges were to be true to the doctrine of originalism, then they would have paid much more attention to the founding fathers original intentions on corporate restrictions, placing them far below their current perch atop the mountain of American politics.

The illegitimate power of corporations combined with their lack of accountability has created real and dramatic tragedy, as illustrated by the afore mentioned examples. But rummaging over the larger scale threats to entire societies, countries, and mankind in general, we see a grotesque, recurrent theme. Much like Frankenstein, corporations are willing to kill, maim, and destroy even their own agents and employees in the name of profit. A recurrent refrain of progressive politics, especially following the infamous Citizens United ruling, is, "Corporations are not people," Mitt Romney's protestation notwithstanding.[8] It might be just as accurate to say, "People that run corporations are not people either."

But corporate CEOs, no matter how lacking in empathy and no matter how detached from the world the rest of us must live in, do not live forever. The damage they can personally do to the rest of us is somewhat constrained by that fact alone. Corporations, however, are imbued with a soulless immortality not shared by the their CEOs. Once created, the monster has a far greater capability to wreak havoc than Dr. Frankenstein himself, and the damage can continue, virtually in perpetuity, until enough torches and pitchforks are gathered and raised to subdue the monster. And monsters there are a plenty, but bearers of torches and pitchforks only a few.

What follows is an unvarnished, unapologetic, heavily-referenced critique of the deadly practices of major corporations, marinated with my own recipe of scientifically placed sarcasm and irreverence, and a dash of prescription medicine on how to save ourselves. Think of it as science, facts, and reality, with an attitude.

P.S. By popular demand, the scratch and sniff version of this book for kids, will not be released anytime soon.

1

The Lead Industry:
We all grew up in Flint, Michigan

"Find out just what any people will quietly submit to and you have the exact measure of the injustice and wrong which will be imposed on them."

—Frederick Douglass

"I coulda been a contender, I could've been somebody" is one of the most famous lines in Hollywood movie history. It is the tearful anguish of Marlon Brando's character, Terry Malloy, that as a boxer, his gifts and talents had been squandered to corruption in the award winning, *On the Waterfront.*[32] For the next several decades, thousands of children in Flint, Michigan might be expressing a similar, heart-breaking lament. Like with Malloy, this squandering will clamp a hard lid on the lives, careers and economic potential of these children. Also like with Malloy, it will be the handiwork of corruption and criminal behavior. Unlike Malloy, it was not self-inflicted, but imposed by others who tossed aside the highest sworn duty of all public officials--to protect the most vulnerable that they serve.

The lead contaminated water scandal in Flint, Michigan, has become a shocking national tragedy. Incompetence, ignorance, arrogance, callousness, lying, and cover-up by an assortment of local, state, and federal officials, ruined the municipal water supply of a city of 100,000 mostly poor African Americans. There is little question that the political impotence of poor people of color emboldened the public officials responsible for the decision to brush aside the risks and consequences to both the people of Flint and to the careers of the officials themselves. Nonetheless, 15 government employees and officials were criminally charged for their actions, and 5 of them face manslaughter charges.[1] The human and economic costs are already substantial. At the very least, it means thousands of Flint children have suffered lead exposure, with likely irreversible brain damage of varying degrees, depending on the amount of exposure.

The events in Flint are a raw political scandal, a public health tragedy, and a profound moral failing of our society. But thanks to the lead industry of decades ago, "We all grew up in Flint, Michigan," in what has been dubbed, "the crime of the 20th century." The number of victims is in the billions, and the economic cost in the tens of trillions. As early as 1983, a report from the UK Royal Commission on Environmental Pollution found that because of the ubiquitous use of lead for several previous centuries, "it is doubtful whether any part of the earth's surface or any form of life remains uncontaminated by anthropogenic lead."[41]

In the toxic substances hall of shame, lead has achieved a level of infamy almost unequaled by anything except perhaps radioactivity. All of us have some lead in our bodies, the result of living in an industrial world, and a sinister lead industry that for decades successfully fought off attempts to curtail products it knew were deadly. But there is no "normal" or "natural" blood level of lead. The CDC and medical organizations like the American Academy of Pediatrics have officially stated that no amount of lead exposure can be considered safe.[2]

Lead has become perhaps the most widely recognized environmental neurotoxin in human history. Lead was mass produced from mines in Spain and Greece around 3,000 B.C. and was first widely used by the Romans, who eventually became the first to suspect its serious health consequences, including madness and death.[3]

Lead was valued because of its softness, durability and malleability and was used for pipes, dishes, cosmetics and coins. Lead pipes have been found still perfectly intact, inscribed with the insignia of Roman emperors. Because of its slightly sweet taste it was routinely added to wines. In fact, lead was used in one-fifth of the 450 recipes in the Roman Apician Cookbook, a collection of 1st through 5th century recipes attributed to gastrophiles associated with Apicius, the famous Roman gourmet.[4]

In medieval Europe, lead was used for roofing, coffins, cisterns, tanks, gutters, pipes, statues and ornaments. It was renowned for its workability for applications like stained glass windows. Lead was a critical component in the development of the printing press, where it was used to produce moveable type.

Although lead is a cumulative poison, there is strong evidence that the brain toxicity of lead is not linearly related to concentration. Small doses likely have greater impact per unit of exposure. Symptoms of "plumbism" or lead poisoning were already apparent as early as the 1st century B.C. Mental incompetence from lead exposure came to be synonymous with the Roman elite, personified by the shockingly imbecilic emperors Caligula, Nero, and Commodus. In retrospect, lead is widely believed to be a significant contributor to the fall of the Roman Empire.

In 370 B.C., Hippocrates wrote about gastrointestinal distress in a person who worked with metals. In 14 B.C., the Roman architect Vitruvius noted poor health in those who worked with lead. In his book, *De Architectura*, he suggested that when running water through leaded pipes, "the lead receives the current of air, the fumes from it occupy the members of the body, and burning them thereupon, rob the limbs of the virtues of the blood. Therefore, it seems that water should not be brought in lead pipes if we desire it to be wholesome."[5] The conclusions drawn by Vitruvius were correct, but well before his time, and his warnings were largely ignored. Dioscorides, a Greek physician who practiced in the 1st century A.D., noticed that exposure to lead could cause paralysis, delirium, swelling and indigestion.

Observations of paralysis in lead-exposed miners were recorded in Europe in the 1600s. American colonists had grown fond of lead's attributes and many applications, but warnings against its side effects were disseminated by none other than Benjamin Franklin. Franklin lamented that his warnings were largely unheeded, and he wrote, "You will observe with Concern how long a useful Truth may be known, and exist, before it is generally receiv'd and practis'd on."[6]

The famous composer Ludwig van Beethoven died in 1827 at age 57, and may have been a victim of lead poisoning. Beethoven had visited countless doctors in search of answers for his poor health, digestive problems, abdominal pain, irritability and depression. An analysis of hair samples showed high levels of lead. Artist Francisco Goya, Queen Elizabeth I, and US President Andrew Jackson were all likely victims of lead toxicity.

Decades later, US medical authorities first described childhood lead toxicity in 1887. In fact, some 19th century paint companies began running newspaper ads touting their "lead-free" paint. Nonetheless, as Franklin had feared, by the 20th century, the US was the world's leading producer and consumer of refined lead.

Ben Franklin warned against the scourge of lead toxicity

Neurotoxin in Your Tank

In 1916, Thomas Midgley, a chemist working for the General Motors Research Corporation was given the assignment of "improving" gasoline, especially eliminating engine "knock." After experimenting with hundreds, and by some accounts even thousands of additives over several years, including melted butter and camphor, Midgley discovered that a lead additive, tetra-ethyl lead (TEL), could eliminate engine knock and increase miles per gallon. Just a smidgen of lead had effectively increased the temperature at which the gasoline ignited, a fuel characteristic which would come to be known as its octane rating.

Over the next few years, as Midgley continued perfecting his additive, he received numerous letters from American and internationally distinguished scientists from around the world, like Robert Wilson of MIT, Reid Hunt of Harvard, Yandell Henderson of Yale (America's foremost expert on poison gases and automotive exhaust) and Dr. Erik Krause of the Institute of Technology,

Potsdam, Germany, all urging him to abandon the "creeping and malicious poison" of TEL. They cited its role in the death of researchers.[7] He was warned it would build up along roadways, in tunnels, and the air where cars trafficked. By 1922 Europe and The League of Nations became sufficiently aware of lead's toxicity that they adopted a resolution banning the use of lead paint for interior use. The US refused to comply with the resolution.

General Motors' President Pierre du Pont was aware of TEL's health dangers. He described it in a letter as "very poisonous if absorbed through the skin, resulting in lead poisoning almost immediately."[8] In December of that year, the US Surgeon General sent a letter to GM warning that leaded gasoline would become a serious threat to the public. All these warnings were dismissed. At the top of the list of lead's attributes was this: it was cheap to refine, always a plus for making a buck. All of the principals involved could have easily chosen any one of numerous safe alternatives instead of lead, like ethyl alcohol. But the oil companies hated ethyl alcohol because it represented serious competition to their business, and because it couldn't be patented, GM knew that it would not add anything to their profitability. Under the name *Ethyl,* leaded gasoline went on the market in 1923. The venality of the conspirators is revealed by the fact that the name *Ethyl* was chosen deliberately because they preferred the public not realize it was, in fact, lead.

In a twist of irony and ultimately hypocrisy, Midgely himself took a prolonged vacation in 1923 because he felt like his health had been affected by his research exposure to lead. In an even deeper ironic turn, Midgely was forced to decline three speaking engagements with the American Chemical Society, which had awarded him a medal for his discovery of TEL.[7] Nonetheless, the next year, he participated in a press conference intended to demonstrate the benignity of Tetra-ethyl lead. He poured it all over his hands, then inhaled its vapors from a bottle for one minute, declaring that he could do this every day without suffering any ill health.

Midgely had plenty of help in the TEL scandal. Prominent among many others were Alfred Sloan and Charles Kettering, who amassed fortunes as the largest shareholders in General Motors and the General Motors Chemical Company, established to produce TEL. You may recognize their names as attached, again ironically, to the famous Memorial Sloan-Kettering Cancer Center in New York.

General Motors, Standard Oil, and DuPont then formed the Ethyl Gasoline Corporation and began pumping of millions of gallons of lead into automobiles to quiet the "ping" of engines and to boost performance. Shortly after Ethyl factories opened for business, the neurologic side effects of lead began to emerge in the workers—confusion, impaired judgment, and a staggering gait--as if they were drunk, but not from alcohol. Their manufacturing facility in New Jersey came to be known as the "House of Butterflies" because workers were often experiencing hallucinating about insects. Despite workers who handled Ethyl getting sick and some of them dying, the Ethyl Corporation, even without the help of modern day Fox News, responded by, of course, blaming the workers for being careless, engaging in "horseplay" with the product, and not taking seriously that their jobs represented "a man's undertaking." The corporate spin was they had made the business mistake of hiring "sissies," not that lead was destroying their employees' brains.

Within 30 days after opening, the TEL plant in Deepwater, New Jersey saw its first death from lead toxicity. Walter Dimock died Oct. 28, 1924, at 1:30 a.m. His brain was riddled with lead and an astonishing amount of lead was found in the rest of the body. New Jersey health authorities responded by ordering Standard Oil to shut down the House of Butterflies.

New York Medical Examiner Charles Norris wanted to take lead out of gasoline as soon as 1924. New York City banned the leaded gasoline, then New Jersey and Philadelphia. But Standard Oil then appealed to President Calvin Coolidge. Coolidge ignored health experts and only consulted industry scientists in assessing the danger of leaded gasoline, setting the precedent for which camp of "experts" would be driving public policy for the next 40 years.

Because of Du Pont's control of local media, the neurologic disorders and deaths among the lead workers was never publicly reported. Within a year, at least 15 workers who worked at the refineries in Ohio and New Jersey that produced the lead additive died. Psychosis with violent insanity, requiring actual straightjackets, usually preceded death. Nearly 300 workers were diagnosed as psychotic, and soon the leaded fuel became known as "loony gas."[5]

In one Standard Oil plant, 80 percent of the staff died or suffered severe lead poisoning. Questioned by reporters in 1924, a company spokesman offered some Standard Oil expert psychoanalysis to reporters, "These men probably went insane because they worked too hard." A few days later, four more workers from the plant died, and 36 others were disabled with permanent neurological damage. According to General Motors, it was all from "working too hard."

The deaths and neurologic catastrophes of its many workers led to Ethyl voluntarily withdrawing TEL from the market in 1925. But in a masterful campaign involving a highly publicized hearing, a docile US Surgeon General committee and claims by industry that the octane boosting and anti-knock properties of Ethyl were an essential, irreplaceable adjunct to modern transportation, the tragic outcomes of Ethyl workers were judged inconclusive as to cause, and the larger issue of whether the public in general should be exposed to leaded gasoline was never addressed. The corporations involved had the audacity to claim The Almighty was on their side, publicly pronouncing that Ethyl was a "Gift from God." If any God was involved, I'm guessing it was the God of the Old Testament playing another one of his devastating practical jokes on people.

While the Surgeon General called for further studies, and even the industry's own paid scientists remained nervous about the use of lead in gasoline, the results of the hearing were broadcast by the venerable New York Times as, "No Danger in Ethyl Gasoline." The lead industry claimed they had been exonerated, and for more than 40 years no further studies were ever done.

For the next 63 years, leaded gasoline was sold all over the world. By 1963, leaded gasoline represented 98 percent of the US gasoline supply. It is likely that at least 68 million US children suffered toxic exposures to lead during that span.[7] I remember when I was a young boy and my father would take me to the gas station with him, the attendant would ask him if he wanted "leaded or unleaded." My dad was a physician, and although he never said why, we always ordered unleaded.

So why was lead still allowed to be added to gasoline despite decades, if not centuries of recognition that it was a potent neurologic poison, and despite its eventual ban from paint? In defiance of even the published research at the time, the U.S. Federal Trade Commission stated that leaded gas was "entirely safe to the health of motorists and the public" and the door was opened for lead to poison the entire globe and all its inhabitants. For over 50 years the Ethyl Corp, General Motors, Standard Oil, Du Pont, and the American

Petroleum Institute obscured, obstructed, and lied about the mounting evidence of a public health catastrophe from tetraethyl lead, aggressively marketing it worldwide and fighting every attempt to regulate or curtail its use.

Dr. Clair Patterson is one of the few heroes that emerges in this saga of corporate greed, government corruption, and tragic outcomes on a global scale. In the 1940s, Dr. Patterson demonstrated the levels of lead in the environment had increased about 1,000 times compared to pre-1923 levels by using ice core samples from Greenland. He then compared modern bone samples to those of 1,600-year-old human remains, and found that modern humans' lead levels were hundreds of times higher. The Ethyl Corporation then dug deeper into their pit of villainy and tried to buy Patterson's silence by offering him lucrative in-house employment. He declined.[8] Patterson was a geochemist who was also part of the original Manhattan Project and probably deserves credit for the original best estimate of the age of the earth—4.55 billion years.

Of course heroes of the common good inevitably make enemies. The lead industry had powerful friends and allies, like a Supreme Court justice, members of the US Public Health Service, and the American Petroleum Institute. For his scientific brilliance and integrity, Patterson was rewarded with a big black ball from government or corporate research funds, and he was dismissed from the National Research Council panel on atmospheric lead contamination. Shilling for industry, the board of trustees at Cal Tech, where he was a professor, tried to get the university to fire him.[7]

By1972, more research compelled the US EPA to issue a ruling to phase out lead in gasoline and was promptly sued by DuPont and the Ethyl Corporation, an action which prolonged tail pipes spewing lead in the US for at least another ten years.

Efforts to legislatively protect the public from lead had begun in Europe in the early 1900s. France, Austria, and Belgium banned lead in white interior paint in 1909. By 1935, Great Britain, Sweden, Greece, Czechoslovakia, Yugoslavia, Poland, Tunisia, Spain, and even Cuba had all banned lead in at least some paints.

But in the United States, where corporate influence held sway, lead continued to spread ubiquitously throughout the American landscape. It came from vehicle emissions and the smokestacks of smelters and paint factories. It leached into drinking water carried by pipes joined with lead solder. It settled in the soil around painted home exteriors where millions of children played and in children's bedrooms as lead-based interior enamel paint deteriorated. It coated play pens and cribs and was chewed on by infants and toddlers. It began to accumulate in urban gardens, making home-grown vegetables an unrecognized, silent health risk.

Other risk factors associated with even higher lead exposure included a paternal manual occupation, residence near highly trafficked roads, living in older buildings, and using hot tap water for cooking. The entire population of the developed world was contaminated by a poison that was being deliberately added to vehicle fuel by the richest companies in the world. The average adult had about one fourth the amount known to produce clinical lead toxicity. It wasn't until 1971 that lead additives to paint were phased out with the passage of the Lead-Based Paint Poisoning Prevention Act. The ban was not complete until 1978.

Even in 1980, lead was still in a myriad of products--gasoline, paint, and hundreds of consumer goods. In 1980, 47 percent of food and soda cans were soldered with lead. In fact, Americans were using ten times more lead per capita than the Romans, according to Jerome O. Nriagu, the world's leading authority on lead poisoning in antiquity. The average American child growing up before the 1980s lost at least 6 IQ points from leaded gasoline and paint.[9]

Much worse for the nation as a whole, that loss of IQ also decreased the percentage of the population qualifying as "intellectually gifted" by about 40 percent, and increased the population of the "mentally challenged" by a similar amount. Numerous studies also showed a tight correlation between blood lead levels and aggressive, anti-social, and criminal behavior.[10]

Dr. Herb Needleman published a landmark paper in the pre-eminent New England Journal of Medicine in 1979, which became the lead industry's waterloo.[11] Needleman had actually worked as a young man in a lead factory and personally observed the mental dullness and hallucinations of the workers in the House of Butterflies. Needleman's study measured the lead content of shed baby teeth, a far more accurate indication of long-term childhood exposure, because within about 30 days, lead leaves the blood to enter the long term storage venues of the teeth and bones. Needleman's study demonstrated a dose-dependent relationship between increasing lead content and a wide range of psychometric measures, including poor organizational ability, lower IQ, distractibility, and impulsive behavior.

Seen as a singular threat to corporate profits, Needleman became the primary target of the lead industry's ire. Shortly after his New England Journal article was published, the industry went on the attack, going after his research and using public relations firms and scientific consultants to undermine his credibility. It was a despicable example of how an industry seeks to warp science and call into question the credibility of those whose research threatens its bottom line, regardless of how much damage that bottom line does to people and society in general.

Industry consultants demanded that the EPA, and later, the Office of Scientific Integrity at the National Institutes of Health, tear apart Needleman's work. And then, in 1991, under pressure from industry consultants, the University of Pittsburgh, in a stroke of pitiful cowardice, formed a committee to evaluate the integrity of his lead studies.

Thankfully, when the [leaded] smoke cleared, the federal government and the university exonerated Needleman and his research. On the national stage, certainly among public health advocates, and even more so in retrospect, Needleman, like Patterson, emerged as a real-life super hero. Nonetheless, industry's counter offensive had exacted a tremendous toll, damaging both Needleman's academic life, much like it had Patterson's, and worse, permanently shackled the field of lead research. Industry's attack sent a chill among academic researchers who might want to challenge the money, muscle, and resolve of highly profitable industries. This same scenario has become a template for manufacturing, chemical, and oil and gas industries to attack academic research, scientists, and institutions that display the courage, integrity, and audacity to challenge industry profits.

In 1980, the National Academy of Sciences reported that leaded gasoline was the greatest source of environmental lead contamination. While lead finally became increasingly targeted as a serious health menace in the United States, the Ethyl Corporation was busy increasing its overseas business 1,000 percent between 1964 and 1981. C.M. Shy of the UNC School of Public Health, in a paper published by the World Health Statistics Quarterly, declared leaded gasoline is "The Mistake of the 20th Century."

Because of industry pushback, including lawsuits, it took the EPA a mere 25 years to complete the phase out of lead from gasoline. In a 1996 announcement heralding the move, EPA administrator at the time Carol Browner said, "The elimination of lead from gas is one of the great environmental achievements of all time."[31] Fair enough. But a more honest characterization would be to admit that it brought a far- too-late end to the monumental, corporate global crime of putting lead in gasoline in the first place. And although the crime and tragedy finally ended, the guilty were never punished.

But there were a few significant gaps left in that "achievement." NASCAR continued to use leaded gasoline until 2008, blanketing their crews and hundreds of thousands of spectators with a fine mist of lead at every event. Even with everything we know now about lead's extreme toxicity, inexplicably and indefensibly, it is still found in aviation fuel for small airplanes and helicopters, known as "avgas." Approximately 150,000 piston engine aircraft across the nation are allowed to use it. For two decades, the EPA has rebuffed petitions from multiple environmental and public health advocacy groups to force elimination of avgas. A 2011 study from Duke University showed elevated levels of lead in the blood of those living up to one kilometer away from an airport.[16] The EPA estimates that 16 million Americans live close to one of 22,000 airports where leaded avgas is routinely used—and three million children go to schools near these airports.[17]

Lead-free avgas is currently being manufactured by Shell. There is simply no reason to continue allowing leaded avgas. Nonetheless, 27 US Senators wrote a letter to the EPA in 2011 urging them to go slow—apparently 25 years isn't slow enough for them—on making a new rule because of the tried and true argument that somehow it would "cost jobs," which is corporate speak for, "it would cost a handful of large campaign donors a lot of money."[18]

By the time lead's reputation became widely sullied, there were two manufacturers left. Both companies tried to hide their activities behind name changes. Ethyl changed its name to Newmarket, and Associated Octel changed its name to Innospec. However, just before that, for their 50th anniversary catalogue, they had the audacity to quote a 1982 letter from a manager to the rest of the company.

"Many funerals have been arranged for lead in petrol–1926, 1943, 1954, 1970, etc.–as I can recall. The grave has been dug, the service arranged, the coffin prepared, the parson and mourners instructed, but the body just would not lie down in the coffin."[12]

As of this writing, Innospec, an American company with most manufacturing sites in Britain, is still selling the deadly toxic tetraethyl lead (TEL) gasoline additive to third-world countries like Iraq, Afghanistan, Algeria, Myanmar, North Korea and Yemen. Reportedly, Innospec recently told shareholders it would seek to "maximise the cash flow" from its declining sales of TEL.[13]

In a true "Lex Luthor" type example of corporate willingness to do unspeakable harm to millions for a few bucks, in 2011, Innospec pleaded guilty in US and British courts to paying massive kickbacks to Iraqi and Indonesian officials to secure lucrative contracts supplying TEL between 2000 and 2008. Innospec was fined millions of dollars by the US and UK governments, and three Innospec executives were sentenced to jail in 2014 for their role in bribing foreign officials to accept their deadly but highly lucrative product. On handing down the sentence, the British judge stated;

"Corruption in this company was endemic, institutionalised and ingrained…but despite being a separate legal entity it is not an automated machine; decisions are made by human minds. None of these defendants would consider themselves in the same category as common criminals who commit crimes of dishonesty or violence….. but the real harm lies in the effect on public life, the effect on community and in particular with this corruption, its effect on the environment. If a company registered or based in the UK engages in bribery of foreign officials it tarnishes the reputation of this country in the international arena."[14]

A report commissioned by the United Nations in 2011 calculated that the yearly global cost of lead in gasoline had reached 1.1 million deaths, 322 million lost IQ points, 60 million crimes committed, and an economic loss of 4 percent of global GDP, or $2.4 trillion.[15] A newer study, to be mentioned later, suggests that number is too low.

Like most other heavy metals, lead is not combustible, it does not degrade, and it cannot be destroyed. Because of its persistence in the environment, the dark legacy of lead exposure will continue indefinitely into the future. The world is now permanently blanketed with this deadly metal, purely for corporate profit. An estimated 7 million tons of lead burned in gasoline in the United States remains in our soil, water, air, and in the bodies of all living creatures, including virtually every human being.[7]

The Enduring Legacy of Toxic Paint

The history of lead's use as a paint additive is as disturbing as the story of lead in gasoline. Lead was added to paint as a coloring agent, to speed up drying, and to make paints more durable and corrosion resistant. The Romans added lead to paint: one reason why the paint on their ruins is still relatively well preserved. In the early 20th century, a house painter would bring with him two buckets—one for the paint substrate, and one with lead powder to mix with it. It was thought that the more lead added to the mix, the better the paint, and therefore, the higher the price. Household paint manufactured before 1950 contained as much as 50 percent lead by weight. A dime-sized chip of such paint contained enough lead to sicken a two-year-old and cause permanent brain damage, a major reason why old homes can be serious health hazards for children.

Three fourths of the homes built before 1978 had at least some leaded paint in or on the house. President George H.W. Bush's dog, Millie, attracted national attention to the dangers of lead poisoning in 1992, when she got sick from breathing lead dust during White House renovations. Lead's sweet taste is one reason small children have been attracted to nibbling on it. Lead is still present in most paints, although at much lower concentrations.

But the disaster that is lead is not just an historical issue. As reported in Bloomberg News,[19] corporations that manufacture paint are still selling oil-based paint with dangerous amounts of lead throughout the developing world where government regulations are lax or nonexistent. Most of the decorative, oil-based paints for sale in 25 low- and middle-income countries contain dangerous amounts of lead. In one 2008-2009 study,[20] the average lead concentration of enamel paint purchased in 10 countries in Africa, Asia, Latin America and Eastern Europe was 23,707 parts per million (ppm). The legal limit for residential use in the US is 90 ppm.

One of the largest paint manufacturers in the world is the US based PPG. Seigneurie, their subsidiary, is the largest paint company in Cameroon. Seigneurie still sells paint with lead concentrations up to 50 percent by weight, 5,500 times the allowable amount in the US.

Mark Taylor, an environmental scientist at Macquarie University in Sydney, Australia says: "The lead in paint limit in the US is 90ppm, so selling paint with up to 500,000 ppm is just incredible. Why would anyone do that? There have been adequate alternatives for decades. When paint contains lead above 90 ppm, it's an indication the metal was added intentionally -- as a pigment, drying agent or corrosion retardant."[21]

Safe alternatives exist for each of these characteristics. Some of the non-lead pigments are more expensive than lead, but coloring is such a small fraction of the total cost of paint, using lead for pigment would not change the manufacturing cost at all. Manufacturers have no good justification for selling the former. Yet the study[22] identified 54 companies that made the paints with excessive lead concentrations, six of which were subsidiaries of US companies and eight of which were subsidiaries of European or Japanese companies. What's more, US entities exported at least 11,000 tons of the pigments necessary to make lead paint in 2012, a trade worth $27 million.

Everyone knows mass migration of manufacturing in the United States to the Pacific rim and other poor countries occurred over the last 30 years of the 20th century. But ironically, American consumers began importing foreign toys, cosmetics, and household products, covered with toxic contaminants like lead, that had been banned in the US. In 2007, when the media exploded with stories of lead in children's toys manufactured in China, the paint was quickly identified as the source. Mattel alone recalled tens of millions of toys.

Although the gun industry has its own chapter in this book for other reasons, it is worth citing them as another guilty party in any examination of public health and lead contamination. Ninety percent of the ten billion rounds of ammunition sold in this country to the military, police, and the public is lead ammunition.[35] Using the familiar "slippery slope" argument, i.e., "first they came for our leaded ammunition, then they came for our guns," then they came for…what, our wives?--the gun industry rabidly opposes any restriction on lead ammunition. No doubt there was a virtually orgasmic celebration at NRA headquarters when disgraced former Interior Secretary Ryan Zinke ceremoniously reversed an Obama ban on the use of lead ammunition and fishing on federal lands on his first day in office.[36]

Dogmatism, stupidity, and irrationality on the issue of leaded ammunition are part of a unique feedback loop among the shooting sport crowd. Insistence on the right to use lead means that those that hunt and frequent gun ranges for target practice are exposing themselves to significant amounts of a toxic substance to which there is no safe exposure, and that has been proven to impair brain function. Shooters are exposed to enough volatilized lead dust and lead fragments that their own health, especially their mental health and judgement, can be the most important casualty of their gun fascination.

Even in a well-ventilated gun range, lead dust drifts onto shooters' hands, arms and clothing, and stays suspended in the air long enough to be inhaled. The more people shooting, the more lead ingested, inhaled, and absorbed by everyone at the range. Studies of shooters at gun ranges show significant spikes in blood lead levels, into the acutely toxic range, that may not return to baseline for as long as 69 days afterwards. The chronic baseline blood levels of shooters is about 67 percent higher than the average non-shooter. And if the range is well ventilated, residential neighbors are eventually exposed as well. When patrons and range employees go home, they can carry enough lead dust with them to contaminate their households and their family members.[37]

There are between 16,000 and 18,000 private gun ranges in the country frequented by the public, and about 1,800 National Guard Armories. In a scathing investigative report of Dec. 2016, the *Oregonian* newspaper found lead contamination was a serious issue for 90 percent of the armories examined, and in half of them, the lead contamination extended throughout the buildings.[40] These buildings are civic landmarks in their communities, where numerous community wide events occur--classes for pregnant mothers on nutrition, with their infants frequently crawling on contaminated floors, weekend drills and shooting practice for part-time soldiers, baby showers, cub scouts, baptisms, ceremonies for deployment, and joyous family reunions. I grew up playing basketball in the local National Guard Armory. The Oregonian stated:

> "Inspectors have found lead dust at alarming levels in armory gyms, drill halls, conference rooms, hallways, stairwells, kitchens, pantries, offices, bathrooms and a day care center...[Lead] contaminated coffee makers, ice makers, refrigerators, dishes, soldiers' uniforms, children's toys, medical supplies, water bottles, carpets, soda machines, bookshelves, fans, furniture, heaters, basketball backboards and a boxing bag. Even a deli meat slicer."[40]

Lead contamination persisted long after the gun ranges themselves were abandoned. In one armory, dozens of ten-year-old boys had a sleep over, and spread out their sleeping bags over a floor where inspectors found lead dust at 650 times the far-too-lax federal "safety" standard.

Lead's health repercussions are the consequence of cumulative exposure. The more frequent the exposure, the more damage to brain function. It becomes a serious occupational hazard for weapons instructors, range employees, police, and defense personnel. Addressing the issue, the Department of Defense lowered their blood lead standard to 20 ug/dl.[38] The current OSHA standards require that an employee stop working if lead levels exceed 60/dl, but they can return to the job if it drops below 40 ug/dl. The reader will see shortly how shockingly absurd these supposedly "safety" standards are.

Lead: A Full Service Toxin

Many heavy metals, like chromium, manganese, molybdenum, nickel, and selenium are toxic at high levels, but are actually required nutrients at lower levels. Not so with lead. There is no known useful biologic purpose of lead at any concentration. Lead mimics the biologic properties of calcium and is stored in the bones and marrow. If you want to know about a pregnant mother's recent lead exposure you can determine that from measuring it in blood. But chronic exposure is manifest in her bones and teeth. During pregnancy and later with lactation, calcium and lead are mobilized from the mother's skeleton, cross the placenta and enter fetal cells. That means a prospective mother's lifetime history of exposure, not merely acute exposure, will impact brain development in all her children. Lead stored in bone becomes an endogenous source of lead throughout a person's lifetime, putting a person at risk long after any external exposure has ended. A diet rich enough in calcium and iron will result in less absorption of lead.

Lead impairs hemoglobin biosynthesis, causing anemia, reduces kidney function, increases blood pressure, is associated with infertility in males by reducing sperm counts and motility, and in females is associated with increased rates of stillbirths, miscarriages, and low birth weight. Lead directly and indirectly alters the function of bone cells, for example, impairing their ability to respond to hormonal stimuli and regulation, and inhibiting the formation of new bone. Lead's greatest impact on public health is its impairment of brain development in children, due to direct exposure from inhalation and ingestion, from breast feeding and from intra-uterine exposure.

The CDC and the American Academy of Pediatrics has stated there is no safe amount of exposure to lead: none.[23,24] Every bit of lead exposure will cause impairment of cognitive abilities. Recent research has established this stark correlation: for every .19 ug/dl of lead in an adolescent's blood, there was a loss of one IQ point. In this study the average lead level was 1.71 ug/dl. Extrapolation suggests that the average loss of IQ was more than 8 points.[25] In the U.S., the average blood lead level of preschoolers is about 1.3 ug/dl, which suggest that even children with no discernible lead exposure are experiencing a loss of about 7 IQ points due to the low levels of lead in their blood.

Other studies show that levels as low 2.0 ug/dl negatively affect a child's test scores.[26] Furthermore, there is strong evidence that, like with other toxins, the relationship between brain damage and lead exposure is supralinear rather than linear, i.e., it is steeper at low doses. This means that "per unit of lead exposure," the toxicity is greater at very low doses.[34] In turn, what that means is; despite significant improvement in lead exposure in the past three decades, children are still suffering significant brain damage from much less exposure.

The new information is especially chilling because in the mid 1970s, the average U.S. preschool child had 15 ug/dl of lead in their blood. Eighty-eight percent of children had a level exceeding 10 µg/dL—which is twice what the CDC considers toxic as of 2012. (Note that the level most recently considered officially "toxic" is still associated with a profound loss of IQ). Among poor black children, the average level was markedly higher--23 µg/dL.[27] That poor minority children suffered so disproportionately had political repercussions. Little political capital was gained for politicians who might champion the "lead cause" because of who were the most conspicuous victims. That certainly contributed to a long delay in regulation of lead in paint and gasoline.

Because no person on earth is fortunate enough to have a zero body burden of lead, every person's intellect has been compromised to some degree.[28] Children are more vulnerable to the neurotoxicity from lead and they absorb a higher percentage of the lead that is swallowed. The microvasculature of a child's developing brain is uniquely susceptible to high-level lead toxicity, characterized by cerebellar hemorrhage, increased permeability of the blood-brain barrier (specialized cells that form the unique architecture of blood vessels in the brain), and blood vessel-related brain swelling.

Lead also affects many of the biologic functions of nerve cells themselves, interfering with critical enzymes, chemicals that act as neurotransmitters, displacing calcium movement in and out of cells, and disrupting the protective coating of nerve axons called myelin. Occupational lead exposure is associated with brain atrophy, permanent white matter lesions, and overall small brain volumes.

Research shows a strong correlation between atmospheric lead levels and violent crime rates (not necessarily white collar crime). A study published in Environmental Research, which used data spanning more than fifty years, reported a "very strong association" between lead levels in children and crime rates twenty years later when they became young adults. The decline in US crime rates which began in the early 1990s fits the pattern with the reduction of leaded gasoline in the early 1970s. Other countries that followed suit saw similar declines, also delayed by twenty years. A Pittsburgh University study showed that juvenile delinquents averaged lead levels four times higher than adolescents with no history of trouble with the law. Some health authorities believe the data suggests that 90 percent of the international drop in crime over the last 20 years can be attributed to reducing lead exposure.[33]

While lead is most infamous for its neurotoxicity, it is also toxic to other organs systems, in particular, the cardiovascular system. The EPA estimated that about 5,000 people a year were dying from lead-related heart disease before it was removed from gasoline. A new study calculates that was a gross underestimation and that even today, as many as 412,000 Americans die annually from heart disease, provoked by past and ongoing lead exposure, almost 20 percent of all deaths,[29] making it almost as deadly as the plague of cigarette smoking.[30]

With the help of profit-obsessed corporations, Thomas Midgley, had made a significant contribution to one of the most sinister public policy mistakes of all time. But Midgley was not done wreaking global havoc and damage that had to be undone by others. The father of leaded gasoline went on to invent chlorofluorocarbons (CFCs), the notorious refrigerants and aerosol propellants which imperiled all of us by destroying a considerable section of the earth's ozone layer before they were banned. Midgely had received numerous awards for his work, but they ultimately belied his place in history. Ironically, his inventions led to his own demise, which happened long before the tragedy of his work became fully apparent. After becoming physically impaired by contracting polio, Midgley invented a machine with motorized pulleys to assist him in getting out of bed or turning over. One day in 1944, as his invention sprang into action, he was strangled to death by the cords and pulleys.[8] There is no comfort in the poetic justice of the tragedy.

Thomas Midgley, inventor of leaded gasoline and the invention that caused his death

Ads for leaded gasoline and leaded paint in the 1920-1930 era

Dr. Herb Needleman Dr. Clair Patterson

Heroes who fought the public health disaster of lead exposure

2

The Pharmaceutical Industry, Part I:

Murder on the Estrogen Express?

"I hope that we shall crush in its birth the aristocracy of our monied corporations which dare already to challenge our government to a trial of strength, and bid defiance to the laws of our country."

—Thomas Jefferson,
The Papers of Thomas Jefferson:
Retirement Series, Volume 10: 1 May 1816 to 18 January 1817

Charles Dodds synthesized Diethylstilbestrol (DES), the first synthetic estrogen, in 1938, but was concerned about the cancer-causing potential of estrogens right from the beginning. Conforming to the British medical community custom at the time that scientific work should be done for the public good, Dodds relinquished his right to a patent. The prevailing custom of the time was that such discoveries should be made available to the public without their having to pay high prices to proprietary pharmaceutical companies. This of course would be considered quaint, naive, absurdly altruistic, and decidedly un-American in today's market.

As early as 1932, research in mice discovered that estrogens induced breast cancer. DES, prescribed primarily to prevent miscarriages, and about five times more potent that naturally occurring estrogen, became the poster child of cancer-causing estrogen when it became obvious decades later that daughters of mothers who took DES during the pregnancy were showing up with markedly increased rates of vaginal and cervical cancers, premature births, spontaneous abortions, still births and neonatal deaths.[1] DES was also later shown to cause birth defects.

The tragic sagas of DES and Thalidomide transformed the medical community's understanding of the uterus—that it was not a sanctuary safe from environmental toxins. Rather, it allowed the human embryo to be exposed and extremely vulnerable to chemical insults. DES was the first recognized human "transplacental" and "transgenerational" carcinogen, meaning that it not only caused cancer in the mother but also eventually the fetus. In other words, it acted as both a seed and ultimately a fertilizer for cancer. DES was also probably the first example of what is now a core tenet of endocrinology and reproductive medicine—endocrine disruption is possible from exposure to external chemicals that mimic or masquerade the effect of endogenous hormones.

"DES daughters" were afflicted with high rates of cancers, several other types of reproductive abnormalities like cervical stenosis and vaginal adenosis, and increased risk for a wide range of adverse pregnancy outcomes. Studies on "DES sons" have been much more limited, but several have demonstrated increased rates of genital abnormalities.[2] What will happen to even further downstream generations of the DES-exposed women is difficult to predict, but worrisome studies in mice have shown an increased susceptibility to tumor formation in the third generation,[3] suggesting the DES grandchildren may also be at increased cancer risk.[4]

Just about every drug company wanted in on the DES money-making machine. It was manufactured by numerous drug companies worldwide and marketed under more than 200 brand names. Somewhere between 2 and 10 million pregnant women world-wide were prescribed DES by pills, injections, suppositories, and creams between 1947 and 1971[10] when the FDA advised physicians to stop prescribing it.

Like other pharmaceuticals produced at that time, DES reached patients with very limited clinical research. And like so often happens with other drugs, once on the market, the mercenary focus of the drug companies led to their marketing it for numerous other disorders and diseases. It rapidly became popular for use in menopausal and post-menopausal symptoms, "chemical castration" for prostate cancer, and suppression of lactation. It was even used to stunt "abnormal height" in girls, and became the forerunner of the morning-after pill for post coital contraception.[5] It was one of the first growth promoters used in livestock--in chicken, sheep, and cattle.[6]

It's not like there wasn't adequate warning about the dangers of estrogens prior to the DES debacle. Several editorials in the Journal of the American Medical Association (JAMA) printed in 1939 urged caution regarding its carcinogenic potential and a thorough investigation of DES prior to approval by the Food and Drug Administration. One editorial stated, "It would be unwise to consider that there is safety in using small doses of estrogens, since it is quite possible that the same harm may be obtained through the use of small doses of estrogen if they are maintained over a long period." Adverse effects of DES on animals had been published[7,8,9] but they were largely disregarded.

Ironically, Dodds' seeming altruism contributed to the DES disaster. Without having to deal with a patent, DES was very inexpensive to produce and highly profitable. Quite simply, there was a lot of money to be made from DES, so caution, science, and patient protection were all enthusiastically thrown to the wind. The pharmaceutical industry dismissed the lack of efficacy of their product and willfully ignored its adverse health effects. Some manufacturers promoted it as an all-purpose elixir for use in all pregnancies.

Defying the warnings in scientific journals, thirteen drug companies originally applied to the FDA for approval to manufacture DES in 1940. But because of warnings in JAMA and other negative research, the FDA turned them down. Then collectively, the companies quietly withdrew their applications. But in the next several months they regrouped and formed what became the first powerful pharmaceutical lobbying group, and, en masse, received FDA approval in 1941.

By 1953, DES was proven to be ineffective for use in preventing adverse pregnancy outcomes. Given the research of the two previous decades sounding the warning on cancer risks, one would think that at least by then, use and production of the drug would have ceased. Not so, given that there was still money to be made. Millions of mothers, sons, and daughters continued to be exposed for almost two more decades.

Ignoring science and the risk to patients by drug companies seeking to profit from DES seems remarkably callous. For thirty years, DES was prescribed for millions of pregnant women and given to millions of livestock despite the evidence of its danger. But that part of the story pales in comparison what followed. In 1971 the

FDA announced to physicians that DES was contraindicated for use in pregnancy, citing cancer risks to the second generation. However it was still given to American livestock until approval for that use was withdrawn in 1979. Regulatory bodies in Europe were much slower to act, so the drug companies happily continued manufacturing and selling the DES cash cow throughout Europe for up to 12 years later. In Mexico, Uganda, and Poland, DES was still being used during pregnancy until the early 1990s. In fact, even after the tragic consequences of use in pregnancy, Eli Lily, the primary manufacturer, did not crease production of the drug until 1997, 26 years after the FDA withdrew approval.

Fountain of Youth, Fountain of Profit

But there were even greater opportunities to make money off estrogen than DES. Hormone Replacement Therapy (HRT) became ubiquitous "therapy" for middle aged women seeking the elusive fountain of youth.

In the19th century it was commonly thought that aging was caused by hormonal deficiency. This assumption was still prevalent many decades later when the estrogen industry exploited this idea to create the "hormone replacement" business. The growingly harsh and unforgiving societal view of aging women, especially middle and upper class Caucasian women, became a gold mine for many industries, not the least of which were drug manufacturers.

Beginning in 1942, with FDA approval, the promotion to patients and physicians of post-menopausal estrogen therapy was a historically unparalleled success for the burgeoning pharmaceutical industry. The first mass-marketed estrogen product was made by Ayerst Labs and made from pregnant female horse urine (who wouldn't want to drink that?), and given the ingenious name Premarin-- (PREgnant MARes's urINe). Drug company marketing credited hormone-replacement therapy with the ability to perpetuate youth and beauty, which continues today.

Between 1960 and 1975, estrogen prescriptions exploded in the United States, increasing nearly 100 percent after the popular press and a best-selling book both claimed that "estrogen deficiency" provoked a long list of diseases related to aging, and that "estrogen replacement" would prevent them. The seductive claim was that it gave women the fountain of youth in pill form, made them happier, more beautiful, and easier to live with. Wow! I'll take a truck load! Better still, let's just spray it over Salt Lake City with dive bombers! Let's add it to the water supply (actually we're pretty much doing that as you'll see later)!

Premarin was a gold mine for Ayerst, but it was worse than snake oil for everyone else. It was cruel, exploitive propaganda. My father, who was a physician, put my mother on Premarin. It likely contributed to her coronary artery disease and ultimately her death.

Doctors' waiting rooms became filled with pamphlets about menopause and estrogen and with articles in women's magazines cheerleading HRT. This was essentially drug company advertising with no finger prints of the drug company left behind.

HRT's heyday was launched by a prominent gynecologist Dr. Robert A. Wilson, who made it his personal crusade to label menopause as a disease and therefore curable by modern medicine, drug therapy in particular. In 1966, Wilson stepped into the limelight with a book entitled *Feminine Forever,* that declared "estrogen deficiency" and menopause as a widespread, but easily curable health threat. His crusade was heavily funded by the drug companies who stood to make billions convincing women they offered eternal youth, happiness, and even sexual vigor, all in one convenient pill.

In his book, Wilson called menopausal women "castrates," stating that "estrogen restores a natural harmony between the rate of aging and life expectancy, a harmony that has been disturbed by the lengthened lifespan of modern women. It is the case of the untreated woman -- the prematurely aging castrate -- that is unnatural." Wilson became influential in leading other doctors to embrace estrogen's off-label benefits with the help of blockbuster ads from the drug industry, free samples sent to doctors, expensive dinners, and consulting fees paid for by the drug makers.

Ayerst Labs hired an elite New York public relations firm to promote the message embedded in *Feminine Forever* for a decade after its publication. Ayerst provided a "service for media" through the Information Center on the Mature Woman. Magazine and newspaper editors frequently published feature articles on menopause with direct help and urging from Ayerst's PR firm. A 1972 film made by Ayerst, *Physiologic and Emotional Basis of Menopause*, asserted that diminished estrogen production during menopause may be the cause of depression, and "The physical alterations that are associated with the menopause may induce emotional changes. When a woman develops hot flashes, sweats, wrinkles on her face, she is quite concerned that she is losing her youth -- that she may indeed be losing her husband." A pill that will keep a woman from losing her husband to a younger woman was nothing less than a Godsend (depending on the husband) and a small price to pay. Even today, this cultural propaganda obviously still lingers in the minds of older women, desperate to halt the ravages of age.

HRT euphoria took a hit, and prescriptions plummeted following the publication of two widely circulated 1975 clinical studies in the New England Journal of Medicine, reporting that estrogen treatment significantly increased a woman's risk of endometrial cancer.

One study showed the difference in risk of developing endometrial cancer between two groups was dramatic: those who used HRT for 5 years or less were 5 times more at risk than non-users, and the risk kept increasing as the duration of the therapy increased, so that those who used estrogen for more than seven years were at 14 times the risk of non-users.

Two weeks after the publication of these studies, Ayerst brought in the artillery for a counter attack, sending a "Dear Doctor" letter to physicians arguing that the articles were "weak studies," claiming that the link between estrogen therapy and cancer was not firmly established. But the presumed safety of HRT took another body blow from more research the following year. In 1976, breast cancer was added to the growing list of diseases that estrogen therapy caused in menopausal women. The first report of a higher risk of breast cancer in estrogen patients was especially alarming for older women, given the poor prognosis at the time for even early-stage breast cancer. Breast cancer is the most common cancer in women, and at the time Monica's cancer was diagnosed (see Prologue), 80 percent of women who got breast cancer eventually died of it.

By 1980, the number of annual estrogen prescriptions had fallen by 50 percent. But the drug makers were not about to accept defeat, or even diminished sales. Pharmaceutical companies learned that they could add progesterone to the estrogen therapy drugs to offset the risks of endometrial cancer. The addition of progesterone to estrogen, known as "opposed therapy," induced a regular bleed, helping to avoid the dangerous buildup of the uterine lining that occurred under estrogen influence, and theoretically, protected it from cancer. But no one could predict exactly what adding progesterone would do to a postmenopausal woman, because it was essentially a foreign hormone to the woman at that point in her life.

HRT prescriptions rose again within a decade. Elizabeth Watkins, author of *The Estrogen Elixir: A History of Hormone Replacement Therapy in America*, connected this sales resurrection to new papers showing that adding progestin to HRT prevented precancerous microscopic pathologic changes. Watkins also reviewed the increasing promotion of estrogen as preventive medicine by researchers, doctors, and drug companies. In the 1980s, drug companies seized on potential long-term benefits as epidemiological evidence emerged for a reduced risk of osteoporotic fractures and a reduction in coronary heart disease.

The rise in sales continued, and in 1992, Premarin was bestowed the honor of the most frequently prescribed drug in the United States, and remained in first or second place for another eight years. Sixty-two million prescriptions for HRT were written in the year 2000 in the US. With considerable boost from the drug companies, ubiquitous prescriptions for HRT were considered good medical practice.

In July 2002, the HRT Titanic hit the iceberg. Government health officials ended a Women's Health Initiative trial on hormone therapy when it was revealing that women given a combination of estrogen and progestin hormones, trade name Prempro, have an increased risk for a deadly trifecta--breast cancer, heart disease and stroke. A subsequent study published months later concluded that women taking estrogen-only pills are more susceptible to ovarian cancer.

The effects of the estrogen/progesterone combination on breast cancer had been unclear through the 1970s. But in the 1990s suspicions about the increased risks of breast cancer due to estrogen and progesterone received research confirmation. For example, a 1989 Swedish study found that after nine years' use, women using estrogen-only hormones had nearly twice the rate of breast cancer compared to women not using hormones, and the risk increased with the addition of progesterone, so that women using a combination of estrogen and progesterone had four times the rate of breast cancer compared to nonusers.

In Sept. 2017, another study was published by the same Women's Health Initiative that showed, as of 2014, women who took HRT had no higher risk of death than a control group that took a placebo. While that was widely heralded as exonerating HRT, the author of the new study wasn't as enthusiastic. The new study did not contradict the findings that HRT did increase the incidences life altering diseases like strokes, heart attacks, blot clots, and cancer.[22]

Cancer and the Pill

The year before Monica was diagnosed with breast cancer a study was published, funded by major international pharmaceutical companies, claiming that oral contraceptives pose no risks of breast cancer or other ill effects. The esteemed Dr. Samuel Epstein, Professor of Environmental Medicine, University of Illinois at Chicago School of Public Health and Chairman of the Cancer Prevention Coalition, issued this press release in response.

> "While the study was alleged to be the largest ever conducted, it was both small scale and insensitive. The study was based on 23,000 healthy women who had "never used" the pill since 1968 and who were subsequently followed up over a 25-year period. The average age of women at termination of the study was only 49, an age when breast cancer is relatively uncommon.
> "In contrast, a 1996 large scale international collaborative analysis of some 54 epidemiological studies, based on over 53,000 women with breast cancer and published in The Lancet in1996, demonstrated that use of the pill starting in adolescence increased risks of breast cancer by 60 percent. These risks are clearly underestimates as reflected by the authors' recognition that 'there is little information about use that ceased more than 20 years ago,' a latency much too short to preclude further major increases in breast cancer rates. Reliance on studies based on such short latencies would have exculpated the carcinogenicity of asbestos, besides the majority of other recognized human carcinogens. Other better designed and well controlled studies have reported much higher risks of breast cancer for women starting use of the pill in their teens or early twenties, especially with use before a full term pregnancy and subsequent prolonged use, and among women with a family history of breast cancer.

> "Moreover, the claim that the current low-dose synthetic ethinyl estradiol pill is much safer than the high-dose mestranol pill used in the 1960's and 1970's is misleading as the former is more potent than the latter, besides being some 40-fold more potent than natural estradiol; additionally, ethinyl estradiol, unlike mestranol, binds to estrogen receptors in the breast. Furthermore, the modern pill is used for much longer periods, often from menarche to menopause, than was the case with the earlier high-dose pills.
>
> "Of related interest, it should be noted that the incidence of estrogen-dependent breast cancers, particularly among post-menopausal women, has increased by 130 per cent from the mid 70's in sharp contrast to only a 27 per cent increase in non-estrogen dependent cancers. This may well be relevant to the risks of the pill as a major source of incremental estrogen exposure. Clearly, unqualified claims on the safety of the current pill reflect interests of the pharmaceutical industry rather than scientifically well-based concerns on women's health."[21]

In a 2009 study[12] of over 50,000 African-American women, epidemiologist Lynn Rosenberg from Boston University, found a 65 percent increase in an especially virulent form of breast cancer among patients who had ever taken the birth-control pill. The risk doubles among those who had used the pill in the previous 5 years and had taken it for longer than a total of 10 years. In 2005, the World Health Organization classified hormonal contraceptives as a group 1 carcinogen, the same class as asbestos and radium.

Dr. Angela Lanfranchi, a surgeon specializing in breast cancer and co-founder of the Breast Cancer Prevention Institute, points to several studies in a public warning about "the pill." A Mayo Clinic meta-analysis found that breast cancer risk increases 50 percent among women taking oral contraceptives four or more years before a full-term pregnancy.[14] The significance of this finding is that this is the most common demographic of birth control pill users in the US. It appears that until pregnancy, breast tissue is more susceptible to cancer until it becomes "stabilized" by the childbearing process. In 2009, another study found that women who took the pill before age 18 have their breast cancer risk nearly quadrupled. An even more ominous analysis from Swedish oncologist Hakan Olsson concludes that pill use before the age of 20 increases a woman's breast cancer risk by more than 1,000 percent, i.e., ten times.

Hormonal contraception pills are essentially the same drug as HRT and carry at least a similar overall increase in breast cancer risk for users, about 25-30 percent. Nonetheless, strong cultural influences help push assumptions that newer, modern birth control pills are safe. In the US at least, birth control pills continue to be touted as harmless and even healthy. WebMD states, "For most women, especially young women, experts say the benefits of birth control pills far outweigh the risk." The American Society for Reproductive Medicine has a page on their website detailing the non-contraceptive benefits of birth control pills, but none of the adverse risks. The National Cancer Institute states (and the American Cancer Society makes a similar statement), "A number of studies suggest that current use of oral contraceptives appears to slightly increase the risk of breast cancer, especially among younger women. However, the risk level goes back to normal 10 years or more after discontinuing oral contraceptive use." My daughter doesn't derive much consolation in having her risk of breast cancer go back to normal after she's already had a double mastectomy. As Dr. Lanfranchi points out, is there any other circumstance where a Group One carcinogen would be prescribed en masse to a large segment of the population?

The safety of birth control pills was again contradicted by the latest and one of the largest studies ever done published in Dec. 2017, involving 1.8 million women from ages 15 to 49, followed for an average of 11 years. Researchers found that after only one year of taking "the pill," risk of breast cancer increased 10 percent. After ten years of use, the risk had increased 40 perent.[15] It is worth noting that this study was even funded by a large manufacturer of hormonal contraception, the Novo Nordisk Foundation.

A Newsweek columnist gushed that the pill "did more good for more people than any other invention of the 20th century."[13] It apparently didn't do that much good for two of my sisters, both of whom had mastectomies after being stricken with breast cancer following several years of birth control pills.

That the incidence of breast cancer in the United States not only stopped rising, but fell 11 percent from 2002 and 2004, subsequent to the dramatic decrease in estrogen sales, shows the extent of the morbidity and mortality caused by propaganda from drug companies--millions of preventable deaths, just in the years that I have been following the estrogen issue.

When a company wants to market a new drug that they know or suspect might be dangerous, it is often a matter of just running the numbers. If fines and lawsuits are likely to be less than projected profits, there's a good chance that drug will be submitted for FDA approval, will get the approval, and will be heavily marketed. Such is the story of Yaz.

Wanting to get in on the lucrative birth control market, Bayer released Yasmin birth control pills in 2001, and later Yaz, both of which contained a new synthetic progesterone, drospirenone. There was no particular advantage in Bayer's new formula, other than it could be used as justification for getting FDA approval in an already crowded market. Bayer promoted Yaz and Yasmin heavily as less likely to cause bloating and weight gain. Soon Yaz became the most popular birth control pill in the US, the result of a Bayer advertising tour de force, "Beyond Birth Control," echoing the original message of the HRT campaign of decades ago—an all-around, life-enhancing elixir—claiming it could not only prevent pregnancy, but make acne disappear, prevent bloating, and minimize the depression and anxiety associated with both premenstrual syndrome (PMS) and the controversial condition of Premenstrual Dysphoric Disorder (PMDD).

Internal Bayer e-mails show officials of Bayer engaged in an extensive Yaz public relations campaign that was "designed to circumvent FDA restrictions on marketing" because they saw the limits "as a threat to the commercial success of Yaz," according to court documents from Yaz lawsuits.[16]

Bayer devised a strategy to hire Judith Reichman, a Los Angeles-based gynecologist who writes a blog about women's health issues, "to engage in off-label promotion" of the Yasmin line of contraceptives. Bayer hired Reichman to the tune of $450,000 just for her willingness to "mention off-label benefits of our products." Coincidently, Dr. Reichman was preparing a soon-to-be published book on women's health. Bayer planned a purchase of 10,000 copies of the book, which contained "off-label claims" about the Yasmin line of contraceptives. The purchase was part of a deliberate strategy to have the book magically appear on the New York Times bestseller list. Bayer also secured a key article in the women's magazine *Allure*, touting extensive benefits of Yaz.

Company e-mails revealed Bayer sales managers promoted to their sales force, use of Yaz as a treatment for all types of PMS when it had only been FDA approved for the most severe form. Despite the FDA forcing Bayer to spend $20 million to correct its unfounded advertising assertions regarding non-contraceptive benefits, Yaz remained the nation's best-selling oral contraceptive and Bayer's second best-selling product, with sales of over $1.5 billion in 2010.

Post marketing, FDA required studies on Yaz, which were conducted by Bayer and supposedly showed Yaz and Yasmin had no higher rates of blood clots than other birth control pills. However, five other non-Bayer funded studies, whose existence was known to Bayer, found something much different, a 50-to-75 percent increased risk for those taking these birth control pills in comparison to others. And one study recorded a 630 percent higher incidence of blood clots.[17] Former FDA commissioner David Kessler accused Bayer of deliberately concealing data about this early on in order to stack the deck for the drugs' approvals.[18]

The FDA called an advisory committee to consider pulling birth control pills containing drospirenone off the market, but the panel voted 15 to 11 that the drugs' benefits outweighed the risks. The Project on Government Oversight (POGO) investigated the advisors on the FDA panel and found that three had research or other financial ties to Bayer, including the panel's chairman. Another advisor had financial connections to manufacturing the generic version of these pills. All four voted for continued approval for Yaz and Yasmin.[19]

The FDA has reported 50 deaths from Yaz between 2004 and 2008, and 11,300 patients have filed Yaz lawsuits against Bayer for outcomes like blood clots, strokes, heart attacks, and gall bladder disease. Bayer has paid $1.5 billion to settle thousands of those lawsuits as of July, 2013.[20]

In 2008, the federal government fined Bayer $97.5 million for orchestrating a blatant kick-back scheme with Liberty Medical Supply, paying bonuses for every patient switched to Bayer diabetic supplies from their competitors. Bayer was forced to enter into an "integrity agreement" (more about that in Chapter 3), with the Department of Health and Human Services. Entering this agreement means Bayer can't be trusted, just like all the other drug companies that were forced to enter those agreements. Bayer has pled guilty to felonies and incurred massive fines—-up to $250 million.

This is what happens in America when a corporation kills or harms us. Despite having all the rights of people, corporations aren't held responsible like people are. We don't throw corporations in jail, and the executives who run those corporations, seldom, if ever, are held accountable.

As of 2012, sales of Yasmin and Yaz had totaled about $6 billion. With all the lawsuits filed this may turn out to be an exceptional case where the profitability of marketing a dangerous drug failed to cover the costs from all the harm done to tens of thousands of patients

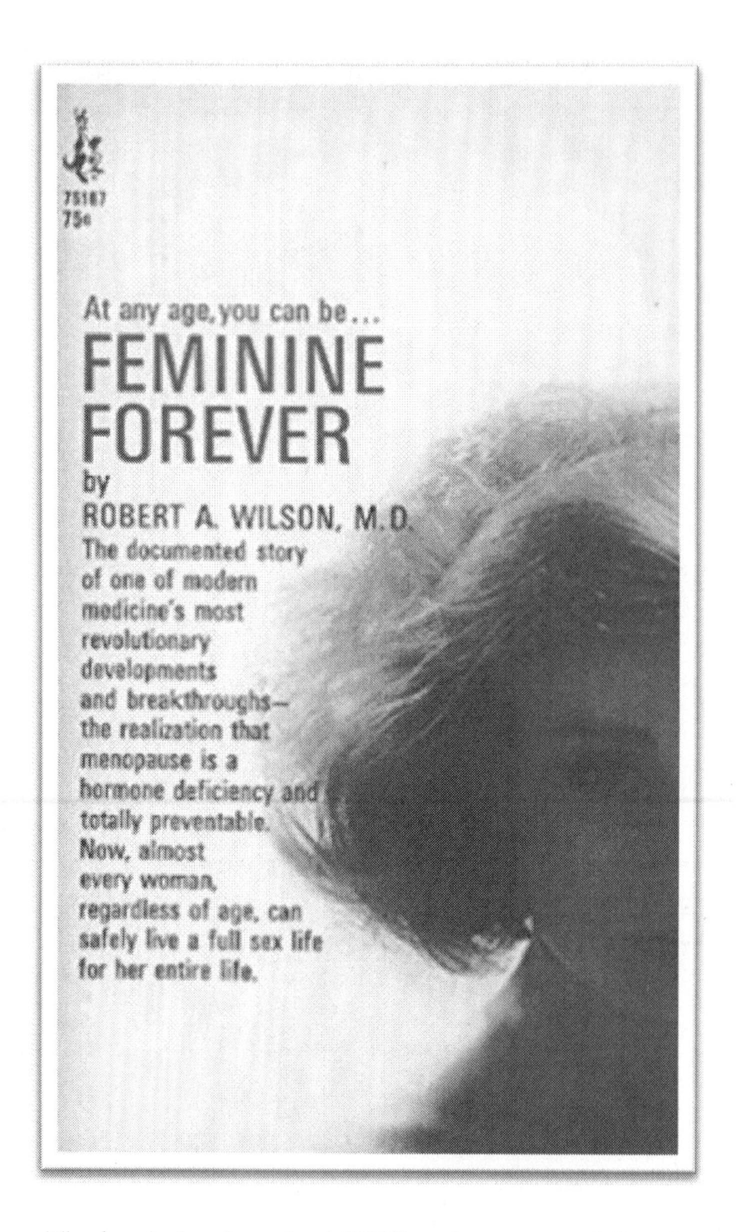

The book that launched HRT as the supposed female fountain of youth

3

The Pharmaceutical Industry, Part II:

Overdosing America

"No one should approach the temple of science with the soul of a money changer."

—-Thomas Browne

The Profit Motive

It is the end of a typical week in the operating room for me. I have provided anesthesia--a varying mixture of unconsciousness, amnesia, pain relief, and life support for 20 operations, including a wide variety of surgical procedures on just about every part of the human body. Organs were taken out, joints replaced, diseases cured, and death likely postponed. All of these patients lived in my city, most came from middle class backgrounds, most were employed and had health insurance. But several of these patients are walking billboards for a crisis that has gripped the entire country. Before they were operated on, they were already addicted to opioids. One was an overt heroin addict, three were addicted to prescription oral narcotics that they had been prescribed by their family doctor or "pain specialist" for various types of chronic pain. One patient was taking 24 different prescription medications. To varying degrees they are all beneficiaries and also victims of the pharmaceutical industry.

This may come as a surprise to some people, although it shouldn't. Pharmaceutical companies are not structured to make us live longer, treat chronic diseases, make us younger, healthier, happier, better looking, or deliver us from sexual impotence, real or imagined. There is no benevolence to their business model. They operate solely to make money. And make money they do, a lot of it. Big Pharma is extremely profitable. And in the course of pursuing profitability, the industry as a whole has shown shocking disregard for human life.

Like the pied piper, Big Pharma has led American doctors to practice a historically unprecedented drug-intensive style of medicine. The drug manufacturers have lured doctors and their patients to an intoxication with the idea that a drug exists for every ailment and discontent, even when changes in lifestyle would be more effective. And they believe that the newest, most expensive brand-name drugs are superior to older drugs or generics, despite those new drugs having a higher rate of adverse drug reactions.

The US is 49th in the world in life expectancy[5] despite Americans taking more prescription drugs per capita than any other country. Spending on prescription drugs more than doubled between 1999 and 2008. Nine out of ten adults over 60 are on a prescription drug,[6] as are one out of every four children and teenagers in the United States.[7]

The CDC notes that the number of Americans taking more than five prescription medications has gone up about 300 percent in the last 20 years. Over one third of Americans over 55 are taking at least five drugs. Almost one in ten were taking more than ten prescription medications.[62] I have seen patients on as many as 25 different prescription drugs, not counting OTC vitamins and supplements. When I was in medical school I was told that by the time a patient is on five different drugs there is virtually a 100 percent chance of having an adverse drug interaction. One doctor observed, "There is no scientific basis for treating older folks with $300+/month meds that have serious side-effects and largely unknown multiple drug interactions."[10]

"Polypharmacy" refers to the use of more medications on a patient than is medically necessary. Several studies have made credible estimates that at least half of older Americans are taking at least one unnecessary prescription drug.[63] There is a well-established connection between polypharmacy and adverse clinical outcomes.

The overuse of prescription drugs carries a staggering price tag. In 2005, it was estimated that adverse drug events were responsible for 4.3 million healthcare visits. Consumer Reports states, "Almost 1.3 million people went to U.S. emergency rooms due to adverse drug effects in 2014, and somewhere between 124,000 and 200,000 died from those events,"[11,62,64] based on data from the CDC and FDA. That makes it the fourth leading cause of death, on par with strokes. About $200 billion is spent annually on all this unnecessary medicine and its consequences.

How did Americans become the biggest pill poppers in the world? If the theme of this book is the stranglehold of corporations over every aspect of our lives, there is no better example than Big Pharma.

Taking advantage of a permissive regulatory landscape, a culture of "what's good for business is good for the rest of us," and good old fashioned bribery—-euphemistically called "campaign contributions" —-Big Pharma adopted a multi-pronged approach involving advertising directly to consumers, infiltrating the world of physicians top to bottom, and assembling a literal attack army of government lobbyists to make sure that state and federal governments serve their interests. From 1998 to 2018, the pharmaceutical industry spent nearly $ 4 billion on government lobbying and campaign contributions, more than any other industry, and double what the fossil fuel corporations spent.[65]

About 20 years ago, the United States became one of only two major countries (the other is New Zealand) where the $600 billion drug industry can advertise directly to consumers. With soothing musical jingles and smiling actors, patients are constantly told to "ask your doctor if [the latest obscenely expensive drug] is right for you." In a world turned on its head by Big Pharma, patients tell their doctors what drugs they should be on.

Much of that is thanks to spending over $6.5 billion on direct-to-consumer advertising and $24 billion spent on various marketing strategies to doctors.[62] For the drug companies, the returns from being able to directly market to consumers and camp out in doctors' offices have been spectacular. For public health, the results have been spectacularly disastrous.

Disease Mongering

While many laud the age of the internet as one that empowers and informs patients and challenges imperious doctors as unique medical authorities, it has also led to patients finding fellow sufferers and the creation of patient advocacy groups that legitimize and institutionalize common, poorly defined symptom complexes as real diseases. And drug companies couldn't be happier about it.

Joe Dumit, Associate Professor of the Anthropology and Science-technology Studies' Programme at the Massachusetts Institute of Technology, has been studying how patients with controversial illnesses like chronic fatigue syndrome or multiple chemical sensitivity syndrome obtain research funding from such institutions as the NIH (National Institute of Health). Dumit found that what made the most difference was the backing of drug companies, who wanted in on the market to treat the disorders. Corporations can make a lot of money convincing basically healthy people that they're not healthy enough, or that they are, in fact, ill, or will shortly become so.

It's tempting to think that patients becoming more of a driving force in their own health care should be a net positive. But at the same time, the door has been opened for the pharmaceutical companies to encroach deeper and deeper into patient care. "Disease mongering" is a term coined by medical journalist Lynn Payer.[59] It describes the strategy of turning the physical and mental vicissitudes of life and the aging process into pathologies, especially ones that can be targeted for drug treatment. Unhappiness, boredom, ugliness, shyness, heart burn, bone thinning, brain thinning, obnoxious personality, and lack of sexual desire can all be turned into a disease and monetized.

In the April 1, 2006, issue of the famed *British Medical Journal*, journalist Ray Moynihan published an article entitled, "Scientists Find New Disease: Motivational Deficiency Disorder."[60] In the article scientists described identifying a new disease whose hallmark symptom was extreme laziness. The article gave the disorder an acronym, MoDeD, motivational deficiency disorder, and claimed that 20 percent of the population suffered from it. It said that extreme cases can lead to death because, "the condition diminishes the motivation to breathe."[60] The diagnosis was made by identifying a motivational rating using a "positron emission tomography." The article said that a biotechnology firm was conducting clinical trials of a drug to treat MoDeD— "indolebant, a canabinoid CB1-receptor antagonist." Of course the entire article was a sarcastic ruse, MoDeD, indolebant, and canbinoid antagonist were all bogus terms, but it was a brilliant mockery of disease mongering.

Vague, ill-defined chronic conditions, like fibromyalgia, have symptoms of fatigue and nonspecific aches and pains, difficult to distinguish from those that essentially normal people have periodically. These chronic "diseases" real or not, have become a gold mine for drug companies in a culture where there is a drug to treat everything that might ever happen to you, including the normal symptoms of aging. And your defense from such maladies is a cabinet full of pills, and a second cabinet full of pills to treat the side effects of the first cabinet full of pills. Whatever happens to you, drug corporations are happy to sell you a pill to fix it, and if your doctor is unaware of the cure, thanks to the drumbeat of direct-to-consumer advertising, you can enlighten him or her about the prescriptions that you need to walk out the door with.

Baldness can cause panic attacks, anxiety, and even job loss according a study funded by Merck, who just happens to make a supposed cure—Propecia. GlaxoSmithKline (GSK) came up with a drug, Lotronex, that treated irritable bowel syndrome (IBS), something that used to describe normal, occasional abdominal cramps. In Vivo Communications helped GSK come up with a multi-year "education program" to change doctors', nurses', pharmacists', and patients' perception of IBS as a "credible, common and concrete disease." [1]

Lotronex was withdrawn from the market because of serious and sometimes fatal adverse drug reactions. But now, according to what has to be the most irritating drug ad campaign of all time, featuring a flaming red haired obnoxious actress dressed up in a flesh colored spandex costume personifying your colon, you can treat IBS with Viberzi. According to its manufacturer, 45 million people suffer from IBS, and plenty of ads for drugs to treat IBS can be found on the website of the Irritable Bowel Syndrome Self Help and Support Group.

Almost 3 million children are on Lipitor, the anti-cholesterol drug that is increasingly suspected of causing a wide range of neurologic pathologies like Lou Gehrig's disease. (It turns out nerve cells require cholesterol to function). More than 20 percent of all American adults are taking at least one drug for "psychiatric" or "behavioral" disorders, as are a shocking one third of all foster children.[8]

Drug companies have convinced elderly women that they will all die from osteoporosis by next Tuesday. A World Health Organization meeting involved in setting guidelines for diagnosing osteoporosis was funded in part by three drug companies who stood to benefit from any treatment campaign.[2] But a typical drug advertised as a woman's best protection from fractures and bone loss, Fosamax (alendronate), only reduced the risk of vertebral fracture from 3.8% to 2.1%. While that can be marketed as a 44 percent relative risk reduction, it is only an absolute risk reduction of 1.7 percent. The cost of Fosamax is about $80/month. A newer drug, Forteo, can be over ten times that amount, $900/month.

If you watch much television you would think that America is collapsing from within--not by government corruption, environmental destruction, or income inequality, but by erectile dysfunction. An advertisement depicts an unhappy couple on the younger side of middle age, lying in bed several feet from each other, with the text: "Erection problems: hard to talk about, easy to treat." Only 3 percent of men in their 40s suffer any kind of real erectile dysfunction,[3] but what normal, red blooded American male can resist the allure of a more rock-hard penis, especially when he assumes every other male is similarly endowed and enhanced?

Rather than researching drugs to treat diseases, disease mongering can take the form of drug companies promoting diseases to fit their existing drugs. Such a market is huge and easily expanded. For example, a senior marketing executive advised his sales representatives on how to expand the use of an addictive, anti-seizure drug, Neurontin (Gabapentin): "Neurontin for pain, Neurontin for monotherapy, Neurontin for bipolar, Neurontin for everything."[4] Neurontin's promotion as an all-purpose tonic is as absurd and as unscientific as Geritol, the sponsor of the 1950s Ed Sullivan Show. Today's drug companies are just as capable of committing fraud as the proverbial 19th century traveling snake oil salesmen.

Pfizer, the maker of Neurontin, pleaded guilty to illegal marketing and agreed to pay $430 million to resolve the associated criminal and civil charges against it. But that was a mere fraction of the $2.7 billion in Neurontin sales in 2003, pumped up by paying "academic experts" to attach their names to articles promoting Neurontin for everything but the kitchen sink—bipolar disease, post-traumatic stress disorder, insomnia, diabetic neuropathy, restless leg syndrome, hot flashes, migraines, tension headaches—and by funding conferences at which all these uses were promoted.

Big Pharma specializes in cannibalizing their competitors' market. If a profitable disease is being treated by a competitor, they will come in and redefine the disorder to suit one of their products.

GlaxoSmithKline was late to the marketplace with the anti-depressant Paxil, so they succeeded in convincing patients that shyness was "social anxiety disorder," and of course, it yields the wonderful acronym—SAD (a bedrock term of President Trump's tweeting).

Paxil was the first medicine approved to treat not just depression, but social anxiety disorder. Never mind that no one has any idea how Paxil may work on SAD, if it does, but a pill to treat SAD has limitless potential. Nervous on a first date? Rub a little Paxil behind your ears with your perfume. Have to give a talk in church? Sprinkle some Paxil on your sacrament wafer. Don't like to raise your hand in class? Smoke 20 mg of Paxil in your e-cigarette. My "social anxiety disorder" occurs only when certain relatives come to visit. So it makes sense to me that I should treat it by putting Paxil in the candy dish for them, and go upstairs and lock myself in the bedroom when they ring the doorbell. In the 1990s GlaxoSmithKline tried to sell the idea that one in every eight Americans has social anxiety disorder.[61]

The new disease and the drug to treat it are forever bound together in the minds of patients, and even physicians. The drug company Roche wanted to get in on the shyness market, so they changed social anxiety disorder into a soul-destroying condition called "social phobia," and they just happen to have a cure, Aurorix. Cymbalta was branded as the drug of choice if a patient had both depression and pain. Are you frustrated having to wait in line, or getting put on hold on the phone.? Then you must have adult Attention Deficit Disorder, and thank God there's Strattera, made by Lilly, to treat that.

In a trend unique to America, the rate of children being diagnosed with bipolar disease has increased forty times between 1994 and 2003.[9] The rush to dope up our children with drugs for anxiety, bipolar disorder and depression can be traced to a federal law passed in 1997, allowing pharmaceutical companies owning the patent for a medicine, an extra six months of exclusive marketing rights by testing its efficacy in children. There is little incentive to do such research on drugs that are not patented, so you can guess which drugs are heavily pushed for use in children. The FDA cannot require manufacturers of a new drug to study its effects in children before approval, and doctors don't need FDA approval to prescribe the same drugs for children that they do for adults. The result is that children's bodies are astutely perceived as potential cash cows for the makers of psychotropic drugs.

Acknowledging at least some controversy over giving children psychotropic drugs, the Harvard Health Publications from the Harvard Medical School published a newsletter entitled, "Should Children Take Antidepressants?" It stated, "The American Academy of Child and Adolescent Psychiatry has established a Pediatric Psychopharmacology Initiative. A group of educators, child psychiatrists, developmental psychologists, and pharmaceutical company representatives will monitor controlled trials, set consistent standards, and promulgate guidelines for researchers and prescribing physicians."

So ask yourself, "Why are pharmaceutical company representatives part of that group? Just how likely are the drug companies to add objectivity, and bring 'non-drug' therapies to such an endeavor?" Americans' recent fascination with doped up zombies is a reality show playing out in front of their own mirrors.

It is currently illegal for a drug to be promoted by advertising before it is available on the market. But drug companies can promote a new disease for which they will just coincidently have a cure. And they often promote the disease for as long as ten years before their "cure-all" drug comes to market. This was the strategy used for such disorders as restless leg syndrome and supposedly low testosterone, given the catchy name "low T" (low testosterone) by a drug company's marketing team. Drug companies spent $153 million on advertisements promoting low T to consumers in 2013. The FDA may control how a drug company advertises their drug, but they are not empowered to control how that drug company can promote a disease.

You have likely seen online or in print, drug company advertisements that include quizzes on asking the reader if they have any of a list of symptoms, with questions like, "Are you a middle-aged man who sometimes feels tired? Has your athletic ability declined at all since your early 20s?" Abbott Laboratories, manufacturers of Androgel, are setting you up for diagnosing yourself as suffering from low T. The maker of Androgel, AbbVie, and five other companies that make testosterone, are now being sued by over 6,000 people who have had predictable and serious adverse outcomes from taking unnecessary and inappropriate testosterone--like heart attacks and strokes.

A quiz[58] for the disorder of "pseudobulbar affect" (PBA), or "emotional incontinence" (sounds dangerous and disgusting doesn't it?), asks you to response to questions like, "Others have told me that I seem to become amused very easily or I seem to become amused about things that really aren't funny," and "I find myself crying very easily." "There are times when I won't be thinking of anything happy or funny at all, but will suddenly be overcome by funny or happy thoughts." Based on questions like that, who is not going to have pseudobulbar affect? Especially when there's such a wonderful scientific name for whatever the hell disease this is, and even more so when you start worrying that other people might think you suffer from, God forbid, "emotional incontinence." Then you start thinking to yourself that if you have emotional incontinence you might have to start wearing an adult diaper around your mouth to prevent embarrassing emotional "leaks." How dreadful!

Well then, if these things happen to you, thank heaven and earth for Nuedexta, which you can take to avoid that catastrophe. Of course, first ask your doctor if Nuedexta is right for you.

To be sure, people suffering from multiple sclerosis, brain tumors, a stroke, ADHD, Parkinson's, or several other neurologic diseases can be emotionally labile, and whether Nuedexta can be helpful in treating PBA from these disorders without significant side effects is not the point. The point is that with this kind of marketing campaign, many people with merely eccentric personalities can be duped into thinking that they have a disease that needs treatment from a drug.

The FDA permits an expedited approval process for drugs intended to treat "rare" diseases, diseases that affect fewer than 200,000 people. But drug companies can manipulate that rule to their advantage by subdividing more common diseases into categories that magically have less than 200,000 victims, and can therefore be granted an expedited rush to market. And they can do the opposite as well. They can push to broaden the criteria for diagnosing a disease and increase the potential customer base for a drug.[57] In fact the diagnosis, definition, and treatment of illness by the medical community is being shoved aside by corporate management of disease.

Crime Does Pay

A standard marketing approach for drug companies seeking to meet "sales quotas" is to send drug reps to doctors and push them to prescribe drugs for off-label use, which, although legal, raises obvious questions of ethics, efficacy, and magnification of side effect risks. The end result is Americans take epilepsy seizure drugs for pain, antipsychotics for the blues, and an antidepressant for knee pain--all because of marketing and sales quotas. This is another variation of "disease mongering."

A deadly example is Cephalon's painkilling fentanyl lollipop, Actiq, which is loaded with fentanyl, the potent pain killer that I use only as a supplement to general anesthesia and for the first few hours of post-operative pain, which are almost always the most painful part of post-operative convalescence. The product was only approved to treat terminal cancer patients in chronic pain who are already on an opioid drug, because life-threatening conditions can occur at any dose in patients without a chronic, build-up of tolerance for narcotics. With the pressure to meet their quotas from supervisors breathing down their necks, Cephalon sales reps were regularly sent to doctors who treated no cancer patients, with free coupons to pass out to patients with common problems like migraines, back pain, and arthritis. A study by Prime Therapeutics found Actiq was prescribed off-label nearly 90 percent of the time. You could go on line right now and get a free coupon for a "reduced price" on your fentanyl lollipop[12] made even easier if you claimed the lollipop was for your pet.

Crime pays for Big Pharma. Like Actiq, Pfizer's Neurontin sales of $2.7 billion in 2003 alone were about 90 percent for off label, i.e., inappropriate if not illegal, use. When Pfizer was fined $2.3 billion for off-label use of four other drugs in the same year, the company entered into a "corporate integrity" agreement with the US Department of Health and Human Services to detect and avoid such problems in future. Pfizer had previously entered into three such agreements in the past decade. That's like offering Bonnie and Clyde a "bank integrity" agreement after their first three bank robberies. Better yet, we could have avoided WWII if we had just offered Hitler a "Europe integrity agreement" after he invaded France and Poland.

As of July, 2012, only one of the top ten drug companies, Roche, was not currently operating under such an "integrity agreement." But before you invite Roche executives over for a congratulatory beer and pleasant conversation, know that over 10 years in the 1990s, top executives at Roche were leading a vitamin cartel that, according to the US Justice Department, was "the most pervasive and harmful criminal antitrust conspiracy ever uncovered."[13] Roche agreed to pay $500 million to settle charges, equivalent to about one year's revenue from its US vitamin business.

Meeting a sales quota is certainly one of the reasons prescription pain killer overdoses are a deadly and explosive epidemic killing 60,000 people a year (as of 2017),[14] nearly eight times more than in 1999, with 500,000 ER visits and costing health care insurers $72.5 billion annually.[15] More on that later.

Stacking the Research Deck and Making America Crazy

The pharmaceutical industry has the largest political lobbying force in the United States, and it pays off like this. If you looked into the medicine cabinet of Americans you would think that the entire country is stark raving, straightjacket-worthy crazy. The highest selling class of drugs in 2009 was anti-psychotics.

Marcia Angell, who for twenty years held the title of editor-in-chief of the New England Journal of Medicine, the world's most prestigious medical journal, wrote in 2009, "The boundaries between academic medicine—medical schools, teaching hospitals, and their faculty—and the pharmaceutical industry have been dissolving since the 1980s, and the important differences between their missions are becoming blurred. Medical research, education, and clinical practice have suffered as a result."[16] The most profitable drugs are usually knock offs of existing profitable drugs of competitors, rather than the result of painstaking innovative breakthroughs.

Drug companies do educate doctors, but only as a means of selling more drugs. Dr. Angell writes, "Drug companies don't have education budgets; they have marketing budgets from which their ostensibly educational activities are funded." My wife commented to me not long ago that while she was in the waiting room to see her doctor, during the 45-minute wait, no less than three drug reps came into the doctor's office to ply their trade. If you go to a psychiatric hospital and ask for a pad of paper, it will likely have at the top the pre-printed names of antidepressant or psychotropic drugs, blurring the line between the care of patients and marketing by the drug companies. It's almost like doctors have their sponsors' logos sewed onto their white coats like NASCAR drivers.

Often, doctors are being spoon fed selected and manipulated information, prompting a belief that drugs are far more effective and safe than they really are, adding to the surge of illicit sales of drugs.

Dr. Angell continues, "Unlike basic medical research, funded mainly by the National Institutes of Health (NIH), most clinical drug trials are funded by the pharmaceutical industry." You don't have to be a genius to see the potential for a conflict of interest. Industry-supported research is far more likely to be favorable to the sponsors' products than is tax-payer, i.e. NIH-funded research. Many drugs were integrated into mainstream medical practice purely on the backs of company-sponsored drug studies without corroboration by independent research. Most published trials funded by drug companies show positive results and, to save money, are conducted overseas—"on sick Russians, homeless Poles, and slum-dwelling Chinese"—in places where regulation is virtually nonexistent.[17] In fact, 90 percent of new drugs approved in 2017 were only tested on overseas populations.[66] A meta-analysis of 1,300 clinical trials showed that drug studies on overseas populations, on average, showed more positive results for the drug companies than those using domestic patients.[67]

For a single drug, the results of many trials are submitted to the FDA. If only one or two are positive, in that they show effectiveness in treating a problem without significant side effects emerging during the trial—the drug is usually approved, even if negative trials far outweigh the positive. Making matters worse, drug studies that show negative results usually are not published at all.

Anti-depressants are at the top of the list of particularly suspect drugs in their therapeutic efficacy. Dr. Angell elaborates, "a review of seventy-four clinical trials of antidepressants found that thirty-seven of thirty-eight positive studies were published. But of the thirty-six negative studies, thirty-three were either not published or published in a form that suggested a positive outcome, perhaps addressing a different symptom."[68]

Reviews of every drug/placebo clinical trial submitted for approval of the six most widely used antidepressants approved by the FDA between 1987 and 1999—Prozac, Paxil, Zoloft, Celexa, Serzone, and Effexor[18]—showed that on average, placebos were 80 percent as effective as every one of the drugs, meaning the drugs were found to be nearly worthless for depression. With favorable results having been published and unfavorable results buried (in this case, within the

FDA), the public and physicians had been duped into thinking they were beneficial.

Trials that might reveal unacceptable side effects can prematurely end after a short-term trial. Like with environmental toxins, the latency period between exposure to dangerous pharmaceuticals and appearance of symptoms or evidence of disease can be decades, and those affected the most can be subsequent generations. The hideous birth defects that appeared in children whose mothers took thalidomide during pregnancy is just one of the starkest examples.

It is hardly unexpected that drug-company-funded clinical trials all too often produce favorable results for their funders. The sponsor's drug may be compared with another drug cleverly administered at an ineffective low dose making the sponsor's drug appear more efficacious by comparison. A drug intended for the elderly can be tested in much younger, healthier people, reducing the chance of exposing adverse side effects. A new drug can be deliberately paired with a placebo instead of an existing competing drug, leaving largely unanswered the question of whether the new drug brings a benefit to patient care and improved health. Those who would excuse this compromised state of affairs claim that attempts to regulate such conflicts of interest will slow medical advances. But that rationalization obscures the truth that drug companies' stacking the deck undermines legitimate medical research and further entrenches a drug intensive practice of medicine.

Many members of the eighteen standing committees of experts that advise the FDA on drug approvals also have financial conflicts of interest with the pharmaceutical industry. Of the 170 contributors to the most recent edition of the American Psychiatric Association's Diagnostic and Statistical Manual of Mental Disorders (DSM-IV), more than half had financial ties to drug companies, including all of the contributing authors to the sections on mood disorders and schizophrenia.[19]

If you ask executives at America's top pharmaceutical corporations about the high costs of prescription drugs, they'll tell you that increasingly high drug prices are needed for research and development of wonderful new, life-saving drugs. But numerous studies debunk those claims. One study, by the group Families USA, found that America's major drug companies are spending more than twice as much on marketing, advertising, and administration than they do on research and development. Furthermore, much of their "research" budget is spent on producing variations of top-selling drugs already on the market—called "me-too" drugs--drugs that bring no additional benefit to patients, whose only purpose is to siphon existing profits away from other drug companies.

Since the 1980s, academic medical centers have struggled for government funding, and drug companies have been more than happy to fill the void. Venerable ivory towers of medical research have allowed themselves to become handmaidens to the drug companies and their business model, doing their bidding in ways that would have been branded as unethical, even corrupt 25 years ago. Individual academic researchers act more like employees of the drug companies, supplying the patients for a study conceived by a drug corporation, then collecting the data specified by the corporate sponsor. Dr. Angell writes that it is commonplace for the company sponsors to "Keep the data, analyze it, write the papers, and decide whether and when and where to submit them for publication. In multi-center trials, researchers may not even be allowed to see all of the data"[69] that eventually may have their name on it.

Beyond grants, academic researchers may now graze at a full smorgasbord of financial ties to the companies that sponsor their work. They can prostitute their name and reputation by becoming company consultants for the products they evaluate, become highly paid speakers at elegant dinners where they cheerlead their clinician colleagues towards writing more prescriptions, enter into patent and royalty arrangements, they may allow their name to be slapped on corporate-conducted research, attach their "brand" to drugs and medical devices at corporate seminars and events, and they may also own stock in the companies whose products they promote.

When nine nationally prominent cardiologists recommended statins be prescribed for cholesterol at a much lower threshold than previous guidelines, essentially doubling the number of people who would be recommended for treatment, only much later, and with only a fraction of the media coverage, did we find out that eight of them had financial ties to the drug companies that stood to benefit.[20]

Perhaps most important, there is no significant source of funding to challenge or provide balance to corporate-sponsored research. Researchers are assumed to be merely truth-seeking scientists with white coats--the universal symbol of altruism, goodness and purity of motive—shining a light through the mysterious world of science that reporters couldn't possibly understand. Conclusions are drawn, and the media usually regurgitates what they are told with virtually no challenging of the study, its funding, design, data, or conclusions.

Dr. Angell concludes, "It is simply no longer possible to believe much of the clinical research that is published, or to rely on the judgment of trusted physicians or authoritative medical guidelines."[68]

Storming the Medical Ivory Towers

The pharmaceutical industry has heavily infiltrated the curriculum in American medical schools, is taking a dominant role in its relationship with the medical profession, and is having a corrupting influence on academic research into its own products.[21]

At Harvard Medical School, the pinnacle of American medicine, where I served as an instructor years ago, of the 8,900 professors and lecturers there in 2009, 1,600 admitted that either they or a family member have had some kind of business link to drug companies, sometimes worth hundreds of thousands of dollars, that could bias their teaching or research.[22]

At every level of academic medicine, including the deans of medical schools, physicians sit on the boards of drug companies, cementing a cozy relationship that doesn't carry the stigma of conflict of interest that it should and would have 30 years ago. Many academic centers are unabashed in catering to drug companies. For example, Harvard's Clinical Research Institute (HCRI), advertised itself as led by people whose "experience gives HCRI an intimate understanding of industry's needs, and knowledge of how best to meet them." Is there a problem with academic medical institutions becoming service centers for the drug industry?

Two thirds of academic medical centers hold stocks in companies that financially sponsor research at the institution. Two thirds of medical school department chairs received departmental income, and almost as many received personal income, from drug companies.

The incestuous relationship between industry and academia is the bastard child of the Bayh-Dole Act of 1980. Before Bayh-Dole, all government funded discoveries became public assets. But now, initially government-funded research will more often end up in a patent owned by a start-up, often formed by the researcher, who then parlays the financial promise of a drug to a larger company, in return for a royalty. The intent of Bayh-Dole was to accelerate shepherding of new technologies towards implementation and product development. But the Act has contributed to the disturbing evolution of academic health centers becoming "the minor leagues" for large pharmaceutical corporations.

Because of Bayh-Dole, drug companies no longer have to do their own initial, innovative research. They can sit back and rely on universities and start-up companies for that. In fact, the larger drug companies now focus on late-stage drug development using licenses, patents, and technology bought from the academic minor leagues.

We know what Big Pharma's agenda is—selling drugs for profit, and the more the better. That is hardly the same as what should be the agenda of doctors, patients, insurers, and tax payers. Academic centers have slid down the slippery slope of allowing those drug companies to dictate their research agenda. Often this "academic" research has no contribution to the public good other than to sustain the profits of a drug company facing lost revenue because a profitable drug is coming off patent. This not only distorts and often corrupts what drugs appear in the market place, it perverts every phase of the practice of health care, from the medical education system itself, to clinical practice, to the kind of medical research employed, and to the financial burden of the entire system.

Dr. Angell continues, "Academic medical research, once a noble profession for those interested in alleviating, curing and preventing disease, has all too often degenerated into a business career, facilitating the interests and profits of Big Pharma."

Cancer research is emblematic of this descent into corporate profit making. If you asked someone if they would rather have their cancer detected and treated or instead prevented in the first place, you know what the answer would be. Can you name a corporation that specializes in and sells to the public "cancer prevention?" Some healthcare networks have realized that prevention improves their bottom line. But beyond that, there is little if any money to be made with such a program or product. Eighty to ninety percent of cancer is environmentally caused, according to the World Health Organization. (Note: environmental includes smoking and diet. About the only cause of cancer that doesn't fall into the category of environmental is "genetic"). So preventing cancer through reducing exposure to culpable environmental contaminates would seem to be where the focus of research should lie. But so long as no one makes money by preventing cancer, in the current healthcare landscape, patients are largely on their own In contrast, there is a lot of money to be made in detecting and treating cancer once it takes hold of a victim. Research on treating and curing cancer is well funded, preventing cancer much less so.

Continuing education after medical school and residency is a requirement for practicing physicians in almost every state. And here, too, the drug industry has pried its foot, a very big foot, into post graduate curriculum.

Pri-Med is one of the highest profile medical-education companies. In partnership with Harvard Medical School, Pri-Med provides post graduate conferences all over the country at minimal to no cost to attendees because of heavy financial support from industry sponsors. The programs feature industry sales-pitch-style symposia combined with free meals, and then, supposedly "industry free" talks by academic faculty during the rest of the day. These "educational pilgrimages" provide a buffet of drug company propaganda while physicians enjoy a free meal, sandwiched between high powered experts from the medical ivory tower. The Harvard name is plastered all over Pri-Med's advertising and promotional materials, and Harvard Medical School and their participating faculty members receive a handsome check from the drug companies. Our standards for objectivity and integrity in the medical industry have been lowered to the point where this obvious conflict of interest has been normalized.

Your Medicine Cabinet: Danger Just One Pill Away

McNeil Consumer Healthcare, a subsidiary of Johnson and Johnson, has advertised Tylenol as "the safest kind of pain reliever" when used as directed. Maybe not. That claim seems at odds with the 150-500 people a year that die from Tylenol toxicity, primarily because it has a very narrow margin of safety. For the decade 2001 through 2010, 1,567 people died from inadvertently taking too much of it. Acetaminophen has caused more deaths than any other over-the-counter pain reliever. Annually 55,000 to 80,000 people end up in the emergency room because of the drug.[23]

Acetaminophen, the active ingredient in Tylenol, is also an ingredient in over 600 other medications taken by one in four Americans every week for colds, headaches, menstrual cramps, hay fever, and various aches and pains. It's in every medicine cabinet in the country.

The recommended dose of acetaminophen is the same as the maximum safe dose—4 gms/day in an adult. This is highly unusual, and obviously dangerous for an OTC drug. Taken over several days, as little as 25 percent above the maximum daily dose—or just two additional extra strength pills a day—can cause liver damage. In fact, acetaminophen is the nation's leading cause of acute liver failure (not be confused with the much different disease, chronic liver failure). Taken all at once, fewer than four times the maximum daily recommended dose can cause death.

When the drug is metabolized by the liver, it is broken down into potentially toxic byproducts. If the liver cannot process those byproducts it can shrivel up and fail abruptly.

McNeil has built Tylenol into a billion-dollar brand, and it is at the center of the controversy. McNeil has developed an antidote that can save lives if given immediately after an acute deliberate overdose. But it is of little help in accidental overdoses, because in those cases the symptoms of impending liver failure take several days to develop, and by then it's too late.

With remarkable success McNeil has repeatedly fought off attempts by the FDA to apply safety warnings, educate the public about risks, restrict dosages, or to otherwise regulate Tylenol's use for over 30 years. In 1977 the FDA convened a panel of experts that advised a warning must urgently be placed on all Tylenol products that it can cause severe liver damage. And presto! Thirty-two years later that actually happened.

Nonetheless, the FDA has still not completed the review of acetaminophen that began back in the 1970s, as part of the agency's larger mandate to assess the safety and efficacy of older medicines. Tylenol has been on the market for 50 years, and more than 30,000 papers have been published on the drug, yet the FDA has backed away from any reform of how it is being marketed and sold.[24]

The story of Vioxx and Celebrex is a microcosm of drug company behavior. When studies on Vioxx and Celebrex became available in 1998 and 1999, many doctors were disappointed. Neither drug alleviated pain any better than the older medicines, and the drugs cost close to $3 per pill. Over-the-counter pain relievers, in contrast, cost pennies a dose.

Merck had known of potential lethal side effects even before launching Vioxx in 1999 but had swept all the research with disturbing results under the rug. Merck suppressed data for three years that proved Vioxx caused an alarming increase in the risk of heart attacks and strokes.[25]

Estimates of deaths caused by Vioxx are as high as 500,000. Merck knew Vioxx was killing people, yet they pressed on with the drug to maximize their profits. Merck's actions fit the legal definition of "negligent homicide." Part of this story is the far too amicable relationship between Merck and the FDA. An FDA scientist who discovered the Vioxx heart connection early on said his FDA bosses forced him to quash information that was potentially damaging to Merck.[26] The most disturbing part of the Vioxx story is that despite paying out billions of dollars in lawsuits, Vioxx still made money for Merck.

Drug companies' ability to make money almost regardless of the efficacy of their product or the coexistence of dangerous side effects, is certainly in part due to their ability to charge virtually any price they want. The cost of drugs in the United States is more than in any other country by far, 40 percent more than Canada on average, the next most expensive nation, and the cost is continuing to increase on average about 10 percent a year, far more than the rate of inflation.

A prescription for Nexium, a drug used to treat acid reflux, costs on average, $18 in Spain. In France its $30, in the UK, $32. But in the United States, a prescription for Nexium costs, on average, $187, six times as much. Lipitor is one of the most commonly prescribed medication in the United States, used to treat high cholesterol. In New Zealand, a prescription for Lipitor costs about $6. And in South Africa and Spain, it costs $11 and $13 respectively. In the United States, a prescription of Lipitor costs, on average, $100.

Several years ago, my neighbor was diagnosed with Chronic Myelogenous Leukemia, (CML). At the time there was no effective treatment for CML, but a drug still in the experimental stages, Gleevac, was showing promise in preliminary trials. Other than controlled trials, the population at large did not have access to Gleevac. I was able convince the business people at Novartis, the maker of Gleevac, to allow my neighbor to be placed on the drug. Seventeen years later, I'm happy to report that he's still alive. Thanks to Gleevac, the ten-year survival rate for CML has climbed from 20 to over 90 percent—a dramatic success story. But Novartis has also increased the price from $30,000 a year, to $92,000 a year, simply because they can. The cost of these leukemia drugs in America is more than twice as high as the average cost in 15 developed nations, and about 28 times the cost in India.

From throughout the world, 100 influential oncologists have launched an initiative to pressure drug companies to reduce the price charged for their chemotherapy drugs. In 2012, 11 of the 12 new-to-market drugs approved by the FDA were priced above $100,000 per-patient per-year.

The exorbitant cost of drugs is much more than a mere pocket change issue. About 20 percent of patients don't fill their prescriptions because of the cost. For a diabetic, access to insulin can be a life or death matter from one day to the next.

Drug Dealers in Suits

On May 30, 2009, in a run-down trailer in Wayne County, Kentucky, 20 month old Kayden Daniels found a coffee cup filled with Liquid Fire, a drain pipe cleaner that was being used by the boy's family to make methamphetamine. Kayden drank it. The Liquid Fire worked just like it does on clogged pipes, eating away at his esophagus and stomach for nearly an hour until Kayden died. His father, Bryan, was charged with murder in the boy's agonizing death.[27]

But if Bryan Daniels was indeed guilty of murder, he likely had help from sources far away from the court room. The role of some of America's largest pharmaceutical corporations in the scourge of methamphetamine use in America, especially poor, rural America, is nothing less than criminal.

It is estimated that there are about 570,000 regular users of methamphetamines in the US.[2] The drug is manufactured by one of two operations: Mexican drug cartels and makeshift, "shake and bake" labs in homes and trailers scattered throughout the country, especially in Kentucky, Missouri, Tennessee, and Indiana. But the home labs get a major assist from Big Pharma.

Meth brings a rush of euphoria, confidence, and boundless energy, and it is highly addictive. It is notorious for wreaking havoc on users, destroying their skin, teeth, heart, their lives, and their families. The trail of devastation reaches every corner of society, leaving ruined families, parents in jail and losing custody of their children, emergency rooms forced to care for injured meth cooks, homes contaminated with deadly chemicals, police departments overwhelmed with cleaning up the mess rather than stopping other crimes, and police personnel riddled with debilitating chronic diseases from repeated exposure from raiding toxic meth labs.

The economics of meth use nationally is an utter disaster. According to a 2009 RAND Corporation study, meth costs the nation anywhere between $16 billion and $48 billion each year.

In recent years, 25 states have attempted to restrict the distribution of pseudoephedrine, the one irreplaceable ingredient for baking meth. The drug industry has viciously fought against curtailing the sale of meth's precursor chemicals since the 1980s at both the federal and the state level. The only states that have succeeded are Oregon and Mississippi. Oklahoma had first started placing pseudoephedrine behind counters and limiting sales, but failed to make it a prescription drug. Nonetheless, Oklahoma saw an immediate drop in the number of labs its officers busted.

But the meth cooks quickly adopted alternative access strategies. They organized groups of people to make the rounds of pharmacies, each buying the maximum amount allowed—a practice known as "smurfing." It was smurfing that then lead to Oregon going a step further and making pseudoephedrine a prescription drug in 2006.

The law worked. Oregon meth labs dropped an estimated 96 percent, and all the attendant social and family tragedies have dropped in proportion. Two years later, Oregon had the most significant decrease in rates of violent crime of any state in the country. Property crimes fell. In fact, by 2009, the overall crime rate in Oregon dropped to the lowest it had been in 40 years.[29]

Mississippi's meth labs dropped 74 percent, and the number of "drug-endangered" children dropped 81 percent after their state legislature passed a similar bill. But Mississippi and Oregon's success was a modest dent in the profits of the pharmaceutical industry, which sells an estimated $605 million worth of pseudoephedrine-based drugs per year, and they responded aggressively to make sure no other states had the audacity to protect their citizens from the meth scourge. Much like the NRA's standard cheer—"Guns don't kill people, people kill people"—Big Pharma's attitude was, "Pseudoephedrine doesn't make meth, people make meth."

As described in the Mother Jones exposé of August, 2013, *Merchants of Meth: How Big Pharma Keeps the Cooks in Business*,[30] the battle then moved to Kentucky in the War to Preserve Big Pharma profits.

Seeing Oregon's success Kentucky law enforcement officials encouraged lawmaker Linda Belcher to file a state bill to make pseudoephedrine available only by prescription. Despite no publicity on the filing of Belcher's bill, almost immediately Kentucky lawmakers were besieged with seemingly angry callers making false claims that they would have go the doctor whenever they felt a little congested.

The angry phone calls were the handiwork of a Washington, D.C., industry association, the Consumer Healthcare Products Association (CHPA), which spent an unprecedented $303,000 in Kentucky in just three weeks on a grass roots campaign of extensive advertising directly to the public and phone banking targeting legislators. The callers followed a script written by an out-of-state PR team.

Using CHPA and a robocall company called Winning Connections, multinational drug giants like Pfizer and Johnson & Johnson had pulled out all the stops, broke lobbying spending records, and beat back cops, doctors, teachers, drug experts, Republican and Democratic lawmakers alike in making sure that Belcher's bill never made it to the floor.

CHPA did its job, twice killing any pseudoephedrine bills in Kentucky. In Alabama, Kansas, Missouri, North Carolina, Oklahoma, West Virginia, and Tennessee, using Facebook, print and radio ads, robo calls, and top-notch local lobbying firms, Big Pharma has made sure the story and the outcome have remained the same. And the number of meth labs in those states have continued growing, with one slight exception.

Eventually, the Kentucky legislature passed a weak, industry-approved measure prohibiting people with prior drug offenses from buying pseudoephedrine for five years. Despite stiff opposition from CHPA, the law also included a reduction in the maximum amount of pseudoephedrine that can be purchased—limits that apparently have slowed the relentless rise in meth labs in the state.

As shameless as Big Pharma has been on methamphetamines, the story of Purdue Pharma's willingness to profit from addicting and killing Americans with narcotics for profit is even more sinister.

Between 1999 and 2010, sales of narcotics like Vicodin, Percocet, and OxyContin quadrupled. Americans consume 80 percent of the narcotics in the entire world, despite having only 5 percent of the world's population. In 2015, 300 million prescriptions were written for narcotics, about one prescription for every man, woman and child in the US. Big Pharma has even seized on the opportunity to openly sell more drugs to treat the side effects of narcotic addiction. A drug for opioid induced constipation (OIC) is now being marketed relentlessly on television. With a good-looking, middle-aged construction worker, a smile on his face, unapologetically talking about treating his chronic back pain with the narcotics prescription given to him "by his doctor," he implies to the viewers that the only problem he has now is OIC. The not very subtle message is that the only problem with being addicted to narcotics is that it "backs you up big time." And hey, Big Pharma can solve that problem, too.

Narcotics can certainly cause OIC, but that can be the least of it. Patients who are given narcotics for chronic pain are less likely to ever return to their previous level of function or go back to work, and are more likely to become addicted, experience sexual impotence, intellectual impairment, and die from overdosing. Chronic narcotic use almost invariably also makes a patient even more sensitive to pain, less able to tolerate even minor painful insults, often leading to a downward spiral of narcotic tolerance and ever increasing dose requirements, sometimes massively increased.

Recently, I was called just after midnight to provide anesthesia for a middle-aged woman, a former athlete, who was writhing in pain from a perforated bowel. She had become a narcotic addict thanks to a family doctor that gave her Percocet for chronic orthopedic pain. She then progressed to IV heroin use. Just outside the operating room she was screaming in pain, not relieved by any narcotics. The perforation of her bowel was most likely caused by a combination of excessive anti-inflammatory drugs, and the OIC from narcotics. She faces numerous more surgeries to drain infections, a likely prolonged stay in the intensive care unit, and in a few weeks, she will be back taking more narcotics and heroin to repeat the cycle. But even she will be luckier than many.

Narcotic overdose deaths have reached epidemic proportions. From 1996 (the year Oxycontin was introduced) to 2016, opioid prescription abuse killed 200,000 people, according to some estimates,[31,32] but the CDC, never one to overstate a health crisis, believes that since 2000 that number is closer to 600,000.[33] And the trend is still accelerating. In 2016 alone, there were 59,000 nationwide. In Ohio, one of the hardest hit states in the country, there were 4,149 fatal drug overdoses in 2016, an average of 12 per day. The CDC estimates that one in five patients going to a physician's office for a pain-related problem were given a prescription for narcotics, and that 11 million people abused narcotics in 2016.[34]

For the first time since the flu pandemic of the early 1900s, life expectancy in the United States has dropped for three consecutive years from 2015 to 2017, and the cause is deaths from narcotic addiction.

Narcotic addiction is unraveling the social fabric of America, crossing all socioeconomic and cultural boundaries, and all age groups. I have taken care of prim and proper looking, wealthy, 70-year old females who are devoted members of the Mormon Church who wouldn't dream of drinking alcohol or even coffee or tea, but are taking six Percocet every day. I have also taken care of desperate, homeless young men with few remaining teeth and abscesses all over their bodies from shooting up heroin. Four out of five new heroin users took their first step towards addiction by inappropriate but legal use of prescription painkillers.

Three factors led to America's opioid crisis: 1) The short-sighted integration of "pain management" into mainstream medical care, 2) Heavy-handed government and business directives to the practice of medicine, and 3) The "anything for a buck" business model of Big Pharma.

It is beyond the scope of this book to review the details of the first two. But briefly, the American Pain Society, a branch of my specialty, anesthesiology, convinced the mainstream medical community to consider a patient's pain so important as to make it the "fifth vital sign." By the mid 1990s, nurses everywhere were asking patients to rate their pain from zero to ten, and just about everyone involved in patient care got swept up in a mindset that patients shouldn't have any pain.

In 1998, the Federation of American Medical Boards (FAMB) issued physicians a misguided reassurance that "in the course of treatment," large doses of narcotics were an acceptable means of reducing patients' pain. In fact, two years later, the Joint Commission, the quasi-governmental hospital rating and oversight body that every hospital is thoroughly beholden to and intimidated by, mandated that every hospital seeking accreditation assess, if not obsess about the degree of pain experienced by every patient they cared for. The Commission allowed Purdue to widely distribute educational materials to hospitals that amounted to almost-sanctioned promotion of OxyContin.[35,36]

The Commission required that healthcare providers ask every patient to rate their own level of pain. It went so far as FAMB pressuring state medical societies to make under treatment of pain punishable as patient abuse or even battery. Inevitably, in that environment, lawsuits were filed for "under treating" pain.[37]

Seeing an opportunity to cut costs by increasing the number of transgressions with which to punish providers, Medicare and Medicaid were quick to jump on the bandwagon. Inadequate treatment of patients' pain was used to increase the number of hoops that hospitals, networks and physicians had to jump through just to maintain their existing reimbursement rates and avoid being sanctioned. They began requiring that hospitals religiously launch patient satisfaction surveys and those surveys were used as a cudgel in hospital and physician reimbursement rates. It takes no stroke of genius to predict that healthcare providers would become eager to narcotize patients rather than defend themselves from poor patient survey scores. With constant patient satisfaction surveys, the arrival of the internet, and direct to consumer advertising of prescription drugs, doctors were no longer able to dictate treatment to patients. Instead, a "customer is always right" atmosphere swept into doctor's offices bringing new pressure to bear for them to hand out pain killers to patients, many of whom came to demand that modern medicine deliver quick and easy pain relief.

But this new regulatory landscape did not develop in a vacuum. In fact in the early 1990s, two relatively new physician organizations, the American Pain Society, and the American Academy of Pain Management played key roles in cheerleading for leaving no pain untreated. But these non-profits groups were happily funded by Purdue Pharma, Johnson & Johnson, and Endo Pharmaceuticals.[38]

Big money was at stake. Prescription narcotic sales were $9.6 billion in 2015. These drug companies seized on the opportunity to profit from the systemwide aversion to pain. But one small company, Purdue Pharma, did it better than anyone else.

The name Sackler is plastered all over institutes and museums on just about every Ivy League university campus. Rooms or wings at the Guggenheim, at the Louvre, at the Tate, at the Met, and many other famous museums carry the last name of their benefactors—Sackler. But those rooms should really be named, The Oxycontin Rooms.

In 1952, three brother psychiatrists, Arthur, Raymond, and Mortimer Sackler, purchased Purdue Pharma. The oldest brother, Arthur, became a trailblazer in pharmaceutical advertising by becoming one of the first to actively court relationships with doctors in clinical practice. They also became the first to market a single drug as a cureall for just about everything that ails you. With this two-pronged approach, they turned the sedative Valium into the first $100 million drug and made themselves enormously wealthy. But they were just getting started.

Purdue then parlayed this strategy into creating their real golden goose, OxyContin. Introduced in 1996, OxyContin was developed as a supposedly time-release formula for a pre-existing narcotic, oxycodone. They had convinced the FDA to grant approval based on flawed studies and claimed that the abuse potential for OxyContin was less than one in a hundred patients. Through an aggressive marketing strategy aimed at increasing narcotic prescriptions in general and OxyContin in particular, Purdue increased its sales 20 times over within only four years. Purdue pushed OxyContin for everybody—for the 30 million sufferers of chronic back and knee pain, women with menstrual problems, anyone with a toothache. In 2001, Purdue spent $200 million promoting OxyContin, and the pay-off was spectacular. Purdue was fined $635 million in 2007 for misleading promotion of OxyContin. But that was mere pocket change for Purdue, given the $31 billion in OxyContin sales from 1996 to 2016.[39]

Getting millions of people hooked on narcotics gave the Sacklers what is likely the largest personal fortunes in the history of Big Pharma.[32] As revealed in an investigative piece by the Associated Press and the Center for Public Integrity,[33] Purdue's main lobbyist, Burt

Rosen, co-founded the very influential Pain Care Forum in about 2005. Purdue describes the Forum as a Washington D.C., policy coalition, but it is clearly a powerful lobbying force that drafts legislation and fights regulations regarding the drug market.

For example, in 2006 the forum conducted "The Epidemic of Pain in America" press conference stating, "Appropriate use of opioid medications like oxycodone is safe and effective and unlikely to cause addiction in people who are under the care of a doctor and who have no history of substance abuse." The briefing came close to being accurate, but left out one important word which would have completely changed the message. It should have been titled, "The Epidemic of Pain Medication in America."

In 2012, the forum widely distributed a report claiming that 100 million people, 40 percent of all adults in the country, suffered from chronic pain. The American Pain Foundation received $150,000 from Purdue to promote the findings in the report through the Pain Care Forum. The foundation closed in 2012 after it was revealed that 90 percent of its funding came from the pharmaceutical industry.

The report was clearly an attempt to preserve opioid sales even as the toll of addiction and abuse was exploding throughout the country. The report got a sympathetic review from the head of the FDA, who cited the report as justification for not doing more to restrain the drugs, and in fact, the FDA did retreat from a planned, more aggressive strategy to restrict the writing of opioid prescriptions. On several occasions the forum spearheaded thousands of public comments to the FDA against the erection of any barriers to opioid access, like the creation of an electronic registry of patient prescriptions.

But about half of the authors to the 364-page report were part of the American Pain Foundation, the American Pain Society or the American Academy of Pain Medicine, all of which are supported by drug company funding.

Healthcare providers by the thousands attended all-expenses-paid Purdue-sponsored conferences at lavish resorts where they were paid and trained to join Purdue's national speaker bureau. Purdue told them all that because OxyContin was time-released it would not cause the euphoria that leads to addiction. But OxyContin certainly did cause a narcotic high, both in patients and in Purdue's stock holders.

Purdue became one of the first to create extensive databases on the prescribing practices of physicians such that they could target those that prescribed the most pain medicine with their arsenal of marketing techniques. This database allowed Purdue to identify hundreds of doctors that were significant outliers in the number of narcotic prescriptions they wrote, indicating recklessness, incompetence, or deliberate drug running. Purdue had no problem with their "overachievers." In turn, Purdue's sales reps reaped bonuses that exceeded their base salary as incentives to ramp up sales.

For a relatively small drug company, Purdue had a massive sales force of 670, who called on almost 100,000 physician prescribers. One of their sales tools would have made a back alley drug dealer proud. They offered free "starter coupons" for OxyContin, giving away up to 30-day supplies for first time users.

The University of Wisconsin School of Medicine and Public Health (UW) came under fire in 2011 when it was revealed that an affiliated Pain and Policy Studies Group that published research and lobbied to defend the broad use of narcotics, had significant financial ties to the drug companies whose sales would suffer from reining in their use. Outside doctors accused the UW Pain Group of acting like little more than a front group for the narcotic drug companies, pushing research that was questionable at best. In doing so, they contributed significantly to the deadly spread of OxyContin nationwide. By far the largest financial contributor to the pain group was Purdue.

David Joranson, a founder of the UW pain group, who, incidentally is not a physician but has only a master's in social work, was a key figure in writing a consensus statement from the American Pain Society and the American Academy of Pain Medicine declaring that opioids were "safe and effective for chronic, non-cancer pain and that the risk of addiction was low."[39]

A co-author of the statement was J. David Haddox, a physician at Emory University School of Medicine, who was at the time being paid as a frequent speaker on behalf of Purdue, and who became a full-time executive for Purdue three years later. Joranson and Haddox also wrote a paper to state medical boards throughout the country warning against regulatory efforts that would obstruct patients' access to opioids and failed to mention the financial strings tying them to Purdue.

One of the most deceptive techniques of drug company peddling is the creation of "patient advocacy organizations" (PAOs). As the names imply, these supposedly nonprofits are often formed on the platform of a particular disorder or disease, i.e., chronic pain, heart disease, or inflammatory bowel disease. There were 1,215 PAOs as of 2015. They lobby and advocate seemingly for the healing or amelioration benefit of patients as victims. Some of them are well known, well respected, and household words, like the American Heart Association, the American Lung Association, and the National Alliance on Mental Illness. Historically, many of these non-profits have enjoyed considerable success in increasing research funding, calling public attention to a disease, and even shaping treatment options. But the pharmaceutical and other industries have jumped on the opportunity to shape the PAOs through funding offers, and even spur the creation of entirely new ones cloaked with the altruism of patient care. Somewhere between 30 and 71 percent of PAOs have a significant financial conflict of interest with the drug industry.[50] As of 2013, the American Heart Association had accepted over $23 million from drug makers.[51] In 2015, drug companies donated $116 million to PAOs, approximately double what they spent on lobbying that year.[53]

The potential for corruption of PAOs is just as real as the potential for corruption of lawmakers. In some cases, PAOs have become little more than merely front organizations for the pharmaceutical industry, whose intent is to manage the disease with a certain drug or class of drugs. Senator Claire McCaskill's investigation of Big Pharma's role in the opioid crisis revealed that Purdue Pharma, Insys Therapeutics, Janssen Pharmaceuticals, Mylan, and Depomed together paid $9 million to 14 non-profit patient groups.[49]

Many pharmaceutical corporations have become simply drug dealers in Brooks Brothers clothes. They blatantly seek to create permanent drug dependency throughout the population, with no hesitation to go after the "children's market." Their business model disregards the collateral damage and human cost of their product. Like a ghetto drug dealer, their innovations are primarily to eliminate their competitors and they achieve their pre-eminence by bribing the establishment, in this case, lawmakers and the medical community, rather than the cops.

Purdue has recently added another layer of exploitation to their sinister business model by capitalizing on the constipation that consumers and addicts often get from narcotics. They recently received FDA approval to market Symproic (naldemedine), a drug for treating opioid-induced constipation. Purdue's website has the audacity to claim, "The launch of Symproic with Purdue Pharma this summer will mark yet another milestone in our commitment to protect the health and well-being of patients we serve."[40]

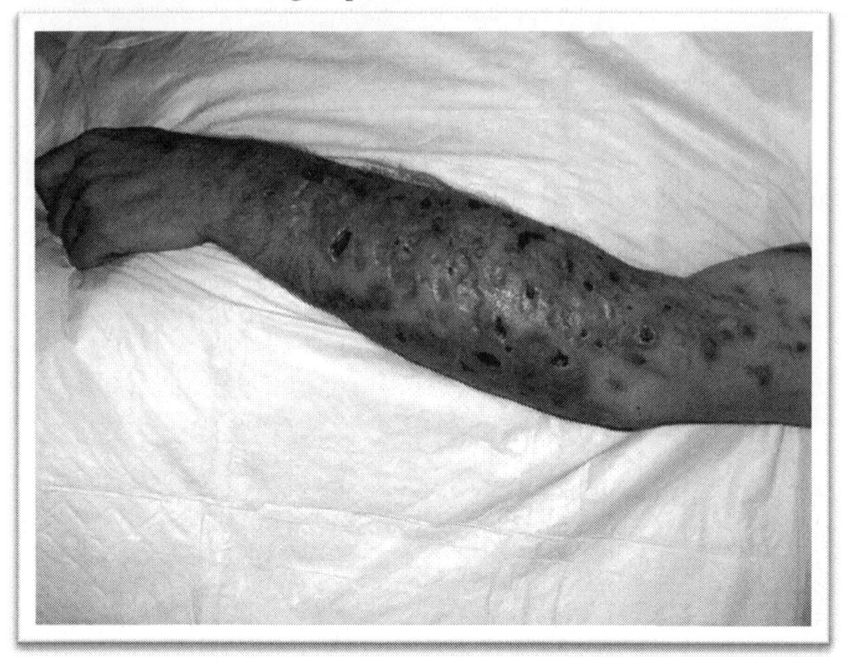

Arm of a heroin addict, riddled with infection and needle scars.

Congress: Driving the Get-Away Car

In the 2000s, "pain clinics" erupted all over the country, run by corrupt doctors and pharmacies. Large and suspicious patterns of narcotic sales by a chain of wholesale distributors, had been identified by DEA officials who tracked them. The DEA could clearly see the evidence that the three largest distributors, CardinalHealth, McKesson, and AmerisourceBergen, were feeding criminal narcotic mills, helping pour millions of pills onto the streets of American towns and cities, creating millions of addicts, and killing people by the tens of thousands. But the foundation of this narcotic empire was Purdue, who had all the detailed information, down to the level of individual prescriptions, on who was actually buying and selling the pills. They could have used that information to help crack down on criminal trafficking, but chose not to.

For years, the DEA had been fining some narcotic distributors and freezing suspicious shipments. Such action had allowed the DEA to act quickly to prevent some drugs from reaching illicit destinations. But there was money to be made, so the drug companies decided to put a stop to that. According to an exposé in the Washington Post,[41] through political action committees, the drug industry funneled $1.5 million to 23 members of Congress who co-sponsored a bill to hamper the DEA's efforts. Republican Congressman Tom Marino received $100,000 and Senator Orrin Hatch, $177,000, to lead the charge in Congress to pass the bill. In total Big Pharma spent $102 million lobbying for favorable legislation from 2014 to 2016.

At the height of the drug epidemic, Congress voted unanimously to strip the DEA of its most powerful deterrent to drug companies suspected of flooding the country with prescription narcotics. Using no less than 46 former DEA officials who had entered the revolving door of government and industry,[42] the drug industry wrote and heavily lobbied for the ironically named, "Ensuring Patient Access and Effective Drug Enforcement Act" that would make it virtually impossible to prosecute unscrupulous narcotic manufacturers, distributors, and chain drug stores. And to seal the deal, they pressured the DEA to fire Joseph Rannazzisi, who for ten years had aggressively led the DEA's division responsible for watching the drug industry.

When President Trump nominated Rep. Marino to be his drug Czar in October, 2017, the hypocrisy led to nationwide outrage over Congress's complicity with the drug industry.

Despite the widespread media coverage of the opioid crisis and the stigma now attached to both patients and physicians, most of the "pain clinics" are still in business, run often by physicians who consider themselves pain specialists. But in my experience, the typical patient that gets their narcotics from a pain clinic is still being sorely mismanaged. Narcotics are still being handed out recklessly for every conceivable type of chronic pain, despite the overwhelming evidence that they are both dangerous and ineffective for treating such pain. I doubt that any of these pain specialists see the all-too-common end result of such mistakes that I see—patients who are riddled with abscesses after shooting up heroin, most of whom arrived at that special brand of hell after being handed prescriptions for narcotics.

Spreading addiction, misery and death worldwide

Because of OxyContin's reputation, domestic sales started to drop in 2011, and by 2016 sales had dropped 40 percent from their peak. Physicians began to understand that OxyContin had unleashed perhaps the worst example of the doctor's creed, primum non nocere, or, "first do no harm." So following the tobacco industry's handbook of shocking ethical failures, the Sacklers have started pursuing an overseas market for OxyContin under an umbrella of international subsidiaries called Mundipharma.[43]

Any excuse that the Sacklers may have conjured up that they didn't know how addictive OxyContin was, can be ceremoniously dismissed as they are "bringing the band back together" and using many of the same tactics that were so successful in the US—running training seminars urging foreign doctors to overcome their reluctance to prescribe narcotics for chronic pain and offering patients discounts on their prescriptions.

Purdue has paid US physicians like Dr. Barry Cole, a Reno psychiatrist and self-proclaimed pain management specialist, and Dr. Joseph Pergolizzi of Florida upwards of a million dollars to conduct seminars overseas on how and why to expand the use of narcotics for chronic pain patients. Dr. Cole is reported to have told an audience of physicians in Veracruz, Mexico, that, "Any side effect is reversible when treatment is discontinued, and there is no permanent damage to the body."[43] He should see some of the bodies I have seen in the operating room.

For the 200,000 and more people in the US that have died from the opioid epidemic as of 2016, it would seem that the claim of reversibility might be a touch false. As a physician who encounters people addicted to narcotics on a daily basis, I can personally attest to the deeply cynical absurdity of that claim.

Mundipharma is focusing its international peddling on well-to-do countries that offer health benefits and have a large middle class. And it is targeting chronic pain patients that are sufficiently healthy to be paying customers, i.e., future addicts, for a long time.

Their marketing campaign in Spain is a good example of the depths to which Mundipharma has been willing to sink to expand the demand for its products. Naked celebrities, actors, pop singers and models urge Spanish TV audiences to stop dismissing chronic back and joint pain as a normal part of life. The campaign is working "miracles," at least for Mundipharma's sales, which have increased 700 percent from 2007 to 2014. And of course, you can't find an adult anywhere in the world that doesn't have some chronic back and joint pain. In pursuit of millions of new customers in Colombia, Mundipharma sent out a news release claiming that 47 percent of the Colombians suffer from the silent epidemic of chronic pain.

A CDC study concluded that about one out of four patients given narcotics for 30 days becomes addicted. In fact, taking narcotics for as little as five days can lead to addiction.[44] Despite the evidence and the proven experience of the opioid crisis in the US, Mundipharma pays hundreds of spokespersons to internationally spread the same lies that they used to get OxyContin in the medicine cabinets of millions of Americans—that chronic narcotic use seldom leads to addiction. One of their marketing videos in China featured a nurse telling a cancer patient, "You will not be addicted if you follow the doctor's instructions." The video was taken down after the Los Angeles Times asked the company to justify it.[43]

Mundipharma's international expansion started in Asia, followed by Latin America, the Middle East, then Africa, and by 2016 has spread their plague of narcotic abuse to 122 foreign countries. Its sales have increased 800 percent over the last five years.

To the extent that chronic pain from terminal cancer is a worldwide health care challenge, cheap, long lasting narcotics like morphine could certainly address it. But multinational drug companies have little incentive to spend much effort distributing cheap morphine. In the U.S. market, Purdue charges hundreds of dollars for a month's supply of OxyContin. Generic morphine is just as effective and costs about 15 cents a day.

At the time of this writing, Purdue has just announced that they are cutting their sales force in half and ending their promotion of narcotics to physician prescribers.[45] Purdue and other drug makers are facing an onslaught of lawsuits for their deceptive marketing practices, from states, counties, and cities whose budgets are being bled dry by all of the sociologic consequences of the opioid epidemic.

Independent of the drug companies, physicians have some of their own responsibility in fostering the narcotic epidemic. Many physicians found it far too easy and convenient to prescribe too many pills rather deal with more phone calls from patients who still complained of pain. In my experience, physician networks and healthcare systems have only recently begun to address the chronic overuse and abuse of narcotics and their role in prescribing them. While late to the game, these developments are hopeful signs that the narcotic epidemic may at least stabilize if not begin to decline.

We Are All Walking Drug Stores

Drug companies spend 19 times more money on marketing drugs than they do on actual drug research and development.[56] Even if you think you are smart enough to avoid being influenced by advertising from an industry determined to overdose us on their menagerie of products, many of us will be unable to avoid their pervasive reach no matter how hard we try. Because of consumers flushing unused drugs down the toilet, or the presence of residues and metabolites in human waste, pharmaceuticals show up as contaminates in the drinking water of about 40 million people.[46]

Approximately 70 percent of pharmaceuticals taken by humans are excreted in the urine,[56] and that means, among other things, Americans have the most expensive urine on the planet. And a lot of that will ultimately end up in waterways because they are not filtered out by municipal sewage treatment plants which were never designed to capture those kinds of compounds.

With every glass of water, you are getting a little of everyone else's Prozac, Cialis, Lipitor, birth control pills, opioids, antibiotics and hormone replacement therapy. A nationwide study by the U.S. Geological Survey in 2000 found low levels of pharmaceuticals—including antibiotics, hormones, contraceptives and steroids—in 80 percent of the rivers and streams sampled. A study in Germany revealed that the most widely prescribed anti-diabetic drug in the world, Metformin, was present in every water sample tested, including tap water, and at concentrations exceeding Germany's environmental safety levels.[47]

The authors of the study stated that Metformin is likely "distributed over a large fraction of the world's potable water sources and oceans." This is likely more than a theoretical public health problem. The federal government doesn't require any testing and hasn't set safety limits for drugs in culinary water. It's basically a free for all. In a 2008 survey by the Associated Press, of the 62 major water providers contacted, the drinking water for only 28 was tested. Of those major metropolitan areas where tests were performed, only Albuquerque, NM, Austin, TX and Virginia Beach, VA had water tests negative for pharmaceuticals.

Rural consumers who draw water from their own wells are not immune from exposure either. Home filtration systems are not capable of filtering out pharmaceuticals and makers of bottled water don't treat or test for them. Furthermore, according to the Natural Resources Defense Council, 25 percent of water sold as "bottled" actually just comes from a tap. Pharmaceuticals also infiltrate aquifers deep underground, the source of 40 percent of the nation's water supply.[48]

More than 100 different pharmaceuticals have been detected in lakes, rivers, reservoirs and streams throughout the world. Studies have detected pharmaceuticals in waters throughout Asia, Australia, Canada and Europe — even in Swiss lakes and the North Sea. Like many other chemicals, many pharmaceuticals are very slow to degrade, persisting essentially unchanged in the environment for up to decades. And there's evidence that adding chlorine, a common process in conventional drinking water treatment plants, makes some pharmaceuticals more toxic.

Mary Buzby, director of environmental technology for drug maker Merck & Co. Inc., said: "There's no doubt about it, pharmaceuticals are being detected in the environment and there is genuine concern that these compounds, in the small concentrations that they're at, could be causing impacts to human health or to aquatic organisms."[71]

Prescription drugs in rivers and streams are damaging wildlife worldwide. Male fish are being feminized, creating egg yolk proteins, a process usually restricted to females. Fish with both ovaries and testes are becoming common place. A recent study by the US Geological Survey of large and small mouth bass at 19 different locations in National Wildlife Refuges found that 60-100 percent of the fish were intersex, i.e. they had female egg cells and testes.[53] While the cause is undoubtedly estrogen-like industrial compounds (see chapter 4) it is also due to the drug Metformin being present in waterways. Anti-depressant drugs are usually the compounds in highest concentration in natural water bodies, and they are profoundly affecting the brains of fish and a real threat to biodiversity. Low dose anti-depressant medication in the Great Lakes are changing fish behavior, including making them abnormally aggressive, even homicidal.[54]

Many of these pharmaceuticals, like other chemical contaminants in the environment, are bioaccumulative, which means concentration in the bodies of an animal steadily increase because the absorption rate of these chemicals exceeds the ability of the organism to metabolize or excrete the chemicals. Pharmaceuticals also are clearly having an adverse impact on sentinel species at the bottom of the food chain such as earth worms and zooplankton.

"We know we are being exposed to other people's drugs through our drinking water, and that can't be good," says Dr. David Carpenter, who directs the Institute for Health and the Environment of the State University of New York at Albany.

Virtually every creature in the animal kingdom, including every human being on earth, is now a living, breathing, mobile, low dose drug store. No one thinks this is a good idea, except the pharmaceutical industry. That's because the degree to which drugs have infiltrated and contaminated all living systems is directly proportional to their profits. And what could possibly be more important than that ?

On a Saturday afternoon in Feb. 2109, well over 100 chanting protesters swarmed the Guggenheim Museum in NYC, an unlikely place for such a demonstration. Then they marched to the Metropolitan Museum and stood on the steps and listened to this refrain. "We have to bring down the Sackler family. They should be in jail, next to El Chapo!"

The Sackler Brothers, Raymond, Arthur, and Mortimer. Now all deceased. Their heirs are multi-millionaires and billionaires.

4

Living in the Chemosphere:
Contaminating Every Corner of the Globe

"The primary purpose of government is to protect those who run the economy from the outrage of injured citizens."

— Derrick Jensen, Endgame,
Vol. 1: The Problem of Civilization

Earth: A Giant Chemical Junkyard

Deep in the heart of the Yasuni National Park in Ecuador live between 200 to 400 Waorani tribesmen. They are "uncontacted"—living in voluntary isolation from the outside world. They became world famous when in 1956 they speared to death five missionaries that had landed their plane in their territory. The Ecuadorean government then made it illegal for outsiders to go in after them for any reason, like "aid" or religious conversion. They are "forest people," whose feet are deformed from a life time of climbing trees, and they are as isolated from modern civilization as any human beings on earth. They hunt with blow darts, wear little to no clothing, and dangle dead monkeys around their necks for adornment. But even they cannot live without being harmed by international corporations. If we were allowed to test them, even the blood of Waorani tribesmen would undoubtedly show contamination with hundreds, perhaps even thousands of the chemicals that permeate the post-industrial, "civilized world."

From the top of Mt. Everest to the Mariana Trench,[1] the deepest part of the Pacific Ocean, from the hair of polar bears to the chromosomes that lie in the nucleus of your newborn baby's brain cells, virtually everything on earth is literally being bathed in harmful chemicals and heavy metals. Taking advantage of this disturbing reality has even found application in the controversial search for energy in underwater formations of methane hydrates. The age of hydrates and sediments that cover them, and therefore the origin of the hydrates, can be determined in part by whether they have traces of the legacy pesticide DDT, which was first used in 1945.

In the late 1970s, chemical and oil companies began to fear that growing public awareness of the toxicity of their manufactured chemicals would start to hurt their business, so naturally they responded with a massive public relations campaign, apologizing for any harm they may have done...I'm just kidding of course. They responded with a massive PR and advertising campaign rather than an apology or re-evaluation of the safety of their products. For instance, Monsanto (see chapter 8) started running print ads showing a young child with a puppy above the slogan, "Without chemicals life would be impossible." Actually without their chemicals, it's something else would be impossible--their profits. Your life on the other hand would be just fine, perhaps a bit more inconvenient, but otherwise fine and probably much healthier.

The term "Chemosphere" currently refers to a famous, futuristic octagon shaped house in Los Angeles designed by architect John Lautner. I think the name more appropriately could be applied to 21st century Earth—the Chemosphere—because our planet is a giant sphere, now drenched in a toxic chemical bath. Multiple studies have confirmed universal contamination of the global environment and virtually all living things with a dizzying array of manmade compounds. Virtually none of these chemicals improves or augments natural biologic systems, and many are toxic to life forms of all types, including humans.

Through the air you breathe, the water you drink, the food you eat, the couch you sit on, and the cell phone you can't put down long enough to read this book, life in modern civilization exposes all of us to approximately 140,000 industrially produced chemicals, 100 times more chemicals than two generations ago.[25] The United States manufacturers or imports 42 billion pounds of chemicals every day. Common fruits and vegetables often contain residues of 60 different pesticides. The average indoor air in homes in the United States harbors 400 chemicals.[3] We are all essentially human guinea pigs in a massive toxicology experiment that we didn't consent to, that has no scientific design or constraints, and it is conducted by people who have little to no concern about how the experiment will affect your health and your life.

The chemical industry wants you to think that there's nothing to worry about. The "science" behind that assumption either doesn't exist, because no tests were ever done, or the science is based on principles that are badly outdated and completely unable to address the risks. WWI became the first large scale demonstration of the power and horror of chemicals to disable, maim and kill army soldiers. The use of tear gas, mustard gas, phosgene and chlorine lead to the labeling of WWI as "the chemists war." In response, the 1925 Geneva Protocol for the Prohibition of the Use of Asphyxiating, Poisonous or Other Gases, and Bacteriological Methods of Warfare was adopted by most major nations. WWI was referred to as the "war to end all wars" in part because of the horror that chemical warfare wrought upon the troops.

But while the use of chemicals was finally being regulated in war, ironically, and simultaneously, a virtual free-for-all was unleashed in the use of chemicals and deadly substances in consumer goods and countless civilian applications.

Lead (see Chapter 1) was almost ubiquitous in drinking water, paint and gasoline, arsenic was added to cosmetics and was the "active ingredient" in the medical tonic, "Fowler's Solution." Chloroform was added to liniments, and "Radium Spray" was used as a bug killer, furniture polish, and disinfectant. Asbestos was added to tooth paste (see Chapter 10), and the most toxic and deadly neurotoxin that we know of, mercury, was a common ingredient added to teething powders for babies and laxatives for adults. War had been cleaned up, but businesses were subjecting everyday civilians to chemical attack. And if you think you are much safer now than post WWI society, think again. The toxins have changed, but they have largely become hidden, and overall your exposure today is likely much worse.

According to studies by the International Agency for Cancer Research and the National Toxicity Program, about 10 percent of manmade chemicals are likely to be carcinogenic. And as stated in the world's most prestigious medical journal, the New England Journal of Medicine, "Every molecule of a carcinogen is presumed to pose a risk."[4] Both heavy metals and many chemicals form covalent bonds with DNA called DNA adducts. This can increase the risk of cancer by activating oncogenes (tumor promoting genes) and blocking anti-tumor genes.

In 1826, the French epicurean Jean Anthelme Brillat-Savarin wrote, "The future of the nations will depend on the manner of how they feed themselves."[5] Little did he know how prescient that was. Americans' naiveté about their food supply is shocking. Many of us assume that the government has a solid system in place to assure food safety. But "safety" is a relative term. Many Americans would think that anything that doesn't kill you within four hours after eating it, is therefore safe. Even if it's something that barely qualifies as food, like for instance *Cheetos*, we tend to assume that while it may not be that nutritious, at least it's not overtly toxic. Moreover, many people haven't the faintest clue how food even appears on grocery store shelves. Seven percent of Americans believe that chocolate milk comes from a "brown cow."[6] Those doing the survey should have asked where people thought strawberry milk came from. Strawberries? Pink cows? Pink cows fed strawberries? Giant cow shaped strawberries? Do British cows have two utters? One that produces strawberries and the other that produces cream? But if you assume that those 7 percent are merely people endowed with an overabundance of stupidity, an even more shocking finding was that 48 percent of people had no idea where chocolate milk came from.

The majority of Americans say that because of pollution, contaminants and safety concerns, they would not buy food that came from China. But about one fifth of the food we eat comes from outside the country, with a significant portion of that coming from China. Knowing the country origin of your food is now more important than ever. But with help from Congress, bought and sold by industrial agriculture corporations, and the World Trade Organization, legislation was passed in Dec. 2106 that eliminated the requirement to label food with its country of origin.

In 2010, researchers examined more than 300 samples from 31 different food types in Dallas, Texas. Every sample tested showed "multiple pesticides."[7] From just the food we eat, we are literally changing who human beings are because we have allowed human bodies to become chemical junkyards. But food is merely one, albeit an important one, source of constant chemical exposure. A seven-year study of 22 healthy adults revealed that human urine carried an average of 2,282 chemicals coming from food, drugs, cosmetics and consumer products of every kind.[8] This information is available online to the public in a database called the Urine Metabolome Database.[68]

Babies are Born "Pre-polluted"

Embryo and fetus are the critical developmental stages through which all future generations must pass. Chemical carcinogens can be as much as 65 times more potent for infants compared to adults and half of a person's life time cancer risk is typically accumulated and set by the age of two. The environment experienced in the womb defines the newborn epigenetic profile, the chemical modifications to DNA we are born with, that can have wide spread implications for disease risk later in life, from cancer to heart, lung, brain, and endocrine diseases.[31]

Furthermore, children are often exposed to more contamination than their parents, and mothers pass on many if not most of these chemicals and heavy metals to their babies in the womb, or infants that they nurse. We should be blunt--babies are born "pre-polluted." In 2005 and again in 2009 the Environmental Working Group commissioned five laboratories to examine the umbilical cord blood of 10 babies and found more than 200 chemicals in each newborn.[9]

In fact, 200 is undoubtedly a very low estimate because one of the many studies addressing this only tested for a few hundred chemicals. Of the ones found in umbilical cord blood, 134 of these compounds are known to cause cancer, 151 birth defects, 154 endocrine disruption, and 158 neurotoxicity. Obviously some of these compounds have more than one effect. Back in 1980, a government scientist concluded that human breast milk was so contaminated with industrial toxins and pesticides that if it were regulated by the USDA is would be banned. Other studies done specifically on the blood of pregnant mothers showed the same disturbing trend. Virtually all pregnant women are walking chemical repositories. Tracking 163 chemicals, 99 percent of pregnant women tested positive for at least 43 different chemicals.[10]

Examining the dietary intake of hundreds of children, researchers concluded that cancer risk threshold levels were exceeded by 100 percent of children for arsenic, dieldrin, DDE, and dioxins. Non-cancer benchmarks were exceeded by more than 95 percent of preschool-age children for acrylamide and by 10 percent of preschool-age children for mercury. Preschool-age children had significantly higher estimated intakes of 6 of 11 toxic compounds compared to school-age children.[11]

Many of these poisonous compounds have been banned for decades, like DDT and PCBs, but because of their persistence in the environment they are still contaminating our bodies. For example, DDT breakdown products have been found in 99 percent of Americans tested,[38] despite the fact that the chemical has been banned in the United States since 1972. It is still used, however, in parts of Asia and Africa. Women whose prenatal exposure placed them in the top 25 percent of the population were almost four times more likely to get breast cancer as adults compared to the lowest exposed group.[39] By comparison, women with the breast cancer gene mutation, BRCA, have a five times increase in their breast cancer risk. In addition, those women with more DDT exposure had more aggressive tumors, and more advanced stages of the disease. This despite studies showing exposure to DDT as an adult does not increase the risk of cancer.

This is even more of a concern given that DDT persists in the environment for decades, is present in the melt water running off of glaciers all over the globe and is re-exposing people throughout the world. Because of the phenomenon of bioaccumulation, people can end up with one million times higher concentrations in their bodies than exist in the ocean, reaching levels comparable those found in countries where DDT is still used for malaria suppression.[40,41]

The history and consequences of DDT use is a perfect microcosm of the larger issue of chemicals contaminating our environment. Jonathan Chevier, an environmental health scientist at Montreal's McGill University says, "There are unforeseen consequences of releasing really large amounts of chemicals into the environment without first understanding the health effects,"[42]

Of the 5,000 chemicals produced and used in high volume in modern society, less than half have any toxicity data available on them, and less than one-fifth have toxicity testing data on developing human organs.[2] This alone is a chilling thought, but virtually all toxicity data that has been generated comes from a testing program called "toxicology risk assessment," the same approach that polluting industries, and unfortunately regulatory bodies, use to assess whether a certain project, like citing of a refinery or incinerator will create a public health risk. It is an antiquated, inadequate and scientifically deeply flawed process.

The science of toxicology is still tethered to a principle first advocated by the father of toxicology, Paracelsus (Theophrastus von Hohenheim), a 15th-16th century Swiss physician, alchemist, and astrologer. His contribution to the Renaissance was the fundamental classic toxicology theory that "the dose makes the poison." In other words, there is a threshold at which low concentrations of toxic substances no longer present a threat to human health and above that threshold, the harm is proportional to the dose. That is still the basis of modern day toxicology and government environmental regulations. However, hundreds, if not thousands of studies have proven that premise to be false for an increasing number of toxins. Some toxins are actually more dangerous at lower concentrations than higher ones, and we have numerous studies showing that the relationship between toxicity and concentration is not linear, i.e. a straight line.

Despite that, the premise is still the basis of toxicology risk assessments, which have become pillars of misunderstanding and misinformation allowing chemical corporations to claim their products are safe and regulatory restraint is unnecessary.

Regulatory toxicology testing emphasized high dose testing on adult animals to discover a threshold below which effects, again primarily cancer, could be identified. The concept of chemicals crossing the placenta and affecting fetal development was an after-thought at best.

Toxicology risk assessments are basically numerical risk assessments. They were first used by the EPA in 1975 under pressure from industry to dismiss health risks of one of the earliest known toxins, vinyl chloride.[26] The EPA felt like they could respond authoritatively if they could calculate the number of cancers that would be caused by exposing the population to vinyl chloride. Federal regulators then started adopting that technique with many other environmental controversies and disputes with industry, and that methodology continues today. The makers of toxic chemicals and industrial polluters are pleased to see it continue because with supposedly hard, mathematical data, they can disguise the risks of their toxins.

This approach however dramatically oversimplifies very complex biologic, physiologic and molecular processes that don't lend themselves to mathematical equations. Furthermore, the assessments do not pass the acid test of scientific reproducibility. Two different "assessors" can come up with widely disparate results given the same input data, especially if they have different motivations[28]—subjectivity rather than objectivity. William Ruckelshaus the former EPA director wrote in 1984, "We should remember that risk assessment can be like the captured spy: If you torture it long enough, it will tell you anything you want to know."[27] In 1991, the National Academy of Sciences wrote, "Risk assessment techniques are highly speculative, and almost all rely on multiple assumptions of fact -- some of which are entirely untestable."[29]

Health risk assessments usually focus on diseases like cancer, but almost completely ignore other health outcomes like infertility, immunosuppression, endocrine disruption, altered behavior and reduced intellectual capacity. The impact on genetic integrity can influence the health of subsequent generations but examining subsequent generations are not usually not part of the research.

Through biological processes many toxic substances accumulate in organisms, concentrate in certain tissues of those organisms, and increase in concentration as they move up the food chain. Humans are at the top of the food chain and because our capacity to break down toxic chemicals is very limited, concentrations in our bodies usually increase over time. Lipophilic (fat-like) toxins like dioxins will concentrate especially in human breast milk. Nursing infants consume 10 to 20 times as much dioxin as the average adult. In just six months of breast feeding a mother can transfer 20 percent of her lifetime accumulation of chemicals like dioxins to her nursing child. If exposure to a certain chemical is deemed not safe for nursing infants, then that chemical cannot be considered "safe" for society.

Classical toxicology assessments assume that the hazard posed by each individual compound, tested out of context and in isolation, can predict the hazard of the entire complex mixtures of chemicals. Obviously no one on earth is exposed to merely one chemical. Furthermore, the potential for toxicity may be much greater than just the additive sum of all these chemicals individually. Specifically, synergistic (multiplicative) effects from chemical interactions are quite possible, if not likely, which would significantly increase the potential for health hazards.

Under this absurd methodology, a safe dose of aspirin, combined with safe doses of ibuprofen, Oxycontin, Toradol, Celebrex, acetaminophen, three glasses of wine, and a pint of Jack Daniels would all be considered individually "safe," but in reality might add up to be lethal. When a supposedly tolerable exposure to mercury is combined with a tolerable exposure to dioxins, to cadmium, lead, and arsenic, etc., etc., the end result can be an intolerable health consequence.

Traditional toxicology assessments wrongly assume that we have a comprehensive understanding of the complexity of biological processes and chemical toxicity when in reality there are vast information gaps. Lack of knowledge cannot be equated with safety; it can only be equated with lack of knowledge. Furthermore, risk assessments never take into account significant genetic differences in the vulnerability to toxins from one person to another. About 15 percent of the population has "multiple chemical sensitivity syndrome" and their response to chemical exposures can be dramatically different than the majority of the population.

Because of the obvious short comings of risk assessments, countries in Europe began embracing a more protective approach to toxic chemical exposures guided by the "precautionary principle." The principle can be expressed this way. When there is reasonable evidence of harm, society is compelled to act to protect the public, rather than wait for scientific proof. This adoption of the precautionary principle was even written into the Maastricht Treaty that formed the European Union. Journalist Peter Montague describes this difference between European and American regulation of chemicals this way. The European model asks, "How much is avoidable?" The American model asks, "How much harm is acceptable?"[30] Let me be even more blunt. The United States approach prioritizes protection of corporate profits rather than public health. The precautionary principle has not been allowed to enter the arena of government regulation under either Democratic or Republican Administrations because of corporate pressure. The result has been stunning regulatory failures.

The failures started with original chemical safety law, the Toxic Substances Control Act (TSCA), passed in 1976, virtually written by the chemical industry. It allowed all 62,000 chemicals that were on the market at the time to remain so, unless the EPA found at some later time that they represented unacceptable risk.

Only almost undeniable health consequences and apparent toxicity after consumer use and wide spread environmental contamination of the products needed to be reported to the EPA. If this seems to you like closing the barn door after the horses have already escaped, bear in mind that before 1976, there was no barn door at all preventing or regulating chemicals that were unleashed upon the public.

Furthermore, the law only allowed testing if information from the chemical makers suggested it was likely to be dangerous. That's like Jack the Ripper on trial for murder with the prosecution only allowed to use evidence provided by Jack the Ripper. Moreover, the law allowed chemical corporations to claim just about everything was a trade secret allowing them to hide critical information from the public, regulators and even health care providers. As a result, the EPA has only tested about 200 of those chemicals.

For chemicals entering the market since then, and there are now 140,000 different chemicals produced by industry, they are also allowed unless the EPA again determines they constitute unreasonable risk.[33] But they only have 90 days to make that determination and hardly ever do they have all the data they need. The end result is virtually all the chemicals made by corporations will make it into the bodies of consumers—from their kale smoothies, from their manscaping body wash, and from the carpet they walk on while they're manscaping with their kale smoothies.

For good reason, asbestos (see Chapter 10) is about as popular as cockroaches. It is one of the few substances that just about everyone recognizes is very dangerous. And you likely think that it has been confined to the museum of toxins. And you would be wrong. Despite it being an undeniable killer of tens of thousands of people, it is still legal to make and sell asbestos. When the EPA tried to ban asbestos in 1991, the asbestos industry went to court, and a judge overruled the ban because they hadn't considered the costs of the ban as required by the deeply flawed TSCA. While asbestos use has plummeted, it is still allowed in brake pads and to manufacture chlorine. The EPA in its entire history has only succeeded in banning eight chemicals--PCBs, dioxins, fully halogenated chlorofluoroalkanes (ozone depleters), hexavalent chromium (the Erin Brockovich chemical), and four metal working chemicals. The last time the EPA banned anything was in 1984.[35]

In June 2016, a bill passed Congress and signed by President Obama, was celebrated as a major overhaul of the anemic TSCA, the Frank R. Lautenberg Chemical Safety for the 21st Century Act. The fact that a Republican controlled Congress passed it tells you a lot about it. The fact that industry supported the bill tells you even more.

The bill took a few steps forward in protecting public health, and several steps back. But that potential progress is now being held hostage by the Trump Administration's eagerness to accommodate the American Chemical Council, the chief chemical industry lobby. The bill gave new authority to the EPA to test potentially toxic chemicals, but Trump's EPA Administrator is demonstrating that the EPA isn't really interested in using that authority.[34]

The absence of data is not proof of safety; it is only proof of ignorance. The most basic information is not available on over 90 percent of the industry inventory of chemical products. The Government Accountability Office says that 95 percent of the information given to the EPA by companies on their new chemicals are wrapped up in confidentiality claims, which makes them unavailable to regulators. Almost all the data that is available is supplied by the manufacturer whose obvious self-interest obscures objectivity. And the cumulative health impact of exposure to all these chemicals is ignored.

Manufacturers don't even have to reveal enough of the chemical mix of their products to allow government agencies to track their spread in the environment or to independently test for toxicity. The TSCA requires manufacturers to submit safety data only if they have it. Most don't, so the EPA is left with computer models to predict whether chemicals will pose health problems, and no one believes that is adequate.

In those rare instances where evidence stacks up implicating a chemical as a serious hazard, industry response follows a well-worn path. "By the time the scientific community catches up to one chemical, industry moves on to another and they go back to their playbook of delay and denial," said Deborah Rice, a former EPA toxicologist who now works for the Maine Center for Disease Control and Prevention.[17]

Regulatory impotence flows from the default assumption that industrial chemicals are safe, unless or until, unequivocally proven unsafe. Essentially human consumers are today's guinea pigs of the chemical industry, just like the previous generation was for DDT, PCBs, asbestos, radiation and lead. Once a dangerous product is put on the market, it is only withdrawn after it has already caused large scale, undeniable damage and death.

Numerous scientific bodies have recently issued official statements and warnings about the public health consequences of the chemicals we are bathed in. In May 2007, 200 of the world's foremost, pediatricians, toxicologists, epidemiologists, and environmental scientists at a worldwide conference issued this warning: "Given the ubiquitous exposure to many environmental toxicants, there needs to be renewed efforts to prevent harm. Such prevention should not await detailed evidence on individual hazards...Toxic exposures to chemical pollutants during these windows of increased susceptibility can cause disease and disability in childhood and across the entire span of human life."[12] The scientists explained that exposure to these ubiquitous chemicals could alter and impair development of critical organs in fetuses and newborns, increasing their chances of developing numerous chronic adult onset diseases, like diabetes, cancer, attention deficit disorders, thyroid damage, and infertility.

In June 2009, the Endocrine Society, comprised of 14,000 hormone researchers and medical specialists in more than 100 countries, warned that "even infinitesimally low levels of exposure [to endocrine-disrupting chemicals] –indeed, any level of exposure at all– may cause endocrine or reproductive abnormalities, particularly if exposure occurs during a critical developmental window. Surprisingly, low doses may even exert more potent effects than higher doses."[32]

In November 2009, the American Medical Association Board of Delegates approved a resolution that called on the federal government to minimize the public's exposure to BPA and other endocrine-disrupting chemicals. The measure was advanced by the Endocrine Society, the American Society for Reproductive Medicine and the American College of Obstetricians and Gynecologists.

In Oct. of 2013, the American College of Obstetricians and Gynecologists and the American Society for Reproductive Medicine, representing well over 50,000 physicians and other health care professionals, issued a joint statement, "Patient exposure to toxic environmental chemicals and other stressors is ubiquitous, and preconception and prenatal exposure to toxic environmental agents can have a profound and lasting effect on reproductive health across the life course." On their website they stated further that, "Reproductive and health problems associated with exposure to toxic environmental agents [include] miscarriage and stillbirth, impaired fetal growth and low birth weight, preterm birth, childhood cancers, birth defects, cognitive/intellectual impairment, thyroid problems"[13]

Endocrine disrupters, or "gender-benders," interfere with or mimic natural hormones, especially estrogen, which are the most potent biologic substances known. What accounts for the difference in physical appearance between Kim Kardashian and Dwayne "the Rock" Johnson are hormones that by volume wouldn't fill a grain of sand.

Endocrine disruptors that have found their way into the global environment from pharmaceuticals, plastics, cleaning products, pesticides, sewage and paint, have had feminizing effects on wildlife. Adding to what I mentioned in Chapter 3 on pharmaceuticals, endocrine disrupting industrial chemicals have undoubtedly contributed to these disturbing observations. Half the male fish in Britain's lowland rivers have been found to be developing eggs in their testes.[45] They are increasingly blamed for a sharp decline in human sperm counts. Although sperm measurement studies are somewhat controversial, one study from throughout the world showed sperm counts have fallen from an average of 150 to just 60 million per milliliter of sperm fluid in five decades. Another study showed a decline in average sperm counts falling further still between 1989 and 2005, from 74 to 50 million per millileter.[14]

While sperm start to die after about seven days of abstinence, it is highly doubtful that in the age of ubiquitous internet porn, lack of sexual activity is the reason for global decline in sperm counts. Many fertility experts believe humans are facing an ongoing and accelerating infertility crisis.

Testicular development begins in utero and continues until about six months after birth, although sperm itself doesn't appear until adolescence. Some studies have shown that pregnant mothers exposed to dioxins give birth to males that show low sperm counts later in adulthood. Male cigarette smokers reduce their sperm count about 15 percent but can largely reverse that trend if they quit smoking. But males whose mothers smoked during the pregnancy, show much greater reductions in their sperm counts, i.e. about 45 percent, with much higher rates of abnormalities, and that the effect is irreversible.[15]

Three decades of studies of cancers in wildlife have shown these are intimately associated with environmental contamination. This is particularly relevant as animals do not engage in poor lifestyle choices that are commonly thought to be the precipitators of poor health—like smoking, drinking beer, eating junk food, and sitting around all day watching NASCAR and Fox News.

An environment choked with carcinogens is of little interest to virtually any large corporation. In May 2010, the usually cautious US President's Cancer Panel, appointed by Pres. George W. Bush, reported that synthetic chemicals can cause "grievous harm" and that the number of cancers for which they are responsible had been "grossly underestimated." The panel advised President Obama "to use the power of your office to remove the carcinogens and other toxins from our food, water, and air that needlessly increase health care costs, cripple our nation's productivity, and devastate American lives."[36]

In their cover letter to President Obama, the panelists wrote, "American people – even before they are born – are bombarded continually with myriad combinations of these dangerous exposures." In particular, the report warns about exposures to chemicals during pregnancy, "when risk of damage seems to be greatest."

Immediately after the report, the American Cancer Society (ACS) issued essentially a rebuttal, criticizing it as speculative, and diverting attention from "more significant" factors like poor lifestyle choices; smoking, overeating and lack of exercise. For the ACS to suggest that Americans can only contemplate one cause of cancer is bewildering. But this "blame the victim" philosophy has been promoted by the ACS for over 40 years, with a trivialization of environmental risks and a determined dismissal of new research.

Dr. Samuel Epstein (see Chapter 2), author of *The Politics of Cancer,* former head of a Congressional committee on cancer, is only one of many critics that argue the ACS "priorities remain fixated on after the fact damage control--screening, diagnosis, and treatment" to the virtual exclusion of cause and prevention.

The ACS has long standing conflicts of interest with a wide range of industries that manufacture chemotherapy drugs, agrichemicals, and radiation therapy equipment. Their spokesman, Dr. Michael Thun, freely admits the society's corporate connections. "The American Cancer Society views relationships with corporations as a source of revenue for cancer prevention," said Dr. Thun. "That can be construed as an inherent conflict of interest, or it can be construed as a pragmatic way to get funding to support cancer control." Or it can be construed as what it is: undermining cancer prevention for their own financial gain--like the ExxonMobil of healthcare.

The plasticizer BPA

Bisphenol A (BPA) was first synthesized in 1891 with the very intention of making a drug that mimicked estrogen. It is little wonder then that when it was ultimately found to be useful in making hard plastic, epoxies and resins, that it behaved as its original research suggested, like estrogen. As a primary substrate for numerous synthetic, polycarbonate plastics, it was commercially produced in the early 1950s. It is a clear solid, making objects almost shatter proof and has become one of the highest volume chemicals in production. It became a ubiquitous part of a wide variety of consumer products from food packaging to cell phones.

When studies began to appear that challenged the presumed benignity of BPA in the early 2000s, images of toxic baby bottles in the mouths of America's infants began to haunt millions of mothers. The BPA emerging controversy stirred a pot filled with politics, health sciences, toxicology and economics. A front page article in the Washington Post in April 2008 read, "US cites fears on chemical in plastics."[43] Americans soon realized that with virtually every can of food and every bottle of water they were swallowing something toxic. With every sales receipt and DVD they handled, and every time they touched their computers and cell phones, they were getting BPA on their hands. In 2015, "BPA's global production was estimated at 5.4 million tones – about a wine bottle's worth of pure BPA for every person on the planet."[53] Predictably, numerous studies showed that between 90 and 100 percent of people tested, in both the United States and Europe, had BPA in their urine. This is all the more remarkable because BPA is broken down quickly in the body and that means that virtually all of us are continuously exposed.

The research on very low dose exposure to BPA showed estrogen mimicking effects with a wide range of poor health outcomes--increased rates of cancer, DNA damage, cognitive and behavioral disorders, and metabolic diseases like glucose intolerance[44]--- all related to endocrine disruption, i.e. hormonal dysregulation. BPA's adverse health effects were the result of its ability to play both offense and defense on the body's hormonal system. It mimicked estrogen, but it also blocked receptors for testosterone.

As this research emerged, Congress passed legislation requiring the EPA to design testing and screening protocols specifically for endocrine disruption. The EPA then asked the National Toxicology Program (NTP) to evaluate the research. In 2001 the NTP released its findings that indeed there was "credible evidence" that low dose exposure to endocrine disrupting chemicals, below the current "safety standards," could cause developmental harm, despite industry sponsored studies that contradicted their assessment. The NTP's recommendation to change testing methods to study low dose exposure and its unwillingness to exonerate BPA was met with immediate alarm and push back by the chemical corporations.

The American Plastics Council then turned to the Harvard Center for Risk Analysis for some comforting research. In case you are impressed, know that this Harvard Center receives financial support from the American Chemistry Council, the Society of the Plastics Industry, Dow Chemical Company, the Business Roundtable, Phillip Morris, and General Electric.[49] The Harvard Center's eventual report[50] was based on a methodology established during a meeting sponsored by the Annapolis Center for Science and Policy, an organization funded by tobacco behemoth Phillip Morris and ExxonMobil, who know a thing or two about throwing up smoke screens to obscure independent research. And the final report itself gave disproportionate weight to two studies, both funded by the American Plastics Council and the Society of the Plastics Industry.

Some of the researchers who felt the Harvard report was a sham, looked at the trail of money funding the research on BPA.[51] Over an eight-year period, from 1997 to 2005, 115 BPA studies examining low dose exposure, conducted by many different laboratories in several different countries, showed extensive changes in prostate and mammary gland development, chromosomal damage, immunosuppression, endocrine disorders, altered brain function and behavioral disorders. In fact, 90 percent of the studies that were funded by the government found some of these disturbing effects at or below previously presumed "safe" doses. Some studies found adverse effects at doses "orders of magnitude" smaller than "safe" doses." None...exactly none of the 11 studies funded by the chemical corporations found any effects.[51]

The federal government started paying attention in 2006. Dozens of national experts met at Chapel Hill, North Carolina, evaluated hundreds of studies, and released the Chapel Hill Consensus Statement which concluded "with certainty" that BPA, at levels typical of which is now found commonly in the human body of Americans, is associated with "organizational changes in the prostate, breast, testis, mammary glands, body size, brain structure and chemistry, and behavior of laboratory animals."[52]

The response of different countries to this granddaddy of reports was illuminating. In 2008, Canada's response was a reflection of the "precautionary principle" mentioned previously, i.e. public health protection should have priority. In contrast, the US Food and Drug Administration (FDA) decided to stick with the chemical industry and declared BPA safe at estimated levels of human exposure. Consumers however didn't buy the "not guilty," and chemical companies began urgently looking for alternatives. Product manufacturers wanted to proclaim that their products, like baby bottles, were "BPA free."

Finally, in a late and tepid response to consumer anxiety over the many studies that showed estrogen mimicking chemicals could increase the risk for developmental disorders in infants, the FDA banned the sale of plastic baby bottles that had BPA as a component in 2012. But as the chemical industry scrambled to find a replacement for BPA their approach was all too predictable. They turned to whatever was the quickest and most likely compounds to duplicate the desired characteristics of BPA.

It doesn't take a genius to predict that compounds that have similar structures would also have similar toxicology profiles. Compounds like BPA's "close cousin," bisphenol S (BPS) and bisphenol PAF (BPAF) for example, were launched allowing the chemical companies to claim that their product was "BPA free" and pretend that they were safe. It has become a musical chairs game of BPA, BPS, BPB, BPC, BPF, BPAF, etc., etc., which is nothing more than a game of "BS" (pardon the slang) because they had no logical or scientific basis for making the claim that any of the BP family were any safer than BPA. If the chemical companies could not be bothered by logic or science they could have looked at a little history. As early as the 1930s British researchers had demonstrated strong estrogen mimicking properties of bisphenol compounds in general, and multiple, specific BPA congeners, like BPB, and BPF.

Inexplicably, but perhaps predictably, regulators bought into the game. For the chemical companies the problem is solved because regulators will consider these BP substitutes as new and entirely "virgin" molecules that can be released upon the public, and then might be studied in depth for decades before they could be ultimately withdrawn, just like BPA, and the cycle repeats itself.

Recent research shows that 81 percent of Americans test positive for one of the new BPA substitutes, BPS, in their urine.[46] Everyone was exposed to BPS before the research on its safety had really begun, and when it did, BPS appeared to be no better, maybe even worse. Researchers found that exposure during fetal development of as little as "pico molar concentrations" (less than one part per trillion) of BPS can change how a cell functions in ways that can lead to "diabetes and obesity, asthma, birth defects or even cancer" later on in life.[46]

Zebrafish are favorites of some researchers because their brain development is similar to humans--more similar to some than others I might add, especially those people with colorful stripes, oversized fins and live under water. In zebrafish exposed to concentrations of BPA similar to what is found in municipal water sources, researchers found a burst of abnormal neuronal growth, even more than what they had found with BPA.[46] The young fish then demonstrated the fish equivalent of hyperactivity in children. Deborah Kurrasch, an assistant professor at the University of Calgary School of medicine, found that doses of BPA 1,000 times lower than the recommended daily allowable dose for humans, can change the growth and migration of neurons to their proper destinations in the brains of embryonic Zebra fish at a developmental stage comparable to the second trimester in humans.[47] This and many other studies compelled the authors to strongly recommend that the entire class of chemicals be withdrawn from the market. And of course that hasn't happened.

A study published in 2011[48] found that of 455 plastic compounds tested, virtually all of them leached estrogen mimicking chemicals, and that leaching was enhanced if placed under the stress of heat, boiling water, microwave radiation or exposed to the ultra-violet radiation of sunlight.[48] A plastic baby bottle can leach as many as 100 chemicals when heated.

Consumers have become familiar with the BPA-free label commonly found now on water bottles, food cans, cash register receipts, baby bottles and pacifiers, and the public has indeed been "pacified," but has also been conned. The subsequent generations of bisphenol compounds are turning out to be just as dangerous as BPA, and the chemical industry still doesn't have to test its products before they end up your blood stream. It is only well after we are all exposed, that evidence about toxicity becomes undeniable, and many lives have already been damaged or ended prematurely, that the regulatory machinery finally gets dusted off and creaks into action. And that's just the way the chemical corporations like it.

Flame Retardants: Making Fire Even More Dangerous

How did all these chemicals infiltrate our lives, indeed the very core of our lives, the very cells that constitute the human body, even the nuclei of cells deeply embedded in our brain stems? The story of flame retardants is emblematic of the pervasiveness of the problem, the motivations of the companies involved, the devastating consequences of the corporate profit imperative, and the abject failure of government oversight. And North American residents' contamination with flame retardants is ten to one hundred times higher than other parts of the world.[16]

Big Tobacco, mentioned in chapter 7, thrived on destroying lives and undermining public health. The more lives destroyed, the more money they made. But they managed to accomplish that feat in ways beyond getting people addicted to cigarettes. Big Tobacco in fact engineered a campaign to soak many of our household items, especially furniture and electronic appliances, in what turned out to be toxic flame retardant chemicals, in a highly successful strategy to divert attention from the fire danger inherent in smoking. A Pultizer Prize winning, six-part series from the Chicago Tribune, *Playing With Fire*,[37]reveals their central role in exposing everyone, including non-smokers, to deadly chemicals.

Decades ago cigarettes first became the object of public condemnation, ironically not so much for smoking related diseases, but for their role in killing people by sparking home fires. Placed firmly in the hot seat, pun definitely intended, tobacco executives decided that pursuing the most obvious solution—creating a "fire-safe" cigarette, one less likely to start a blaze—might hurt their marketing and appeal to consumers. So the industry insisted it couldn't make a fire-safe cigarette that would still appeal to smokers and instead shifted attention to the couches and chairs that were going up in flames. Of course that makes as much sense as blaming the existence of forests for forest fires, something that legislators in my home state of Utah have actually done.

With advocates for burn victims and firefighters pushing for changes to cigarettes, Big Tobacco changed the playing field entirely by launching an aggressive and cunning campaign to get firefighting organizations to adopt tobacco's cause as their own. The industry poured millions of dollars into the effort, doling out grants to fire safety groups and hiring consultants to court them. One executive of the Tobacco Institute boasted, "Many of our former adversaries in the fire service defend us, support us and carry forth our federal legislation as their own." State and federal lawmakers had tried unsuccessfully since the 1920s to enact laws requiring fire safe cigarettes. Ultimately regulators in the pivotal state of California succumbed to the Big Tobacco strategy of making everything but the cigarette fireproof. In 1975 the state of California passed a law known as Technical Bulletin 117 (TB117), which required polyurethane foam to resist an open flame, launching one of the biggest mistakes in public policy in the history of the US. Almost all furniture manufacturers found it too difficult to comply to more than one standard, so by default California's ill-advised flame resistant standard became the de facto standard of the entire nation.

Hiring a former fire marshal lobbyist, the three companies that made flame retardants, Albemarle, Chemtura, and Israeli Chemicals, eventually succeeded in a campaign to create worldwide standards requiring that the plastic casings of electronics resist a candle flame and posted internet videos comparing the flammability of computer monitors with and without flame retardants.

Achieving scientific consensus on any issue or controversy is usually a slow process, coming as the result of extensive peer reviewed studies, and eventually requiring the buy in of the majority of the world's relevant scientific experts. Changing public policy based on that scientific consensus is a subsequent and equally slow process, often times dragging out for decades, even after consensus among the real experts is reached. Most laws involving environmental protection have followed this painstaking and unfortunate scenario.

But this scenario was turned on its head for flame retardants. Despite the fact that independent testing showed flame retardants didn't work, the chemicals did virtually nothing to prevent fires in foam covered furniture or appliances, the foundation was laid for an entire industry to be created to keep us safe by drenching us and everything we touch in chemical additives. The scientific evidence supporting this public policy was virtually non-existent. Even the authors of the only two studies cited as supporting flame retardant efficacy have publicly stated their research was misrepresented. The evidence suggesting the hazards of flame retardants steadily mounted almost from the day they were introduced. Even the manufacturer sponsored studies raised serious concern. In 1977, a flame retardant brominated tris was banned from use in children's pajamas after researchers showed that it could damage DNA in animals (see chapter 16). But once a chemical is released to the marketplace, it is virtually impossible to remove it because of either the limited powers of the EPA, the enormous power of lobbyists, or simply apathy rather than political will.

"Despite consumer demands and even policy actions by states to phase out some of the worst flame-retardant chemicals, the chemical industry has not responded by producing safer chemicals, they're simply substituting other equally toxic chemicals," said Steve Taylor, program manager at the Environmental Health Strategy Center. "So we're stuck in a game of chemical Whac-A-Mole."[18]

Flame retardants are semi-volatile organic compounds. They are not embedded in the manufacturing process, but are sprayed on afterward. Therefore, they routinely escape as vapor or airborne particles that will land on surfaces or settle in dust. Friction and heat generated through normal use of a product, like furniture for example, can provoke their release. The cushions of a typical living room couch can contain up to two pounds of flame retardant chemicals that are in the same family as banned pesticides like DDT. Sitting or lying on the couch releases into your indoor air, a plume of flame retardant chemicals that is either inhaled or absorbed through your skin. However, most human exposure to flame retardants comes from ingesting or inhaling large amounts of contaminated household dust, rather than from people's diet or what they absorb through their skin.[19]

Blood levels of flame retardants doubled in adults every two to five years between 1970 and 2004. A typical American baby is born with the highest recorded concentrations of flame retardants among infants in the world. A study by the Environmental Working Group revealed that toddlers average five times higher levels in their blood than their mothers do.[20]

Essentially worthless flame retardant chemicals eventually permeated everything on the planet and they are largely impervious to degradation. But their lack of efficacy just scratches the surface of the absurdity and perniciousness of it all. The US government allowed one generation after another of flame retardants onto the market with minimal to nonexistent assessment of the potential health risks. Many of the chemicals bioaccumulate (they are absorbed into the body at a higher rate than they are eliminated), and biomagnify (the concentration increases as you move up the food chain). Humans reside at the top of the food chain, especially a nursing infant, and that's why even small concentrations of toxic compounds distributed throughout the environment, can translate into serious public health threats. Dr. Arlene Blum, UC Berkeley chemist and executive director of the Green Science Policy Institute, calls flame retardants the asbestos of our time.[69]

Firefighters have markedly increased rates of cancer. Female firefighters in California between ages 40 and 50 have a rate of breast cancer six times higher than the national average. Firefighters have increased rates of cancers of the respiratory, digestive, and urinary systems,[21] and significantly increased risk of brain cancer, leukemia, non-Hodgkin's lymphoma, multiple myeloma.[22]

Fires release common pollution components like particulate matter, recently declared the most important environmental cause of cancer. But other, more unique toxic compounds are present in smoke from fires, some of which ironically are the flame retardants that were supposed to make fires less dangerous.

Polybrominated biphenyl ethers, or PBDEs, have been the most commonly applied flame retardants. Studies in laboratory animals and humans have linked them to thyroid disruption, memory and learning problems, delayed mental and physical development, lower IQ, premature puberty and reduced fertility. The largest study of children and flame retardants, led by Brenda Eskenazi, director of Berkeley's Center for Environmental Research and Children's Health, showed that children with higher exposures to PBDEs in utero or in infancy have impaired physical coordination, attention and IQ. Since tris flame retardants were removed from children's pajamas in the 1970s, more than 3,000 peer-reviewed studies have documented the ability of similar classes of flame retardants to accumulate in humans and/or show adverse health effects.

Other flame retardants have been linked to cancer. Firefighters have been shown to have levels of the PBDE in their blood three times higher than the general population. Worse still, PBDE react with other chemicals during a fire and yield even more toxic byproducts, like dioxins and furans. Firefighters can have levels of those toxins hundreds of times higher than the general population. One firefighter described the toxicity unleashed by the average residential house fire as, "It's Love Canal, and it's on fire." Foam treated with PBDE gives off more carbon monoxide, soot, and smoke than untreated foam, and it's those three things, not burns, that are most likely to kill someone in a fire.

It may come as no surprise that the American Chemistry Council, the industry trade group, has come out defending flame retardants as necessary for public safety and dismissing studies linking them to adverse health outcomes. Only a few of these flame retardants have been taken off the market because of research suggesting serious health dangers. For example, chlorinated Tris (TDCPP), which replaced brominated Tris after it was voluntarily withdrawn, was itself voluntarily removed from children's pajamas in the 1970s because of studies linking it to cancer. But despite its danger, chlorinated Tris was not banned, and a recent study showed that 40 percent of couches manufactured between 1985 and 2010 still contained chlorinated Tris.[23]

A report, *Hidden Hazards in the Nursery*, from the Washington Toxics Coalition and Safer States and Maryland Public Interest Research Group (PIRG), found toxic flame retardants in 17 out of 20 new baby and children's products tested. Bassinet pads, nursing pillows, diaper changing pads, and car seats were some of the items tested. The most prevalent flame retardant found was chlorinated Tris. California recently classified chlorinated Tris as a carcinogen, especially for kidney cancer, and evidence links the chemical to neurotoxicity as well as hormone disruption. The Centers for Disease Control and Prevention estimate that 90 percent of Americans harbor flame retardants in their bodies. I wouldn't be surprised to see their manufacturers try and spin this as a good thing, and market them for "heart burn."

Because regulations in California played a pivotal role in launching the flame retardant debacle, California became the focal point of the efforts to undo the damage. According to an investigative report by Environmental Health News, the chemical industry, specifically Albemarle, Chemtura, Tosoh, and Israeli Chemicals spent at least $23.2 million to lobby California officials and donate to campaigns from 2007 through 2011, so that fire safety regulations that launched their highly profitable, highly hazardous, and totally worthless flame retardant business would be left untouched. As the threat of regulation changes gained momentum, these companies hired the public-relations firm Burson-Marsteller, which has served clients ranging from foreign military juntas to tobacco and pharmaceutical companies to those responsible for the Bhopal Disaster (see chapter 13) and the Blackwater private military fiasco.

Burson-Marsteller are experts at creating "astroturf" groups, for the purpose of ginning up the false perception of public support for a point of view that would otherwise be unpopular, which is exactly what they did for flame retardant manufacturers. They even paid a physician to fabricate emotional stories about tragically burned children to beat back any thought of regulatory change.

The *Playing with Fire* series revealed that the doctor's testimony was false and that a chemical industry front group had paid him $240,000 for his help. In 2014, the physician, Dr. David Heimbach, who was facing disciplinary charges in Washington State, surrendered his license to practice medicine.

Kaiser Permanente, the country's largest HMO system, concluded the evidence against flame retardants is sufficiently strong that it announced it will stop purchasing furniture treated with flame retardants. Kaiser released this statement, "Chemicals used as flame retardants have been linked to reproductive problems, developmental delays and cancer, among other problems. Concern over the health impacts to children, pregnant women and the general public has been growing in recent years, as scientific studies have documented the dangers of exposure."[70]

In response, pretending all the damning evidence didn't exist, the North American Flame Retardant Alliance (NAFRA) of the American Chemistry Council asked Kaiser to reconsider its decision.

"The use of flame retardants in upholstered furniture can help prevent fires from starting and/or slow the rate at which small fires become big fires, providing valuable time for persons to escape danger," Cal Dooley, the council's president and chief executive, wrote in a June 17, 2014 letter to Kaiser.

In Nov. 2013, Gov. Jerry Brown changed California's regulations so that furniture manufacturers were no longer required to use flame retardants, but did not out law them. Because it will be decades before all flame retardant furniture is removed from homes and offices, this tragedy will continue to bring ill health, cancer and death to thousands, if not millions, for many years into the future. Meanwhile, the retardant manufacturers enjoyed record profits as late as 2011 and global revenues of approximately $5 billion.

Despite California's forward move on the issue of flame retardants, a recent investigation showed that virtually no furniture salespersons in California knew whether their furniture had flame retardants in them or not, and no manufacturers' labels indicated whether they were present. Predictably, a bill recently introduced in the California legislature to require disclosure by manufacturers is being fiercely opposed by the American Home Furnishings Alliance, the North American Home Furnishings Association and the Polyurethane Foam Association. Beyond these and other industry groups, it is being fought by the National Federation of Independent Businesses (NFIB) and the California Chamber of Commerce.[24]

The scandal of flame retardants touches multiple corporations, trade associations and business groups. They are guilty of willfully ignoring the science that their source of profits is killing and harming thousands of firefighters and the public at large. They have become accomplices of the corporate masters of death—the Tobacco Industry.

Scotch Taping a Toxic Legacy

In 2005, chemical giant DuPont was given the largest corporate fine in the history of the EPA, $16.5 million, for a 40 year long cover up of serious health hazards of one of their signature products, a compound known as C8 (Teflon). As part of the settlement with the EPA, DuPont voluntarily agreed to phase out manufacturing of C8 compounds over the next nine years. Given that Teflon sales that year had reached $1 billion, and given that many of those health consequences were fatal, that settlement reeked of about as much justice as offering your neighborhood axe murderer a deal where he slows down his murdering of people over the next several years, ultimately agrees to start using a different axe, and pays a $50 fine.

The same year, DuPont entered into a $300 million class action settlement with 70,000 people who lived near its manufacturing plant in Parkersburg, W. Va. where it made C8 and dumped its waste into the local air and water. DuPont refused to accept any responsibility or admit any guilt, and the fine and settlement barely scratched the surface of DuPont's perfidy. Neither has it done much to abate the global health hazard of this family of toxic compounds given the broken system that utterly fails to protect public health from powerful corporations.

Teflon was supposed to be the poster child of the "miracle of modern chemistry"—a coating surface that does things that are almost too good to be true. It created a "miraculous" surface that made cookware truly non-sticking. Close cousins of Teflon were the main ingredient in stain and water repellant fabrics for clothes, carpets and furniture. Similar chemicals were made for fire-fighting foam, dental floss, fast food wrapping, pizza boxes, microwave popcorn, ski wax, and over 3,000 products that consumers were exposed to routinely. These miracle surfaces led to wonderful profits for Fortune 500 companies like 3M and DuPont. But there was a dark side to these products that can hardly be described as miraculous. It turns out the entire Teflon family are the chemicals from hell. And no one took more advantage of weak federal regulation than the makers of these wretched chemicals.

In the world of industrial chemistry, compounds that are composed of eight carbon atoms in a running chain are, unimaginably, known as C8. Perfluoroalkyl substances, like PFOS and PFOA (perfluorooctanoic acid), are C8 substances that have found extensive modern day utility because they are extremely resistant to break down anywhere--in the environment or in the plant and animal kingdoms, including humans. Longer chained fluorinated alkyls also tend to degrade until they reach a derivative state of eight carbons atoms, and at that point there is essentially no further break down. This means that even if all C8 compounds were banned, there would still be increasing levels in the environment. They are like "Terminator" compounds, you can never get rid of them. If in the last 20 years you have eaten anything cooked in non-stick cookware, eaten any fast food, worn any clothes, sat on any furniture, walked on any carpet, or flossed your teeth like you're supposed to, then you have C8 chemicals in your body that will never leave. In America, 99.7 percent of people have C8 compounds in their blood.[54]

Few of us are aware of our contamination, and virtually no one became contaminated through their own volition. These compounds are found in newborn babies, breast milk, umbilical cord blood, every type of wildlife ever tested, from polar bears to Asian tigers. One of the main companies that marketed and sold C8 compounds had put a safety threshold of PFOA in water for human consumption at 1 ppb (one part per billion). By 2003, tests showed that the blood of the average adult American had four to five times that amount. The average blood sample of people who lived in the Ohio Valley where DuPont's main Teflon manufacturing plant was located was 83 ppb. The median level for people living closest to the plant whose drinking water was contaminated by the plant was an astonishing 224 ppb. In 2009, the EPA set a preliminary limit of 0.4 parts per billion for short term exposure in drinking water. It was never finalized during the Obama Administration, and there is obviously no reason to think it will be under a Trump EPA.

For the first several decades after their discovery, the 3M corporation, perhaps most famous for Scotch Tape, was the main manufacture of C8 chemicals. When 3M was attempting to study exposure to Teflon in the late 1990s, they had intended to use as controls, "clean" blood samples from unexposed patients found in blood banks. Much to their own surprise, the only blood samples they were able to find that weren't contaminated with those Teflon chemicals was preserved blood from soldiers who died in the Korean War before the chemicals were placed on the market. The company then realized that they had basically contaminated every person in the country, if not the world.[55] Shortly after, 3M issued a nondescript press release announcing that they would begin phasing out production of PFOS because of "new data" that showed low levels of the chemical had been widely found in the environment and in people.[56]

To the casual observer this may have looked like a responsible thing to do--a corporation volunteering to sacrifice their business as a precautionary gesture to protect public health. But as we all know, looks can be deceiving.

In 2010 the state of Minnesota filed a lawsuit against 3M because it turns out the company knew 40 years ago that their C8 chemicals were accumulating in humans and in fish. Within a few years after that their own research began showing toxicity to the immune system of humans. The suit claimed that 3M "acted with a deliberate disregard for the high risk of injury to the citizens and wildlife of Minnesota."[57]

After seeking $5 billion in damages, in Feb. 2018 Minnesota settled with 3M for $850 million. Documents revealed by the Minnesota Attorney General showed numerous examples over 40 years, of 3M altering or concealing their own research, and painting over the human disease potential of C8 and their other products with a deceptive veneer of safety. Moreover, according to an exhaustive report by *The Intercept*, it appears that 3M abandoned or failed to pursue other research that should have been precipitated by their own studies showing alarming results.[57]

As other chemical companies have done, 3M also heavily promoted research by academics who were funded by 3M, like John Giesy, a professor at the University of Saskatchewan, and a fellow of the Royal Society of Canada. Giesy claims in his biography to have published "more than 1,100 peer-reviewed articles" and is essentially the world's top scientist in his field. Giesy was accused by the Attorney General of Minnesota of accepting at least $2 million from 3M to stage manage the science intended to investigate the potential toxicity of C8 compounds while he worked at Michigan State University.[58]

The Attorney General accused Giesy specifically of portraying himself to his scientific colleagues as an "independent" editor of scientific journals, while insidiously; accepting payments from 3M, referring to himself as part of the 3M team in private communications, working to prevent the publishing of studies showing toxicity of 3M chemicals, sharing unfavorable, confidential research manuscripts with 3M employees, trying to erase any paper trail connecting him to 3M, and telling 3M that he would "buy favors" from other scientists on behalf of 3M.

Despite spending the bulk of his career as a professor at public universities, Giesy found some way to amass a net worth of about $20 million. The Attorney General's complaint stated, "This massive wealth results at least in part from his long-term involvement with 3M for the purpose of suppressing independent scientific research."[59]

Not surprisingly, Giesy, 3M, and the University of Saskatchewan all dispute this characterization of Giesy and his work. After reviewing the documents and Giesy's response, University of Manitoba academic ethicist Arthur Schafer concluded that Giesy's response is unconvincing and that the allegations against him are "the most serious he's ever seen."[59]

3M Hands the Ball Off to DuPont

While eight different companies were involved in the production and sale of C8 compounds, DuPont was by far the biggest player. Only few years after DuPont launched their business strategy that lead to contaminating the world with leaded gasoline (highlighted in chapter 1), they began their next chapter of toxic contamination of humanity with their C8 compounds. In the 1930s DuPont hired an in-house physician, pathologist Wilhelm Hueper, to investigate what appeared to be an epidemic of bladder cancers among its workers. When Hueper reached the point of publishing research that implicated DuPont's products in the cancer, the company prohibited him from publishing the research and fired him in 1937.[60]

Documents unsealed during lawsuits involving 3,500 individual claims revealed that DuPont became aware of possible toxicity of C8 compounds as early as 1954. By 1961 DuPont's own researchers had confirmed that toxicity. Over the next several decades DuPont in-house scientists found that C8 compounds caused "enlargement of rats' testes, adrenal glands, livers, and kidneys."[60]

A 1965 DuPont study of rats found that just a single dose of the chemical could have a prolonged effect. Two months after such an exposure the rats showed livers that were still about three times larger than normal. C8 compounds' resistance to degradation in the environment had parallels within the human body. They bind to proteins in plasma and are carried throughout the body, penetrating every human organ.

In the 1970s DuPont's studies showed that workers were accumulating C8 in their blood, they were showing increased rates of endocrine disorders, and their liver tests were more likely to be abnormal. Rhesus monkeys fed C8 compounds became chronically ill and died. DuPont's own studies showed that there was no safe level of exposure to C8 compounds in animals. But the company did not alert any government regulators. By the 1980s, animal studies showed higher rates of birth defects in animals and a few of Dupont's pregnant women employees gave birth to babies with facial birth defects, prompting the company to move its female workers out of the Teflon division.

Confronting the fact that C8 was toxic started presenting waste problems by the 1960s. So DuPont first buried 200 drums of it on the banks of the Ohio River not far from their plant. Then they came up with the brilliant idea of loading the drums onto barges and dumping them into the ocean, and they continued to do that until 1965.[60]

DuPont then started disposing of its C8 waste in unlined landfills and ponds, sending it up company smoke stacks and just dumping it into the Ohio River. By 1984, Dupont's own tests revealed C8/PFOA in the drinking water of communities down-stream. Nonetheless, the corporation continued to increase production and told no one outside the company of the pollution of municipal water or the health risks revealed from their in-house studies.[61]

DuPont's in-house researchers had proven that PFOA could cross the placenta and expose fetuses to the chemical. They also knew that a 3M study in rats found the same facial birth defects as were found the babies of some of their pregnant women employees. But DuPont didn't want to find any more damning evidence or deal with the implications of the evidence they already had. So DuPont terminated their studies and did not report their findings to the EPA despite being required to do so by law.[62]

Instead of doing anything to reduce their emissions or reconsidering the basic safety of the chemicals, DuPont responded by doubling down on C8, ramping up their use of the product and keeping quiet about what they knew of its toxicity.

In 1989, DuPont learned the employees at their West Virginia plant had an abnormal number of leukemia deaths. Later that year they found an abnormally high number of kidney cancers within their male employees. Still they said nothing.

For over 50 years DuPont studied and documented the health effects of C8 compounds in animals and humans--the testicular tumors, enlarged livers, kidney cancers, and the endocrine disorders. In 1991 DuPont knew that livestock were drinking from water down-stream from one of their plants that they had contaminated with C8 compounds at a level of 100 times greater than the internal "safety" limit the company had set for drinking water.[63]

In 1999 DuPont and 3M worked together on a study in monkeys in the hopes of establishing an exposure dose that could be classified as safe. Instead they found that even the lowest possible doses of C8 compounds were making the monkeys sick and concluded that there was not a safe level of exposure.[63] The results of the study prompted 3M to give up on manufacturing C8 compounds. But not DuPont. Studies showing the toxicity of C8 compounds were buried by the companies that made and sold them for at least 20 years.[64]

The most recent studies on C8 compounds have tied them to an extremely broad range of health effects, including cancers of the prostate and ovary, lymphomas, infertility, arthritis, ADHD syndrome and immune disorders in children. An independent panel of epidemiologic experts that was formed as part of the settlement between DuPont and the 70,000 members of the nearby communities, studied the possible connection between C8 exposure and human diseases. After following 70,000 people over seven years, and producing 35 peer reviewed studies, the panel found probable links between exposure to these chemicals and 55 different diseases. In 2006 the EPA confirmed that PFOA is a likely human carcinogen.

In 2016, the EPA released an official, "voluntary" advisory for PFOA and PFOS exposure. In it they warned against concentrations in drinking water exceeding 70 parts per trillion.[65] That concentration is equivalent to 70 grains of sand in an Olympic-size swimming pool. A study from The Department of Health and Human Services concluded that number is six times too high for infants and nursing mothers, but that study is currently being suppressed by the Trump Administration.[62] Other reviews of the data indicate that current guidelines are "orders of magnitude" too high for human exposure.[66]

If there is any good news in this story, the blood levels of C8 compounds in the U.S. population is falling. But the bad news is that, once again, we are all victims of chemical "whack-a-mole." Like with so many other compounds, the law allows DuPont and other chemical companies to replace C8 compounds with similar chemicals with only minor changes, whose toxic potential has not been tested, and which may turn out to be just as dangerous to human health, maybe even more so. And in 20 years we'll repeat exactly the same cycle

Rest assured that cycle has already begun. DuPont is marketing a second generation C8 type compound, a fluoropolymer they call GenX. DuPont filed 16 reports with the EPA showing evidence of numerous health effects from GenX in animals, "including changes in the size and weight of animals' livers and kidneys, alterations to their immune responses and cholesterol levels, weight gain, reproductive problems, and cancer."[67] In other words the same package of outcomes as with C8 compounds.

A report sent by DuPont to the EPA shows that the biodegradation of GenX is essentially zero, the same as C8 compounds. Chemours, the chemical company spun off from DuPont in 2015, produces and sells GenX, releasing it into the environment in unknown quantities because, like with tens of thousands of other compounds, there is no current government regulation to stop it, minimize it, track it, or prevent human exposure to it.

The sordid tale of the Teflon family of C8 compounds has much larger implications than their own toxicity, and the corporate behavior of 3M and DuPont. It is an illustration of the capture of the regulatory process by the very industries that are supposed to be subject to regulations, and the extent to which corporations will endanger the lives and health of populations large and small to protect profits, and the all too common complicity and corruption of academic personnel and institutions.

Children gleefully playing behind a truck spraying DDT in the 1950s.

5

Dirty Energy Death Star:
Writing Your Personal Obituary

"If another country came in here and blew up our mountains
and poisoned our water, we'd go to war.
But a company can do it."

—Paula Jean Swearengin
2018 Democratic candidate for US Senate
West Virginia

The name "London fog," has a certain allure to it, historically evoking mystery (ala Sherlock Holmes), danger (ala Jack the Ripper), style and sophistication (ala the famous clothing line). The name "London Smog" is much more pernicious, but is far more descriptively accurate. London was the site of the most famous air pollution event of the modern world, the Great London Smog of 1952.

On Friday, Dec. 5, 1952, a meteorological "inversion" began setting up over London--cold air was trapped by a high pressure system of warm air above. Emissions from public diesel buses, industrial smoke stacks, and coal from home furnaces quickly became concentrated near ground level. A 30-mile-wide patch of smog became so thick that people reported not even being able to see their feet as they walked. The city became paralyzed, transportation ground to a halt. A slippery black "ooze" coated sidewalks, walls, clothes, and faces. Indoor movie theaters closed because the smog obscured the screens. But the Great London Smog was much more than an eerie, aesthetic affliction, it was deadly.

The event lasted less than five days, but deaths sky rocketed within a matter of hours. Initially, 4,000 people died, most from pneumonia and bronchitis and most of the victims were at both ends of the age spectrum.[1] But mortality rates didn't return to normal for six months, and looking back, epidemiologists estimate that the five day event cost 12,000 Londoners their lives, and 150,000 were hospitalized.[2] The four day event is regarded as the sentinel air pollution event of 20th century civilization, leading to the shaping of public opinion and the first of government regulations.

The great London Smog of 1952 was the sentinel air pollution event of modern civilization

Pictures, movies and dramatizations of the event almost seem like fiction. But the event was all too real, and is still impacting lives today. On a smaller scale, similar short term, deadly pollution nightmares occurred in Belgium in 1930, and Donora, Pennsylvania in 1948. More recent studies of the aftermath of the 1952 Great London Smog greatly expand our understanding of the residual health consequences of the event and of air pollution in general. Researchers looked at the rates of lung disease among the survivors who were either in utero, or in the first year of life during the five-day event.[3] Sixty years later, those who were exposed to just that weekend of smog, reported an increased rate of asthma during childhood and adolescence of 20% compared to unexposed controls. The presence of asthma that persisted into adulthood was 8-10% higher.

This study is only one of hundreds that demonstrate even short term air pollution can have life-long consequences if the exposure occurs during critical developmental windows of fetal life. Other consequences of in utero exposure will be dealt with later in this chapter.

My organization, Utah Physicians for a Healthy Environment (UPHE), was formed as a response to a dense, month long winter inversion in January 2007 that blanketed Salt Lake City, much like the Great London Smog of 1952. What made it even worse was the lack of attention the inversion received from the media, lawmakers, the public, or even health professionals. As physicians, the original members of UPHE knew that this build-up of pollution was a health hazard, but even what we knew only scratched the surface of what medical research had already established. A silent killer, like the destroying angel depicted in the movie the Ten Commandments, had descended over the urban areas of the state, but no one was even talking about it. It was apparent to us that virtually no one realized the depth of community wide danger that was unfolding. And that fact was hammered home a few years later.

Shooting the Messengers

Just about the polar opposite of London, is Vernal, Utah, a town of 10,000 in rural Utah, surrounded by sage brush, vast expanses of beautiful deserts, mesas splattered with a color palate of soft pastels, and lonely plateaus. Vernal is a blue collar town. There are no strip joints in Vernal, and not a lot of bars. A lot of people go to church on Sunday. Like most of Utah, it's a Mormon town that trumpets family values. Drilling is one of those family values. Bumper stickers in Vernal say, "Honk if you love drilling." Vernal politicians certainly do. Utah state lawmakers do, their solid Republican Congressional delegation does. With jobs, increased tax base, new community recreation centers, burgeoning store fronts on main street, people with money to spend--what's not to like? Well, perhaps dead newborns.

Donna Young was a midwife in Vernal with 20 years' experience managing home births in Idaho and Utah. She lives in the Uinta Basin, the heart of the fossil fuel drilling frenzy in Eastern Utah. On May 8, 2013 she had the first still birth of her career. At the funeral service a few days later, she noted what seemed like an extraordinary number of infant graves with recent dates at the cemetery. She decided to investigate.

She didn't get any help from local authorities, but eventually information gleaned from obituaries and mortuaries revealed 15 cases of infant mortality (most of them still born, or death shortly after birth), in 2013. Looking back to 2010 revealed a modest upward trend, but then a huge spike in 2013. This is sparsely populated rural Utah. The entire county's population is about 35,000 people, so we're talking small numbers. But per capita this was a perinatal death rate six times the national average. It is all the more concerning because national infant mortality rates have been dropping slowly and steadily for almost 50 years, including about 10 percent in the last decade.

What is going on in Utah's Uinta Basin to explain newborn babies dying? An abrupt surge in teenage mothers, drug or alcohol use? No evidence of that. Is there a genetic explanation? Genes don't change that quickly. Is there a sudden onset of medical incompetence by the area's health care providers? No reason to think so. That leaves one other possibility. Is there something happening in the environment? As a matter of fact, yes.

Major cities with pollution problems have either high ozone, like Los Angeles, or high particulate pollution, like Salt Lake City, depending on the time of year. But the Uinta Basin has both simultaneously, making it unique and probably the most polluted part of the state, and one of the most polluted parts of the Western United States. Studies suggest that the two may act synergistically to impair human health. Add to that high levels of the by-products of oil and gas drilling, storing, processing and transporting—diesel emissions and hazardous compounds like benzene, toluene, and naphthene--and you have a uniquely toxic air pollution brew in Vernal.

A study by the University of Colorado[5] in fact showed that during the height of the drilling activity in the area in 2013, the levels of VOCs in the atmosphere were as high as what would be produced by 100 million cars, eight times more cars than are registered in the Los Angeles Basin. That is indeed a pollution nightmare. Wherever you have a pollution nightmare, if you look for it, you will find a public health nightmare.

In 2014 Donna Young approached UPHE with her observations about Vernal's infant mortality. We reviewed the literature, became concerned about it ourselves, and eventually approached the local and national media with our concerns. The "Vernal newborn death controversy" was covered by Newsweek,[6,7] the LA Times,[8] Denver Post,[9] Rolling Stone,[10,11] Bill Moyers,[12] and regional newspapers.[13] I appeared on MSNBC to discuss the possible causes and implications. I soon became a target of the fossil fuel industry, and state officials. I have learned many times over, that in a very conservative state, with extraction industries wielding great economic and political clout, impugning the safety, wisdom, integrity, consequences, legality or public interest of those businesses can very quickly make you an "enemy of the people."[14]

Donna learned an even harder lesson. She was targeted by the community's power brokers as whistle blowers often are. She received a threatening "legal" letter from the local hospital. She was told by one of the local doctors that everyone wants to take her down "politically" and ruin her career. She also received ominous, anonymous phone calls threatening her physically. Someone tried to poison livestock on her ranch. She began carrying a gun to protect herself.

Undaunted and somewhat naive, I decided that several of our UPHE doctors would travel the 150 miles to Vernal to conduct a town hall on how pollution harms public health, and newborns in particular. Although our fears of possible physical violence thankfully went unrealized, the public turn-out was sparse at best, with almost more media in attendance than towns people. Donna came accompanied by a body guard.

After our presentation I had the "honor" of experiencing a fracking waste pond, up close and personal. Our intent was to take measurements of the concentration of VOCs in the center of town and near the fracking operations. It turned into a real eye-opener, perhaps more appropriately, a "nose opener." On a clear spring morning with virtually no wind, temperatures about 65° F, accompanied by a film crew and a journalist from the *Rolling Stone*, Paul Solotaroff, we hiked about a half mile from the highway to a fracking waste water pond.

When we came within about 400 yards of the pond we noticed that there were no employees on the site. At about 200 yards away it became obvious why. We were stunned, almost physically overcome, by the powerful fumes. We covered our noses and mouths with cloths and masks but to no avail. Our plan was to hang air sample canisters on the fence surrounding the pond and wait the 30 minutes required for them to fill up. No one else was volunteering to take one step closer, so I covered my face, and ran as fast as I could to the fence and ran back, trying to not take a breath during the "storming of the pond." Paul told me a month later that after visiting the fracking pond he acutely lost his sense of smell, and it hadn't come back.

We visited another site that we were told was another fracking waste water operation just a few miles from Vernal. From a distance it looked to us like a farm, in that an agricultural sprinkling system was mounted on huge slow rolling wheels, moving pulsating sprinklers throwing water into the air. When we came within about 500 yards of the site, it became apparent this was no farm, and that was not just water being sprayed into the air, it was fracking waste water. The smell was the same and just as noxious as what overcame us at the first pond. It wouldn't have taken much wind to blow that misted, toxic waste water all over Vernal itself. Vernal residents commented to us that they could frequently smell that fracking pond several miles away in the center of town. About the time we had scrambled through some barbed wire to hang our sampling canisters, a menacing looking guard in a truck came by to shoe us off the property. We got our air samples anyway.

Air pollution from fossil fuel combustion, be it from smoke stacks, tail pipes, or fracking, or the personal air pollution of cigarette smoking, is a combination of three types of toxins. 1. Gases, like ozone, carbon monoxide, nitrogen and sulfur oxides (affectionately known as NOx and SOx), and VOCs (volatile organic compounds like benzene). 2. Particulate pollution, which includes carbon based "soot," dust, and liquid droplets formed in the atmosphere from precursors like NOx and SOx, and ammonium compounds. 3. Chemicals usually attached to those particles. Inhalation of any of these three components is treated by the body as an act of invasion, which is repelled by the body by generating a cascade of inflammatory chemicals to fight back. It's much like defenders of a medieval castle launching lighted arrows into the surrounding forest to repeal invading hordes. And like lighted arrows, the inflammatory chemicals may drive off the invaders, but a forest on fire will come back to haunt the defenders of the castle as well. The inflammatory response narrows arteries and reduces blood flow. But the particles themselves, once inhaled, can be delivered by the blood stream to virtually any cell in the body, and any critical organ. They have been found to infiltrate the kidney, liver, the lining of the blood vessels, the brain and the placenta of a pregnant mother. They preferentially accumulate in parts of the blood vessels that are already narrowed and atherosclerotic. Once inside a cell, the smallest particles can penetrate the nucleus of the cell where the chromosome lie, and interfere with the functioning of chromosomes.[15]

Diseases of virtually every organ system can follow. Strokes, heart attacks, every type of lung disease, cognitive impairment, cancer, accelerated aging, and sudden death, including infant mortality, all occur at higher rates among people exposed to air pollution. In the case of a pregnant mother, the placenta, perhaps the most vascular of human organs, is compromised for the same reason, and it should be easily understood that pregnancy complications and impaired fetal development—think birth defects, miscarriages, and still births--can be the result. Many epidemiological studies show that to be the case.

That increased infant mortality in the Uinta Basin could be the result of the increased air pollution is certainly consistent with well over 160 medical studies. It is not only plausible, but very likely. Although it is still early in the arc of medical research, other studies show harm to infant health specifically from proximity to fracking operations[16] and other types of fossil fuel operations, like living downwind of a coal fired power plant,[17] or near mountain top removal coal mining.[18]

Our physicians group also came to learn that two Vernal families who lived almost next door to each other, had infants with the same congenital disorder called tracheomalacia, which is a fancy term for a cartilage defect of the trachea that can cause it to collapse and obstruct normal breathing. It can be a life threatening disorder, especially in young infants, and especially so in the setting of a respiratory infection common to infants. Fortunately, most patients improve as they grow older. We were unable to get any of the local physicians to help investigate this, but one of the mothers found out that the pediatric clinic in Vernal was seeing 32 patients with that diagnosis, which given the size of the local population and the number of patients seen in that clinic, was about 7-8 times the number of patients you would expect with a birth defect that affects one in 2,000 babies.

The Vernal, Utah drama is also a metaphor for a much larger battle with global implications on two fronts. First, fossil fuel combustion wherever it occurs, has serious, immediate and long term health consequences from the pollution emitted as described above. Let's be more blunt—it kills people: a lot of them. Second, its role in the climate crisis is also killing people now, and many more in the future, and we'll address that it the next chapter.

While the medical research on air pollution is much more complete in 2019 than it was 20 years ago, it is no longer new science. But in the early 1990s the idea that air pollution actually killed people was new, and it was in large part brought to the public arena by researchers like Dr. C. Arden Pope, an economics professor at Brigham Young University.

Pope is handsome, middle aged, but looks younger with a full head of dark hair, and enjoys the physical fitness and physique of someone who exercises a lot. As such, he has reason to be concerned about the air he breathes while he exercises. Pope's personal life, demeanor, research, and career is above reproach, just like his physicality. He is unflappable, but avoids the spot light, and is hardly a self-promoter. He grew up on a farm and has often said, "I didn't know the word environmentalist ever stood on its own. I only heard 'darned' environmentalist." He talks in a southern Idaho accent he calls, "muddy." [19]

Pope's groundbreaking research on particulate pollution in 1988, examined the rate of hospitalizations for respiratory complications of children in Provo, Utah, his home town, where a steel mill had closed down. He found that those rates had dropped in half compared to the rates when the steel mill was in operation. The head of clinical research at the EPA at the time said, "particles [particulate pollution] did not hit our radar screen until [Pope's Geneva Steel paper] came out. I mean, no one even thought about them." Up to that time researchers had focused on ozone. But Pope's study laid the ground work for a shift to particulate pollution, which is now known to overshadow ozone as a toxin and is now widely regarded as the most pervasive environmental health scourge.[20]

Pope's results were not surprising to anyone who lived near Provo. Before the steel mill shut down, Geneva would blanket Provo and Salt Lake City in a suffocating layer of brown soup that would often completely obscure the mountains on Salt Lake City's west side and rival the Great London Smog. Even without research to confirm it, locals knew something that ugly just had to be unhealthy, if not deadly.

Pope's results unleashed the wrath of the owners of the steel mill, two high profile pillars of the community, brothers Joe and Chris Cannon, and much of the city, because the steel mill was the largest employer of the county. For much of the community when there were jobs and money to made, the truth could be very unwelcome—a story reminiscent of Donna Young's. But this was just the beginning for Pope trying to do research in the cross hairs of polluting industries. He collaborated with other world class researchers like Doug Dockery and Joel Schwartz on epidemiologic papers, one that followed 8,000 people and a second one that followed 500,000 people, that yielded results that left even the researchers stunned. Both studies showed, unequivocally, very low concentrations of particles, at the levels that hung over most US cities, were actually killing people, a lot of them. These papers became the bedrock of the EPA's clean air standards and their enforcement, with enormous public policy implications.

Big polluters were furious. Rather than accept the science and their culpability and responsibility, an extraordinary coalition of 500 industry groups launched a coordinated attack not just on the science, but the scientists themselves. When the EPA eventually sided with Pope and his colleagues, industry immediately went to court to stop the EPA from saving lives. Finally, in 2001, the US Supreme Court ruled in favor of the EPA, and the new standards went into effect. Despite being vindicated, Pope would be the last person to gloat.

All three of the types of pollution released by the combustion of fossil fuels, wood, or biomass--gases, particles, and chemicals-- are toxic to human health. Inhaling air pollution has basically the same systemic health consequences as cigarette smoking, only to a lesser degree--unless you're doing your inhaling in Beijing, China, then eliminate the "lesser." A world renowned group of researchers released a report in 2017 in the premier British medical journal concluding that air pollution is the number one environmental cause of human deaths, killing more people than road accidents, violent crime, wars, and fires combined. Worldwide it is responsible for one in six premature deaths, approximately 9 million total, and inflicts an economic loss of $4.6 trillion, 6.2 percent of the global economic output. In the most severely affected countries, pollution is responsible for one in four deaths.[21] The World Health Organization (WHO) has calculated that nine out of ten people in the world breath enough air pollution to harm their health.[22] Over half of the world's population live in cities, and many of the largest cities in the world have "achieved" levels of air pollution that are almost unimaginable. And it's not just the meccas of pollution in China and India. London is drifting back towards the toxic days of the 1952 smog event, as is Paris and the other major cities of Europe.

About 210,000 premature deaths occur annually in the US due to air pollution.[23,24] According to this study, 200,000 deaths could be attributed to particulate pollution, and 10,000 could be caused by ozone. Another study suggested that ozone is a more sinister health menace. Through its interference with healthy fetal development and placental viability, ozone has been estimated to cause 8,000 still births in the United States alone.[25]

So while it's inarguable that fossil fuels once provided the energy to launch modern civilization, it is also inarguable that today, with clean energy alternatives, they have become a deadly anachronism, and the purveyors of fossil fuels and their handmaidens in government are not just willing "Merchants of Doubt,"[26] but also merchants of death. The health consequences of air pollution are indeed staggering, and they extend far beyond premature death. Lobbyists for the dirty energy cabal of coal, oil and gas companies have been virtually living in the offices of their allies in Congress ever since the 2010 "Citizens United" Supreme Court ruling allowing suitcases of "campaign contributions" to unravel the protections of the Clean Air and Clean Water Acts.

The remainder of this chapter is a detailed, scientific review of the extent of the health consequences fossil fuel corporations are willing to subject all of us to, in order to maintain and enhance their bottom line. In an era of short attention spans some may find it mundane. But in this case, mundane may save your life.

Correcting the Myths

Peabody Coal and ExxonMobil want you to think that the air you breathe is clean enough. These companies profit from misconceptions about air pollution prevalent among the public, politicians, and regulators. Somewhat surprisingly, lack of familiarity with the medical science is even the norm among health professionals. Let's unravel these misconceptions.

1. Contrary to the very existence of federal air quality standards, there is no safe level of air pollution.[27,28] Even concentrations well below the EPA's National Ambient Air Quality Standards (NAAQS) for particulate pollution and ozone are not safe. Specifically, we know they still have a strong association with premature death. You may ask, and rightly so, "What is the point of having the standards?" The answer is simply that they are far better than nothing. The EPA has an advisory board, the CASAC (Clean Air Scientific Advisory Committee), that traditionally is composed of some of the nation's best air pollution researchers and experts. But regrettably, the EPA often doesn't follow their recommendations, and under both Democratic and Republican Administrations, polluting industries and their allies in Congress often succeed in applying enough overt and behind the scenes political pressure to keep the NAAQS weaker than the science alone would dictate.

Not only is there no safe level of air pollution, but the relationship between pollution and the most serious health outcomes, like sudden death, is not linear, it is hyperbolic. Plotting the graph between those two shows a steeper curve at low doses. In other words, per unit of air pollution, there is an even stronger correlation with disease consequences at low doses.

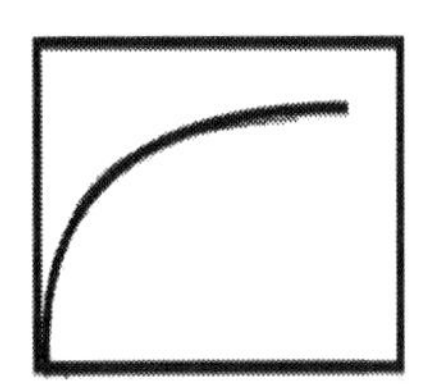

An example of a hyperbolic curve. Air pollution concentration is plotted on the horizontal axis, and the health consequences, like sudden death, are plotted on the vertical axis

Another clinical example of that curve is: smoking just one cigarette a day is half the risk of smoking a full pack of 20 cigarettes.[399] That same principle applies to community air pollution, which further undermines the idea that a little bit is acceptable.

2. Virtually everyone is affected whether or not they have overt symptoms. For most people it will be shorter life spans than they otherwise would have enjoyed. A pregnant mother is unlikely to have

any symptoms even if her baby is being harmed. In an elderly adult often the symptoms are manifest only after the damage has been done, like with a heart attack or a stroke, or the evidence of cancer has emerged.

3. Although everyone is affected there are genetic differences in disease susceptibility provoked by air pollution. Moreover, the coexistence of other diseases, like diabetes, can magnify the potential for other disorders.

4. Even short term air pollution matters, the London Smog weekend is the quintessential example.

5. Air pollution averages don't tell the whole story. Sometimes averages don't tell any of the story. Imagine a house in Oklahoma's tornado alley destroyed in seconds by a tornado, with wind speeds of 240 mph. The 24-hr average of the wind speed at the house's front door was normal, despite the house being destroyed by the wind. We often hear resistance from lawmakers using the argument that our average air pollution is "fine," or within government standards, therefore, "Not to worry." The average air pollution over London in 1952 was likely "normal," but thousands of people still succumbed to one particular smog event.

6. Timing is critical. Brief exposure during the first trimester of intra-uterine life has a much different impact on one's life compared to the same brief exposure at middle age or even during adolescence.

7. What's important is not the amount of pollution emitted into the air, but what and how much pollution a person or community of people actually inhale. In fact, what's inhaled is not as important as what ends up inside the cells of your body. Obviously these three are related, but not as closely you might think. The name of this concept is "intake fraction," and it takes into account microenvironments, discrepancies in the characteristics of different types of pollution, and how many people actually inhale the pollution emitted.[29] Regulators focus on how much pollution is emitted, but ideally the intake fraction would receive much more consideration than it currently does.

8. Not all pollution is created equal. The same concentrations of particulate pollution can have much different consequences depending on the type and source of the pollution, the size of the particles and the chemicals that may be attached to them. Lebron James is a basketball player, considered the world's best. Twelve-year-old Lebaron James from Sandy, Utah is also a basketball player, in that he

shoots baskets in his driveway on the weekends. Just because both are basketball players doesn't mean they have an equal chance of winning an NBA title. The same is true of air pollution particles. They can be counted the same, but the damage they do can be much different.

Let's follow the journey of a pollution particle that has the most potential to cause harm. The particle is inhaled, reaching the tiny air sacs in the lungs, the alveoli, then picked up by the blood stream, and distributed to just about any cell in the body, including those that are part of the most critical organs of the body, like the heart, lungs and brain, and in the case of a pregnant mother, the placenta. The particle can then penetrate the cell membrane, enter the cell, and then enter the nucleus of the cell where the chromosomes lie. At each point along this journey, smaller particles have an advantage in ultimately reaching the nucleus of the cell. An ultrafine particle (0.1 micron and smaller) has a much greater potential for toxicity than a fine particle (2.5 microns, which about 1/30th of the diameter of a human hair). When PM2.5 is captured on a filter at a government monitor, only the weight of the particles is measured. A few larger particles will weigh as much as hundreds of smaller particles, and both will contribute equally to what is announced as the pollution level (PM2.5) in your area. Yet the hundreds of smaller particles can do much more damage because of the journey just described, a crucial fact not addressed by EPA or state regulations.

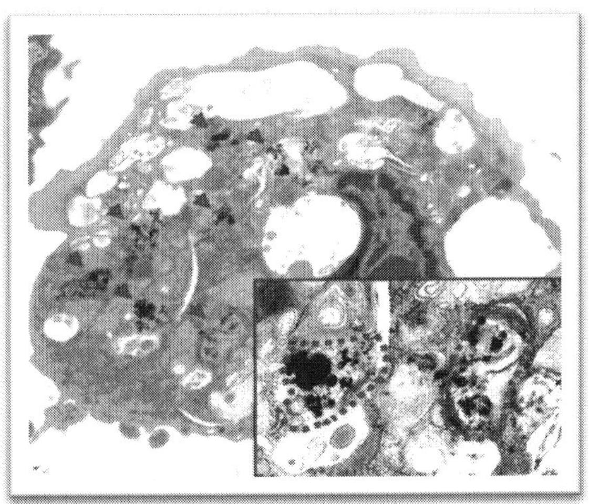

Electron micrograph showing a human cell that has been "invaded" by air pollution particles (arrows pointing to them).

9. Microenvironments matter. Pollution is not evenly distributed throughout a community. Local pollution sources, like industrial smoke stacks, corridors of heavy traffic, wood burning stoves and restaurants, gravel pits, and commercial diesel truck rest stops can create dramatically different levels of pollution within even a few blocks. But again this fact is largely ignored in strategies for achieving compliance with clean air standards.

The Most Basic Biologic Need

If you are an outdoors type, you have probably heard this survival adage recited about the number three. You can survive three minutes without air, three days without water, three weeks without food, and three months without sex. My postscript to that is, after three months without sex, to hell with the food and water. But the adage does convey that breathing clean air is the most basic biologic need of any living creature.

To begin a discussion of air pollution's disease consequences we should start at the microscopic level. Similar to the physiologic consequences of cigarette smoke, both ozone and particulate pollution precipitate a low grade, systemic inflammation leading to small artery narrowing, and enhancement of clotting mechanisms. Chemical markers of these changes are found even in young healthy adults, breathing pollution.

Once inhaled, a typical PM2.5 particle can migrate to the smallest airways of the lungs, translocate to the tiny arterioles, be picked up by the blood stream, and delivered throughout the body to major organs like the heart, brain, liver, spleen and kidneys.[30] It's like an incredibly efficient microscopic Fed Ex delivery network.

Evidence of pollution particles ending up in the urine of children was recently revealed.[31] As with cigarette smoke the effect can be almost immediate. In fact, within 15 minutes those particles appear in the blood and urine of human subjects and have been found to be still present in the body at least three months later,[32] like an unwelcome house guest that won't ever leave. Chronic exposure to even low concentrations of pollution are associated with an acceleration of atherosclerosis and significant arteriolar narrowing and stiffness. Moreover, those particles preferentially accumulate within the lining of blood vessels, at sites where inflammation, narrowing and atherosclerotic plaque already exist, exacerbating the inflammation and reducing blood flow.

Some, but not all of the inhaled particles will dissolve or disintegrate in body tissues over time, eventually being eliminated in urine, stool, and the lymph system. But even then, they may leave a toxic calling card, including fragments of heavy metals and harmful chemicals that were attached to the particles. But if a person is continually exposed to air pollution, then one is never really rid of the contamination.

Air pollution causes average blood pressure to increase, including within minutes of exposure, increasing the stiffness of arteries, with an effect on all organs. Increases in blood pressure are found even in newborns and children.[33,34,35,36,37,38,39,40,41,42,43,44] Over time, the inflammation increases the thickness of the lining of blood vessels, and impairs their integrity, especially in the microscopic vessels, narrowing the diameter of blood vessels throughout the arterial system, leading to rigidity (the "lead pipe syndrome"), which decreases blood flow, and accelerates the growth of atherosclerosis as part of the aging process itself.[45,46,47,48,49,50,51,52,53] And these changes are found even in the blood vessels of young, other-wise healthy adults.[54]

As suggested by the finding of pollution particles within the blood stream almost immediately after inhalation, pollution episodes lasting only a few days, like the Great London Smog, can exact a toll on our health by quickly and dramatically ramping up body wide inflammation.[55] Conversely, home air filtration units have been shown to reduce pollution particles inside the body, as well as the inflammation that comes with it, albeit with some delay.[56] Furthermore air pollution exposure alters the lipid profile of the blood, decreasing helpful high density lipids (HDLs).[57,58,59,60]

Air Pollution: You Are What Your Grandparents Inhaled, Your Grandchildren Will Be What You Inhale

With these biochemical and physiologic studies as a back drop, providing compelling explanations and plausible biologic pathways, it is no surprise that an enormous body of epidemiologic research shows that air pollution contributes to a list of clinical diseases and poor health outcomes every bit as long as what is attributed to smoking cigarettes. At the top of that list is four of the five leading causes of death: heart attacks, strokes, respiratory disease and cancer.

As of 2017, the EPA's annual standard for fine particulate pollution (PM2.5) was 12 ug/m3 (micrograms per cubic meter of air). The presumption of regulators has been that level of pollution is "safe" for human exposure. However, a growing body research beginning at least by 2006 shows in fact that the relationship between particulate pollution and death continues well below 12 ug/m3. In fact, as mentioned above, there is no safe level of air pollution, and that relationship continues right down to the lowest limits of pollution that we can measure. Landmark studies have shown that for every 1 ug/m3 PM 2.5 of chronic exposure, deaths from all causes increase approximately 1 percent.[61,62] So at 12 ug/m3, community death rates are increased about 12 percent. Additional deaths accrue from acute pollution episodes not reflected in annual averages--think London Smog, or Utah's winter inversions.

Moreover, ozone exposure causes additional mortality, increasing death rates about 1 percent for every 10 ppb. ozone. The EPA's National Ambient Air Quality Standard, i.e. presumed "safe level," for ozone is 70 ppb. Ozone levels that comply with that standard increase US mortality an additional 7 percent. Combining "safe levels" of PM 2.5 and "safe levels" of ozone, increases mortality rates about 17 percent. That doesn't really sound very safe does it? This is a perfect illustration that the EPA's regulatory thresholds, while better than nothing, are more arbitrary and political calculations than they are medical science. This is clearly not "safe" air pollution, any more than driving a little drunk is still safe to get behind the wheel.

There has been the suspicion that most deaths related to air pollution are explained by frail, elderly people with heart and lung disease dying shortly before they would have otherwise--the "harvesting" effect. No doubt this subset of the population is at highest risk from air pollution mortality. But new studies show that the average mortality victim of air pollution is about 79 years old, but loses ten to eleven years of their life expectancy.[63,64]

Another fascinating study demonstrates that the increased mortality effect of air pollution persists for decades. The air pollution you breathed while you were watching the Brady Bunch in the 1970s, is still increasing your risk of a premature death today.[65] Episodes of air pollution much shorter than the Great London Smog event are capable of triggering heart attacks and strokes, likely by rendering already vulnerable atherosclerotic plaques unstable and therefore friable, and by increasing blood vessel clotting susceptibility. One study showed that pollution typical of heavy traffic congestion can triple the risk of a heart attack within an hour after exposure,[66] and women had an even higher risk. It is worth noting that the air pollution in the cabin of a car stuck in traffic is about twice as high as outside the car.[67] Another study found that a very modest spike in PM 2.5, between 15 and 40 ug/m3, was associated with an increase in the risk of stroke of 34 percent within 24 hours. The risk was found to be greatest within 12 to 14 hours.[68]

There are other mechanisms by which air pollution can trigger sudden death. Numerous studies show that air pollution can disturb the electrical signaling within the heart, causing dangerous and even fatal rhythms.[69,70,71,72,73,74,75,76,77,78,79] Blood clots originating in the legs and pelvis can migrate to the heart and lungs causing acute lung dysfunction or even death. Because of air pollution's enhancement of blood clotting mechanisms, there is a remarkable correlation with rates of deep vein thrombosis, and the correlation is even stronger for diabetics.[80,81,82,83,84]

Enlarged muscles may be an advantage if you're in your speedo, on the beach, trying to attract the attention of teenage girls. But the only thing an enlarged heart muscle will attract is a hospital bed. When the blood vessels are narrowed and thickened, the heart has a more difficult time pumping against that resistance and over time causes the heart to enlarge, which is one of the reasons why high blood pressure must be treated. Because of pollution's effect on the blood vessels, people living near heavy traffic develop an oversized muscle mass of the right side of the heart.[85] Numerous clinical and laboratory studies have shown that air pollution also decreases the performance of heart muscle, with consequences ranging from decreased cardiac performance in athletes to increased hospitalizations and death for patients with heart failure.[86,87,88,89,90,91]

A study in animals found that even in utero exposure to particulate pollution was associated with impaired performance of the primary pumping mechanism of the heart later on in adulthood.[92] Other long term health effects of air pollution during critical stages of fetal development will be explored in more detail later in this chapter.

If you flattened out an average pair of human adult lungs with their 500 million alveoli--the tiny air sacs where gas exchange occurs--the resulting surface area would equal that of a full-size tennis court at Wimbledon. Now imagine twenty tennis courts, which represents an average rate of 20 breaths per minute for an adult at rest. How much air pollution do you think can land on the surface of twenty tennis courts? How much pollution could land on the surface of 29,000 tennis courts, which reflects the number of breaths that a person takes in 24 hours, even more if you're breathing hard from exercise. Not every alveoli participates in every breath, so let's cut that down to 15,000 tennis courts. The amount of pollution that can land on 15,000 tennis courts is the amount of pollution that your lungs inhale every day. And your lungs are the gateway to most of the pollution that enters the rest of your body. Your lungs are literally a warehouse sized sponge, absorbing and storing the pollution from fossil fuel and wood combustion.

Air pollution's adverse effect on the lungs is perhaps the most intuitive consequence for the lay public. Invariably, media reports of air pollution events, policy debates, and research focus on how air pollution affects the sensation of shortness of breath and asthma attacks. Oh, that it were that simple. Air pollution decreases lung function in otherwise healthy individuals, exacerbates virtually all pulmonary diseases, contributes to respiratory infections, [93, 94, 95, 96, 97, 98, 99, 100, 101, 102] and plays an aggravating, and likely causative role in reactive airways disease (asthma).[103]

Inhalation of particles can lead directly to scarring of the lungs.[104] Pollution changes the composition of, and increases the population of the bacterial flora and immunoglobulin that inhabit the airways.[105] It is associated with increased rates of hospitalization and death from respiratory diseases of virtually every type, from neonates to the elderly.[106, 107, 108, 109,1 10, 111, 112, 113, 114, 115, 116] Even brief exposure to ozone and particulate matter reduces lung function even in young healthy adults, and the reduction can last for a week after the pollution exposure has ended.[117, 118, 119, 120,1 21] Long term exposure to ozone causes an increase in overall mortality in addition to that from particulate matter, and most of the ozone-related mortality is from respiratory disease.[122, 123]

Particulate matter pollution also causes increases in respiratory mortality.[124] Air pollution can permanently inhibit lung growth in children, [125, 126, 127] preventing them from ever achieving their full adult lung capacity, which would normally occur at the approximate age of 20. I often ask audiences that include parents of young children and prospective parents if they have aspirations that their children will be world class athletes. Most of them raise their hands. Their "great (if unrealistic) expectations" can be so charming. And then I tell them the bad news. Most children who grow up breathing typical urban pollution have probably lost enough of their potential lung capacity they have little chance of becoming an Olympic athlete in an endurance sport. That they should instead shoot for bowling or chess is not what parents want to hear.

After reaching their peak, the average person loses about 1% of their lung capacity with every year of aging. If someone never achieves their full lung capacity, then that decline begins from a shorter peak, and there are few things that correlate more with life expectancy than lung capacity. That means a likely earlier death.

In fact, prenatal exposure can reduce fetal lung development, impairing lifelong lung function, and increasing susceptibility to respiratory infections later in life.[128,129,130,131,132,133,134]

There are very limited opportunities to compensate for defective or inadequate organ development in utero. There is no second chance at normal embryonic brain formation. I mentioned in Chapter 4 that babies are essential born "pre-polluted" from chemical contamination of the mother's environment. While air pollution is obviously not the only source of toxins for a pregnant mother, it is certainly one of the most important, most ubiquitous, and one about which there is little a mother can do to reduce her exposure risk.

It was once thought that the placenta acted as a barrier that shielded a human embryo from most environmental toxins. Now we know, unfortunately, that is not the case.[135, 136, 137] For decades physicians have known enough to advise women who may be pregnant to avoid almost all medicines, drugs, alcohol, and cigarette smoke. With this information in hand, that precaution should expand widely to exposure to other potential toxins in air, water, food, consumer goods, cleaning supplies, and personal-care products.

Every year about 136,000,000 babies are born worldwide. Most of those mothers breathe enough air pollution at some time during the pregnancy, even if the exposure is short lived, that the health of those babies has been diminished, in many cases with life-long consequences.

In fact, the greatest impact of air pollution on public health might very well be how exposure of the pregnant mother adversely affects fetal development. Particulate matter and the chemicals and heavy metals attached to them, like lead, mercury, dioxins and polycyclic aromatic hydrocarbons (PAHs) can cross the placenta throughout pregnancy, and the systemic inflammatory process triggered by ozone can interfere with critical organ development, suppress the immune system, and set the stage for diseases later in life.[138, 139, 140, 141] For some toxins, like mercury, levels in fetal blood generally exceed those of the mother.

Even worse, multiple generations can be harmed by a single exposure. Performance of a computer requires proper functioning hardware and software. If either is defective, the computer will not function as designed. In the world of microbiology, genes operate like a computer's hardware, and chemical attachments to genes, "epigenetics," function like a computer's software, telling the hardware what to perform. Generally, it takes less of an environmental toxin to precipitate epigenetic dysfunction than genetic damage, a term often used for actual breakage of the DNA's covalent bonds. But just like genetic damage can be passed on to future progeny, so can epigenetic dysfunction.

A pregnant mother as well as her fetus can be harmed by air pollution. If the fetus is a female, her eggs are also being formed in utero and their DNA is being developed, making a third generation vulnerable from a single exposure. Although sperm itself is not produced until adolescence, testicular development begins in utero, and so both a male and female fetus can pass on to subsequent generations a wide variety of increased risk for disease, morbidity and reproductive disorders.[142, 143] For example, three generations of mice show the chemical markers associated with asthma merely by exposing the first generation during pregnancy.[144] The mechanism of intergenerational disease transfer is alteration of "gene expression," i.e., epigenetic changes impairing the immune system of the animals. Air pollution exposure of a pregnant mother can cause epigenetic changes in both the mother and the fetus that can be just as disruptive as genetic damage.[145, 146, 147, 148, 149, 150, 151, 152, 153] Exposure of a sperm-producing male can impair sperm quality and function, fragment sperm DNA, and trigger aneuploidy (wrong number of chromosomes), ultimately risking genetic integrity of any progeny.[154,

[155, 156, 157, 158] And, not surprisingly then, air pollution is associated with decreased fertility.[388, 389] I suppose someone might seize on this information, and in a stroke of genius, start a campaign to reduce teen pregnancies by making them breathe truckloads of air pollution. However, with smoking and now "vaping,"(see Chapter 7), teenagers have already figured out a way to expose themselves to more than enough pollution.

Genes play a critical role and maintaining virtually every aspect of good health. For example, a person has genes that promote tumor growth and genes that suppress it. If either of these genes are turned on or off inappropriately, a function of epigenetics, the risk of cancer will rise. Exposure even to brief episodes of pollution at critical stages in the development of the human embryo can cause a person to experience an increased likelihood of multiple chronic diseases including those of the heart, lungs, immune system, and brain and even obesity, diabetes, cancer, and shortened life expectancy. One common denominator is likely air pollution's adverse influence on epigenetics.[159, 160, 161, 162, 163, 164, 165, 166] Ultra fine particulate pollution can affect development of the fetal brain in numerous ways, one of which is changing the epigenetics of neuroprotective genes on nerve cells.[167]

Changes in mitochondrial DNA (mtDNA) can serve as a marker of cumulative oxidative stress. Increased PM2.5 during the third trimester of pregnancy is associated with decreased mtDNA content suggesting heightened sensitivity to this kind of biological damage in a fetus.[168, 169]

Chemical attachments to DNA, like methyl groups (a carbon atom with three hydrogen atoms), can alter the activity of the involved DNA segment. Changes in the methylation of DNA is one avenue of triggering adverse epigenetic changes. Placental mitochondrial methylation of DNA is positively associated with maternal air pollution, which is inversely associated with mitochondrial DNA content. This is likely a reflection of air pollution causing mitochondrial death.[170]

Telomeres are repeating sequences of DNA at the ends of chromosomes that act much like the end caps of shoe laces, protecting the chromosomes and keeping them from unraveling. Every time a cell divides it loses a little of its telomere length, ultimately limiting the number of divisions. The down side is that it limits a cell's lifespan and prevents us from achieving immortality. (Regrets to those buying fountain of youth schemes on late night infomercials). The upside is that telomere attrition is thought to protect against the unlimited proliferation of cancer cells. Telomere shortening corresponds with aging, even further with age-related diseases, chronic inflammation, and oxidative stress, and is associated with early death. On the other hand, a large study of almost 65,000 people over 22 years found that genetically long telomeres were also associated with a higher risk of cancer deaths.[171]

However, even in patients eventually diagnosed with cancer, a pattern of premature telomere shortening was found in the years preceding the cancer diagnosis. Then the accelerated telomere shrinkage stopped three to four years prior to the cancer diagnosis.[172]

In fact, testing for the length of telomeres has been advocated and commercialized as a marker of how rapidly cells are aging, and therefore a predictor of health risks and life expectancy years in advance of old age. An enzyme, telomerase reverse transcriptase (TRT) acts as a protector of telomeres. If an organism has strong TRT, then it can live long after it should otherwise have died. Speaking of organisms that should have died long ago, Keith Richards must have incredible TRT.

What do telomeres have to do with air pollution? Air pollution shortens telomeres, accelerating the aging process.[173, 174, 175, 176]

It may surprise many people that the initial length of telomeres at birth is largely a function of environmental factors. Inheritance plays only a minor role. More specifically placental telomere length is highly correlated with fetal telomere length, which in turn is highly correlated with adult telomere length. So the length of placental telomeres plays a significant role in determining a person's life expectancy. Air pollution shortens the length of placental telomeres, leaving a strong imprint on life expectancy even at life's earliest stage.[177]

A brand new study has found that air pollution can damage mammalian germ cells in prospective parents, such that air pollution inhaled by the parents even before conception, can impair the health of the second generation, even if that second generation is never exposed.[400]

Genetic damage, epigenetic changes, and shortened telomeres are not the only way that air pollution can threaten a fetus. As mentioned previously, the signature physiologic consequence of air pollution is a low-grade arterial inflammation, arteriolar narrowing, and increased vulnerability to clot formation. This affects blood flow throughout the body and to all organs.

During lectures I have often asked audiences what is their favorite organ. It's a loaded question, but the range of answers is fascinating and entertaining, especially when men are forced to answer honestly, which, predictably, they never do. But the best answer to that question is the organ that no one reading this book still has, and that is the placenta (apologies to any males that still have an umbilical cord attached to their mother). Any impairment of placental function can have life-long consequences by compromising fetal development of every critical organ, or the viability of the baby itself.

Approaching the issue from multiple angles, researchers have documented air pollution's connection to impaired placental arterial formation, reduced placental blood flow, and even an increase in the cross section of vessels on the fetal side of the placenta, a compensatory response to reduced blood flow on the mother side of the placenta.[178, 179, 180, 181, 182]

It takes little imagination to grasp that if blood flow to the developing embryo or fetus is compromised, then biologic stress can be the end result. That biologic stress has been confirmed by a wide range of studies. Air pollution during pregnancy precipitates the chemical markers of inflammation, a prelude to pregnancy complications in the near term and chronic disease vulnerability later in life.[183]

C-reactive protein is a blood marker for systemic inflammation. The risk of high C-reactive protein levels during pregnancy, measured before the third trimester, is elevated significantly with air pollution exposure to the mother.[184] For every 10 ug/m3 of PM2.5 (less than the EPA's standard for annual average), the risk of intrauterine inflammation (IUI) increased 240 percent. IUI contributes to, or is a mechanism for multiple types of pregnancy complications.[185]

The inflammatory response, vascular insult, and chromosomal perturbations are the likely common pathways for the wide range of adverse clinical outcomes with air pollution exposure, including those related to pregnancy. Premature birth, defined as a birth before 37 weeks' gestation, is one of most common pregnancy complications. It affects 12.5 percent of babies born in the US and has increased 20 percent in the last 15 years. Prematurity interrupts pregnancy prior to full development of important organ systems (brain, lungs, liver, etc.)

It is the leading cause of death near delivery in otherwise normal newborns, and a major contributor to respiratory failure, gastrointestinal problems, inadequate brain development, seizures, bleeding into the brain, jaundice, prolonged ICU stays, and SIDS once they leave the hospital. Premature babies that succeed in avoiding or overcoming these serious consequences are still left with increased lifetime risks for cerebral palsy, loss of hearing and vision, impaired cognition, behavioral disorders, chronic lung disease, hypertension, heart disease, and diabetes.

Numerous studies link outdoor air pollution to prematurity. One large study linked 2.7 million pre-term births per year in the US, 18 percent of all pre-term births, to particulate pollution.[186] The 9/11 dust cloud from the collapse of the Twin Towers in 2001, was shown to be associated with significantly higher rates of premature birth and low birth weight in the babies of pregnant women in Manhattan, nearest the site. The study's authors stated, "the impacts are especially pronounced for fetuses exposed in the first trimester, and for male fetuses. We estimate that in this group, exposure to the dust cloud more than doubled the probability of premature delivery and had similarly large effects on the probability of low birth weight." This is more evidence that even short-term air pollution exposure can affect the developing fetus, and therefore, lifelong health.[187] Even one- to two- day episodes of air pollution can trigger premature births.[188]

Low birth weight syndrome (LBW) and intra-uterine growth retardation (IUGR) refer to smaller than normal babies carried to term. This affects about 8 percent of newborns and the list of short-term and long-term consequences are very similar, including higher rates of still births. Air pollution exposure significantly increases the rates of LBW and IUGR.[189, 190, 191, 192, 193, 194, 195, 196, 197, 198] Even incense burning during pregnancy is associated with smaller weight babies with smaller head circumference.[199] Smoke is bad for babies (and everyone else), period, whether it comes from cigarettes, vaping, fireplaces, candles, barbecues, tail pipes, or smoke stacks.

Normal thyroid function is critical to fetal growth and development, especially for the fetal brain. Particulate pollution is associated with reduced thyroid hormone, even at levels below the EPA's standards.[200] Brain Derived Neurotrophic Factor (BDNF) is a protein that augments the growth, maturation, and differentiation of neurons, much like a "Brain Miracle Grow," but you can't just pick it up at Home Depot. BDNF is an important contributor to fetal brain development, in concert with thyroid hormone. The fetus benefits from gene expression for BDNF coming from the placenta, which is also reduced with pollution exposure.[201]

Knowing that air pollution affects the blood vessels, it is no surprise to find that it has a strong association with hypertension of pregnancy and pre-eclampsia.[202,203,204,205,206,207,208,209] With this disorder both the mother and baby are at risk, and it increases the risk of premature birth over 200 percent. Premature rupture of membranes is another important pregnancy complication that predisposes both the mother and baby to increased risk, including infection and hemorrhaging. Membrane rupture has been less studied but appears to be correlated with air pollution as well.[210,211]

Gestational diabetes is another complication that puts both mother and baby at risk. It occurs in up to 14 percent of pregnancies, and the incidence is increased with both PM 2.5 and ozone exposure.[212,213] Given air pollution's capability of delaying and impairing fetal development, it would seem logical to find increased rates of birth defects. Various congenital and nervous system defects are found with pregnant mother exposed to more air pollution, especially to air toxics like benzene.[214,215,216,217,218,219,220,221,222]

Given all the other research on air pollution and adverse pregnancy outcomes, one would expect that research might specifically show that stillbirths are associated with exposure. Unfortunately, this is indeed the case.[223,224,225,226]

A study of over 220,000 births from across the US showed that both chronic and acute ozone exposure during the week prior to delivery increased the risk of stillbirth.[227] The authors concluded that about 8,000 stillbirths per year in the US could be caused by ozone exposure. For the week prior to delivery, the correlation was an 8-10 percent increase in stillbirths for every 10 ppb ozone. The EPA standard is currently 70 ppb. This study is more evidence of the profound inadequacy of the standard in protecting public health. Miscarriages, a fetal death before 20 weeks' gestation, are more difficult to study but have also been shown to increase with exposure to traffic pollution.[228]

A few studies suggest that the primary culprit in particulate air pollution's adverse effect on pregnancy outcomes is PAHs, which are often attached to air pollution particles, rather than the particles themselves. More evidence that not all air pollution is created equal, and we should be paying much more attention to those sources that create high levels of PAH pollution--wood smoke, diesel exhaust, and industrial pollution.[229, 230]

Air pollution particles can find their way into virtually any cell in the body, and when they do, an immune response is generated. The inflammation affects the brain, causing neuronal damage, neuronal loss, loss of brain mass, cortical stress measured by EEG, enhancement of Alzheimer-type abnormal filamentous proteins, and cerebrovascular damage. Many of these changes can be found in children and young adults.[231, 232, 233, 234, 235, 236] The inflammation can cause specialized cells that line blood vessels in the brain, called the "blood-brain barrier," to lose their normal function of acting as a barrier to foreign particles, chemicals, and the body's own molecules of inflammation from entering the brain.[237, 238, 239]

The immune response to the inflammation can also include the release of antibodies to nerves themselves. Reduced blood flow can be the result of this inflammatory process and anatomic evidence of blood and oxygen deprived areas of the brain appear on MRI scans as spots of White Matter Hyperintensities (WMH), found throughout the brain, but primarily in the pre-frontal cortex. The WMH interfere with nerve to nerve signaling and impair brain function, like brain "dead zones." WMH used to be considered an expected hallmark of advanced aging, now we know otherwise.[240]

The presence of WMH doubles the risk of dementia and triples the risk of a stroke. They impair physical coordination, increase the risk of depression, and are inversely associated with intelligence. These foreboding WMH are found even in children and young adults exposed to high levels of air pollution and seldom found in children breathing clean air.[241, 242, 243]

Air pollution particles and their attached heavy metals and toxic chemicals can actually end up inside the brain from two routes. One is through inhalation by the lungs, translocation to the capillaries in the lungs, then distributed by the blood stream with greater access afforded by this disruption of the brain's normally protective blood vessel barrier. But another is through the nose. Pollution particles can attach themselves to the lining of the nose, then to the olfactory nerve fibers in the nose, and then, like a conveyer belt, migrate directly back to the brain stem itself. [244, 245]

Toxic, nano-sized particles from high temperature fossil fuel combustion called "magnetites" have been found at autopsies in people as young as three years old.[246] People with higher concentrations of these metallic nanoparticles are known to be at higher risk for Alzheimer's, and the kind of brain damage these magnetites can cause is consistent with the disease. These particles, compromised of iron oxide, platinum, nickel, and cobalt, can originate from industrial, vehicle or other sources of pollution. The researchers found "millions of these particles per gram of brain tissue" after studying numerous autopsies. The lead study author said these results are "dreadfully shocking."[247] Magnetites were responsible for an average of 1/100th of the weight of the brains examined. Think of magnetites as an energy zapping, havoc wreaking "kryptonite" on the human brain.

We have already begun the discussion of pollution and fetal organ development, but it is such an important issue that more detail is warranted. Remarkable research in both humans and animals has shown that prenatal exposure to pollution harms the architecture of the brain. In mice, prenatal exposure to diesel exhaust results in impaired mentation later on when they are adults.[249] In humans, there is a linear relationship between the amount of PAH exposure of a pregnant mother and the loss of brain white matter, primarily on the left hemisphere. As yet we don't know the reason the left hemisphere suffers the most loss. The researchers found no safe level of PAH exposure.[250]

The loss of white matter, brain volume and PAH exposure, in turn, correlated directly with cognitive loss and behavioral disorders measured later on in childhood.[251, 252, 253, 254]

Numerous other studies show a link between pregnant mothers' exposure to the full range of air pollution components--PAHs, traffic pollution, coal combustion emissions, carbon monoxide, benzene, and nitrogen oxides--and decreased intelligence measured later on in childhood.[255, 256, 257, 258, 259, 260, 261, 262, 263, 264, 265, 266, 267] Animal studies examining the effect of pollution exposure during what would correspond to the first month of human life, show loss of brain mass around the ventricles, pockets of cerebrospinal fluid in the center of the brain, causing abnormal enlargement of the ventricles. This anatomical anomaly, in turn, is associated with schizophrenia, autism, attention deficit disorder, developmental delays, and cognition handicaps.[268] White matter loss in the elderly also corresponds to air pollution exposure.[269]

Alzheimer's and Parkinson's disease are characterized anatomically by abnormal brain architecture, loss of brain volume, aberrant biochemical and neurotransmitter function and the deposition of abnormal protein tangles and plaques in the brain, like "tau" and beta amyloid. Air pollution is associated with all of these kinds of brain abnormalities even in children, [270, 271, 272, 273, 274, 275] and is likely a contributor to the growing worldwide epidemic of Alzheimer's, responsible for about 20 percent of the disease according to one study.[276]

The latest research examined the brains of 203 people, ranging in age from 11 months old to 40 years old. At autopsy (causes of death were usually trauma), every single brain but one showed those abnormal Alzheimer proteins, even in the 11-month-old. And the amount of these abnormal proteins was proportional to the amount of air pollution where the subjects lived. The principle author, probably the world's expert on this type of research, said, "Alzheimer's disease hallmarks start in childhood in polluted environments, and we must implement effective preventative measures early. It is useless to take reactive actions decades later." [248]

Well over 150 clinical studies confirm that pollution exposure is associated with almost the full range of neurologic disorders throughout the age spectrum, including lower intelligence, diminished motor function, attention deficit and behavioral problems, accelerated dementia, memory and cognitive loss in elderly adults, higher rates of strokes, relapses in multiple sclerosis, impaired olfactory sense, Parkinson's, and other neurodegenerative diseases, anxiety, and depression.[277, 278, 279, 280, 281, 282, 283, 284, 285, 286, 287]

Air pollution is associated with higher risk for developing and seeking treatment for virtually all mental disorders, and a much higher mortality risk for those with mental health and behavioral disorders, including suicide.[394, 395, 396, 397]

Recently, a criminal defense attorney reached out to me because she was wondering if her defendant might have developed homicidal urges because of the air pollution he grew up with in his childhood. I am slow to support that theory. However, the question may not be as much grasping for straws as it may seem. We do know that chronic traumatic encephalopathy (CTE), the brain disease found in 99 percent of NFL players studied at autopsy,[288] is characterized by many of the same kind of pathologic findings as is present in Alzheimer's, is linked to air pollution, and is associated with loss of impulse control, anti-social aggression, and violent behavior. The WMH mentioned above are predominantly found in the prefrontal cortex, an area of the brain impaired in people who manifest poor decision making ability, aggression, and antisocial behavior.[289] Another fascinating study showed that in addition to increasing depression and suicidal tendencies, air pollution is associated with an increased tendency to criminal activity and unethical behavior.[386] So maybe the attorney was on to something.

The same vascular changes that are responsible for heart attacks are also responsible for pollution-related strokes. Numerous studies show significantly higher rates of strokes with chronic and acute exposure, including within as little as hours after the onset of a spike in pollution.[290, 291, 292, 293, 294, 295, 296, 297]

Children are more vulnerable to air pollution than adults. They inhale more pollution because of a higher respiratory rate compared to their body mass, and they have a higher heart rate which combine to disseminate more pollution particles throughout the body. The protective barriers from specialized cells lining the lungs, blood vessels of the brain, the nose, and the GI tract are less well developed than those of an adult. These factors increase a child's exposure, subjecting them to a chronic state of environmental stress, provoking dysregulation of genes that control inflammation, immune response, cell viability and communication between brain cells. At the same time, because of the "in progress" developmental status of the fetal, infant, and childhood brain and nervous system, the damage can be much greater.

Childhood brain disorders associated with pollution have been well documented by a robust body of experimental, clinical, epidemiologic, and pathology research. The impact on brain function can be almost immediate. One study showed that attention span in school children was impaired by the air pollution they breathed that day on the way to school.[398]

Although autism carries a unique and alarming stigma, and it can be profoundly life altering, it is only one point on the wide spectrum of childhood and developmental disorders of the brain that have a strong connection to environmental neurotoxins. Air pollution is one of the important delivery mechanism for those toxins. The connection between autism and air pollution grows steadily stronger. [298, 299, 300, 301, 302, 303, 304, 305, 306, 307, 308, 309, 310, 311, 312, 313, 314] That should be of particular concern to states like Utah and New Jersey, because Utah has had the highest rates of autism of any state in the country until being recently eclipsed by New Jersey. According to the CDC in 2012, one of out of every 32 Utah boys has been diagnosed with some form of autism.[315] Autism will be discussed more thoroughly in subsequent chapters.

No one is more infamous for their air pollution than China, even though some cities in India and other parts of the world are often under just as dense blankets of smog. It didn't help that in November 2013, an eight-year-old Chinese girl became the youngest known victim of lung cancer, with no known risk factor other than their toxic air. The WHO declared air pollution the most important environmental cause of cancer, more important than second-hand cigarette smoke. They placed it the Group I category, the same as asbestos and ionizing radiation. The most common cancers associated with air pollution are the same as with smoking—lung and bladder cancer—with about 15 percent of all lung cancer deaths coming from air pollution. However, for just about every type of cancer—lung, bladder, breast, prostate, ovarian, cervical, brain, eye, nasal, pharyngeal, laryngeal, esophageal, liver, stomach, pancreatic cancer, and childhood leukemia—there is at least one study that has found a statistical connection to air pollution. [316, 317, 318, 319, 320, 321, 322, 323, 324, 325, 326, 327, 328, 329, 330, 331, 332, 333, 334]

For breast cancer, the most common cancer in women, there are now several studies showing a connection to air pollution. For a woman that lives in a typical American city, her chance of breast cancer is increased about 250 percent from air pollution.[335] To put that in context, smoking increases the chance of lung cancer about 1,500 percent.

Notably, a study shows that victims of childhood acute myeloid leukemia have significantly higher levels of nanoparticles from air pollution in their blood.[336] Pre-natal pollution exposure is associated with increased rates of multiple childhood cancers like brain cancer.[337, 338, 339, 340] Furthermore, air pollution is associated with decreased survival in patients with all types of cancer, especially breast cancer.[341, 342, 343, 344, 345]

Plausible explanations for air pollution's carcinogenic potential at the molecular level involve many of the same pathways for other morbidities--inflammation, immunosuppression, oxidative stress, epigenetic changes, and defective repair mechanisms and replication of DNA.[346]

There are now more than 60 studies that show a connection between air pollution and metabolic disorders like type II diabetes, and a few that have found a connection to type I diabetes, and metabolic syndrome, i.e., obesity. Moreover, it's a two-way street because diabetics are more vulnerable to all the aforementioned adverse outcomes with pollution exposure.[347, 348, 349, 350, 351, 352, 353, 354, 355] Long-term and short-term air pollution exposure decreases insulin sensitivity, even in children.[356, 357, 358, 359, 360, 361] Chronic inflammation, like with so many other morbidities, is the likely common denominator as it can interfere with function of the insulin-producing beta cells in the pancreas.[362,363,364,365,366,367,368]

There is much national hand wringing and scolding of Americans for becoming a nation afflicted with obesity. In my own practice of medicine over the last 38 years, I have seen first-hand the steadily increasing girth of our population. Indeed, gastric bypass operations have become an industry in and of themselves, including now even being offered to teenagers. While there are many factors involved in this trend, one factor seldom discussed is the relationship to environmental contaminants, including air pollution. Air pollution promotes obesity and metabolic syndrome, and the mechanism is almost undoubtedly its provocation of chronic inflammation.[369] Prenatal pollution exposure has a particularly strong association with childhood obesity.[370, 371]

Particulate pollution is ingested from air, food, water, and hand-to-mouth activity. Upon arrival to the GI tract, like an army of tiny invading hordes, the particles can alter the composition and function of bacterial flora, increase bowel inflammation, affect digestion, and make the intestinal barrier more permeable allowing greater passage of bacteria from intestine into the blood stream.[372]

Other epidemiologic research shows the clinical connection between air pollution and several increasingly common bowel disorders—inflammatory bowel disease, irritable bowel syndrome,[373, 374, 375, 376] and appendicitis, including the risk of perforated appendicitis.[377, 378] Miscellaneous consequences of air pollution exposure include an increase in rates of infant mortality and SIDS,[379, 380] lupus,[381] juvenile arthritis,[382, 383] sleep apnea,[384] and reduced kidney function.[385, 387]

For some people, what's going on inside their body and how they feel isn't what's important. How they look on the outside is the only thing that matters. If you're such a person, rest assured that air pollution is still not your friend. Like the inside of you, air pollution accelerates aging on the outside as well. It depletes anti-oxidants from your skin, making you look older, with more pigment spots and more wrinkles than you should have for your age. Baby boomers in America are spending about $100 billion a year trying to look a little younger than they are. Rather than waste your time and money on surgery, Botox, waxing, chemical peels, massages, facials, moisturizers, make up, truck -loads of magic potions, and who knows what else, you might look better by just putting a good air purifier in your bedroom.[390, 391,392]

Not only will clean air make you look better, it will also make you feel better. Researchers found a significant inverse correlation between air pollution and happiness. Strangely, they also found the converse, unhappiness leads to more air pollution, in other words, a vicious cycle.[393] I venture to guess that reading about all the health consequences of air pollution can bring even more unhappiness. But despite the aphorism, I don't believe that ignorance is bliss.

For several decades the cabal of polluters—dirty-energy industries, manufacturing-groups, chemical companies, auto makers, and the U.S. Chamber of Commerce---have been eager to deny you the right to breathe clean air. Under the Trump Administration, they have turned the federal government into their own private candy store with pipeline approvals, a bonanza of drilling permits, carving up national parks and precious landscapes for mineral and fossil fuel extraction, and rolling back every imaginable rule and regulation. The air pollution they are determined to force feed you is a full service public health disaster. They are willingly punching your ticket to a list of diseases as long as an encyclopedia and claiming the right to shorten not only your life, but that of your children and grandchildren.

While the fossil fuel industry deserves the lion's share of attention when it comes to the air pollution that city dwellers inhale, in some cities and many rural areas, corporations that make wood-burning appliances are actually a more serious threat to public health. Wood smoke is likely the most toxic type of pollution the average person ever inhales, and neighbors who live near a household that burns wood in a stove or a wood boiler (OWB) can be subjected to shocking levels of pollution. One study showed that an OWB can create outdoor concentrations of particular air pollution up to 8,800 ug/m^3 at 50 feet away. To put that in context, Beijing, China's worst day of 2018, showed a level of 288, a mere fraction of the hazardous levels to which neighbors of OWBs are regularly subjected.

The corporations that make and sell wood stoves and boilers, Central Boiler, and members of the Hearth, Patio, and Barbeque Association (HPBA) share something in common with Peabody Coal, ExxonMobil, and Big Tobacco. Just like these giant corporations, their products are serious health hazards, yet their company concerns are only with selling more product, while obfuscating and denying the disease and premature death they are responsible for.

In 2014 I served as an expert witness for a family suing their neighbor over being subjected to this level of pollution from an OWB. The picture above, taken by the plaintiffs, speaks for itself.

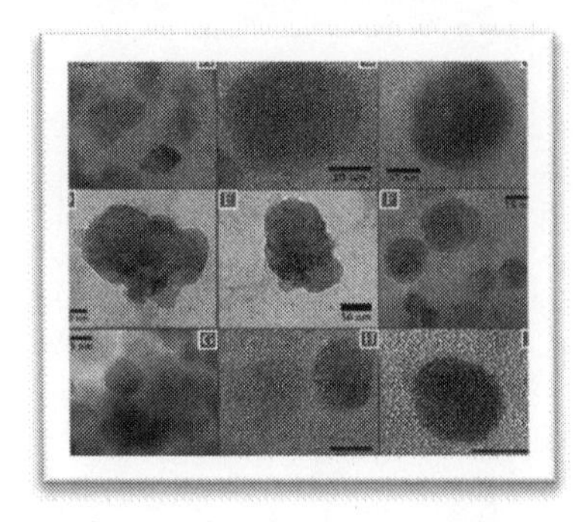

Electron microscopic view of tiny pollution particles extracted from human brains--millions per gram of brain tissue. Reprinted with permission of the PNAS
https://doi.org/10.1073/pnas.1605941113

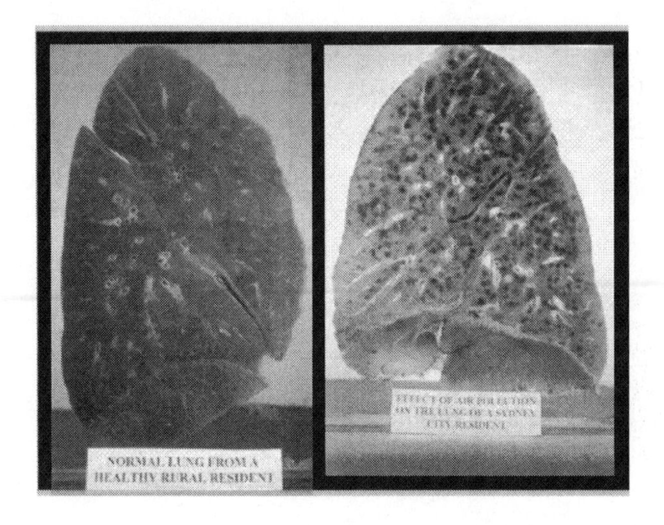

The lung on the left is from a healthy person who lived in a rural area of Australia, breathing relatively clean air. The lung on the right is from a non-smoker who spent their life living in an urban area, Sydney, Australia, which is not particularly polluted for a major city.

Air pollution can damage DNA such that prenatal (intra-uterine) exposure, and even pre-conception exposure of germ cells (sperm and ova) can precipitate chronic disease and shorten life expectancy.

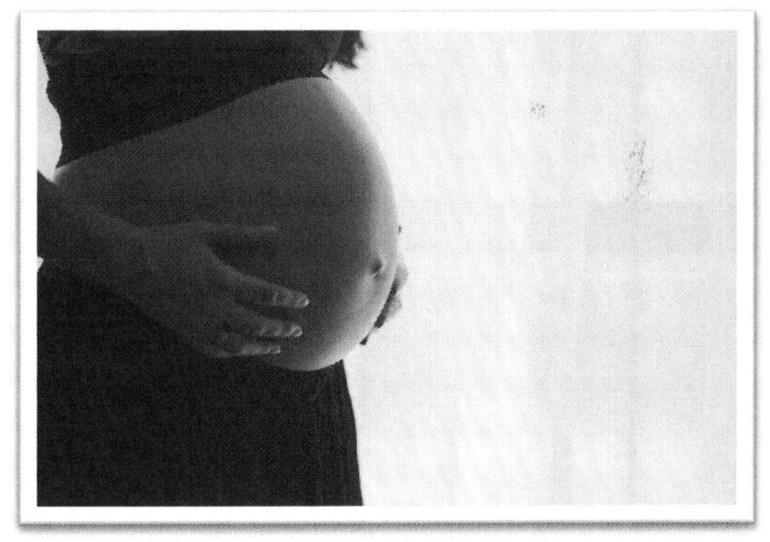

Air pollution particles have been found in human placentas, suggesting a pathway for harming fetal development.

Scenes from what is probably the largest air pollution protest rally in the modern history of the United States, at the State Capitol in Salt Lake City, Utah, January 25, 2014, attracting about 5,000 protesters.

6

The Dirty Energy Death Star:
Writing the Planet's Obituary

"Global warming isn't real because I was cold today. Also great news! World hunger is over because I just ate."

—Steven Colbert
Comedian, TV talk show host

"Conservation may be a sign of personal virtue but it is not a sufficient basis for a sound, comprehensive energy policy"

—Former Vice President
and Halliburton Executive
Dick Cheney

"Changes to weather patterns that move crop production areas around–we'll adapt to that. It's an engineering problem, and it has engineering solutions."

—Rex Tillerson
former CEO of Exxon-Mobil
former Sec. of State

Global Warming: A Brief History

The concept of "resource curse," also known as the "paradox of plenty" refers to the phenomenon of countries with abundant natural resources, like fossil fuels, minerals, and forests, having less economic equality, less democracy, less government stability, and worse environmental protection, worse development outcomes than countries with fewer natural resources. It's certainly a reflection of an imbalance of power between transnational corporations that exploit those resources and immature, ineffectual, and often corrupt governments. But the concept of resource curse might be even more applicable to the entire planet, which is having its natural resources exploited with almost the exact same outcome on a global scale.

The earth has just recently experienced its 406th straight month of global temperatures above the 20th century average.[79, 110] At this writing the temperature is a record setting 105° F in Salt Lake City, no relief is expected for a week. Western forests are being obliterated by drought, pine beetle infestations, and wildfires. Reservoirs are only half full. Meanwhile the north shore of the Hawaiian island of Kauai was just pounded with a record-breaking 49 inches (over 4 ft) of rain in 24 hours. The summer is just getting started, and I'm taking time off from watching TV reruns of my favorite show on the Sarah Palin Channel to write about something slightly more important---the climate crisis, i.e., the likely end of human civilization (although I'll admit that having Sarah Palin have her own TV channel is the clearest sign of the end of human civilization).

Rest assured, no matter how concerned you are about the climate crisis, it is not enough. I deliberately use the term "climate crisis" rather than "climate change," which is still the popular lexicon. "Global warming" was used repeatedly by George W. Bush in his speeches in 2001. The issue was, at the time, an Achilles heel of the political fortunes going forward for G.W. But in a 2002 confidential memo from the now notorious political strategist and linguist Frank Luntz, consulting for the Republican Party, Luntz concluded that the term "global warming" was frightening to people. He warned that the Republican Party had, "lost the environmental communications battle" and urges its politicians to encourage the public in the view that there is no scientific consensus on the dangers of greenhouse gases."[1] Luntz went on: "The scientific debate is closing [against us] but not yet closed. There is still a window of opportunity to challenge the science." Luntz specifically advocated replacing the term global warming, with the less frightening sounding "climate change."

The strategy succeeded, perhaps beyond Luntz's and the Republican Party's wildest and most cynical dreams. Politicians from both parties, the media, the public, and even scientists had been duped into using the more benign sounding phrase, and public concern about the issue receded proportionately.

I reject the term, global warming as well, because the consequences of humans loading the atmosphere with greenhouse gases is much more than just global warming, and parts of the world may actually experience periodic catastrophic global cooling.

To dispute the science of global warming takes an astonishing degree of denial of history as well as science. It is willful ignorance or fraud. The physics of the greenhouse gas phenomenon is as elementary as gravity. The roots of the science can be traced back to many individual observers and scientists as early as the 19th century. Chief among them was the mathematician and physicist, and Secretary of the French Academy of Sciences, Joseph Fourier. Around 1824 Fourier, calculated that the earth would be significantly colder (as in uninhabitable) if it were only heated by solar radiation from the sun. He was the first to propose what is essentially the greenhouse effect, postulating that atmospheric gases could present a barrier to the escape of heat from solar radiation.

Eunice Newton Foote had a paper presented at the American Association for the Advancement of Science meeting in 1856 showing that CO_2 at the earth's surface increased warming by the sun. John Tyndall expanded on Fourier's work in 1859 and found that methane and CO_2 were effective in blocking the rebound of infrared radiation from the earth's surface[59] (which otherwise would normally happen at night and corresponds to atmospheric warming's being more pronounced at night).

Other researchers were studying global warming by 1882. By the late 1890s Svante Arrhenius, a Swedish Professor of Physics, went so far as to calculate that doubling atmospheric CO_2 would provoke warming of the earth by 5-6°C.[6] He won the Nobel Prize in Chemistry in 1903.

These scientific advances in the understanding of greenhouse gases did not go unnoticed by the media of the time. Historian Jeff Nichols found that between 1883 and 1912, numerous articles in such newspapers as the New York Times, The Philadelphia Inquirer, and the Kansas City Star were published discussing a rise in atmospheric carbon dioxide that would affect the climate.[85]

be insufficient for him. Were MALTHUS alive now he would take great pleasure in asserting that the combustion of coal increases twice as rapidly as the human race, and that there is every reason to believe that the rate of combustion will before very long be still greater than it now is. When we reflect that all the gases given off by burning coal enter and contaminate the atmosphere, and that the latter is a constant quantity while the former is steadily increasing, we gain an idea of the danger which threatens us.

Excerpt from an 1883 New York Times article

In the magazine *Popular Mechanics*, an article by Francis Molina was titled, "The Remarkable Weather of 1911," and the subtitle was, "The Effect of the Combustion of Coal on the Climate--What Scientists Predict for the Future"

REMARKABLE WEATHER OF 1911

The Effect of the Combustion of Coal on the Climate — What Scientists Predict for the Future

By FRANCIS MOLENA

THE year 1911 will long be remembered for the violence of its weather. The spring opened mild and delightful, but in June a torrid wave of unparalleled severity swept over the country. The cities baked and gasped for breath, while the burning sun and hot winds withered the corn and cost the farmers a million dollars a day. A little later England was scorched and France and Germany sweltered. The mercury went above 100 deg. in

The mean temperature of every month except November was above the average of that of the 40 years covered by the records of the United States Weather Bureau. The average daily excess was from four to six degrees.

With only one month out of twelve below normal, one may well ask if the climate is not changing and getting warmer. There is a general impression among older men that the good old-fashioned winters in which

In that article were these observations: "The mean temperature of every month except November was above average of that of the 40 years covered by the records of the United States Weather Bureau. The average daily excess was from four to six degrees.... With only one month out of twelve below normal, one may well ask if the climate is not changing and getting warmer.... It has been found that if the air contained more carbon dioxide, which is the product of the combustion of coal or vegetable material, the temperature would be somewhat higher." The article was published in March of 1912. Now, it didn't get everything right. It predicted that it would take hundreds, if not thousands of years before the warming became a crisis.

In the 1930s, an English engineer, G.S. Callendar, compiled probably the best meteorological data from around the world at the time and announced that mean global temperatures had risen between 1890 and 1935 by about 0.5° C.[86] From his temperature data he resurrected the theory about warming due to greenhouse gases that had remained dormant for about 30 years.

In 1960 Charles David Keeling demonstrated that atmospheric levels of CO_2 were, in fact, rising. The arrival of computer modeling allowed more scientists to refine the work of Arrhenius and yielded more accurate predictions of temperature rises. By the 1950s, climate scientists were becoming concerned among themselves about greenhouse gases, and by the 1960s and 1970s they were advising politicians and policy makers in the US.

In 1965 the President's Council of Advisors on Science and Technology warned President Lyndon Johnson in a lengthy report[2] that fossil fuel combustion was causing CO_2 to accumulate in the atmosphere and was creating a dangerous geophysical experiment causing perturbation of the climate. If continued, it would cause disruption of a hospitable climate to the detriment of human civilization. Even in 1965, scientists had enough information to debunk many of the talking points used by climate deniers today. In retrospect, the report was amazingly prescient for its time. Along with global warming, the report predicted shrinking of polar ice caps, rise of sea levels, and warming and increased acidity of ocean waters. President Johnson referred to that report in a speech before Congress in November 1965. "This generation has altered the composition of the atmosphere on a global scale through a steady increase in carbon dioxide from the burning of fossil fuels."[5] This is likely the first mention of the issue by a major politician.

Robert M. White, the first administrator of the National Oceanic and Atmospheric Administration and later president of the National Academy of Engineering said, "We now understand that industrial wastes, such as carbon dioxide released during the burning of fossil fuels, can have consequences for climate that pose a considerable threat to future society."[11]

The JASON Committee is an exclusive and secretive group of elite American scientists,[12] funded by the Pentagon, to give advice to the US government on a wide variety of issues. It's kind of like an Opus Dei cult of brilliant scientific nerds, with mandatory costumes of tweed jackets, pants that don't fit, and wild and crazy hair. (Full disclosure, I let my membership lapse when they started charging dues in Bitcoin).

In 1979 the JASON Committee predicted that by 2035, CO_2 in the atmosphere would double, raising global temperatures 4 to 6°F, and warming at the poles of 10 to 12 degrees F. Their prediction has been very prophetic. The JASON Committee was the forerunner of the Intergovernmental Panel on Climate Change (IPCC).

The fossil fuel companies themselves, engaged in their own research on the greenhouse gas phenomenon. Exxon had even gone to the extent of equipping a state-of-the-art supertanker with technology to measure CO_2 levels in the ocean so they could learn about its role in the carbon cycle. Internal documents prove that they were warned about the consequences of fossil fuel combustion in the late 1970s by their own scientists. Exxon, the American Petroleum Institute, and nearly every other multinational oil and gas company were doing and sharing research, and they formed an industry task force on the climate implications of fossil fuel combustion, CO_2 emissions, and the greenhouse gas phenomenon. They all knew about human-caused global warming and its catastrophic implications. Their research even allowed them to predict when the first clear climate consequences would be manifest.[13]

At one of these industry task force meetings in 1980, Stanford Professor John Laurmann went further than the JASON Committee, telling the task force that his models estimated a 5°Celsius rise by 2067, with "globally catastrophic effects."[14]

In May of 1981, Exxon's scientists weighed in with a report to their management estimating that "global temperatures will increase by 3°

Celsius with the doubling of the carbon dioxide emissions in the atmosphere, which could cause catastrophic impacts as early as the first half of the 21st century."[7] Exxon had even begun adopting estimates of atmospheric CO_2 into its corporate planning, and throughout the 1980s the company acknowledged the cause and effect of fossil fuel combustion and the climate crisis.

The Empire Strikes Back

By the 1990s, top-level management at Exxon and the other oil majors did an about face, and rather than burden themselves with altruism, social responsibility, or accountability, they embarked on the seminal example of pure psychopathic corporate behavior and began a campaign of fighting any attempts to curb the climate crisis, especially the Kyoto Protocol. In 1989, the same year that the United Nations formed the Intergovernmental Panel on Climate Change (IPCC), Exxon, BP, and Shell formed the ironically named Global Climate Coalition (GCC). Instead of acknowledging the science or any of their own culpability for playing a dominant role in the looming global catastrophe, GCC embarked on a plan to undermine public and lawmaker acceptance of the science and vigorously lobby against any public policy that would reduce greenhouse gases and/or diminish their bottom line. After all, what's the survival of the human race when there are quarterly profits to be made?[8]

As the science continued to mount, predicting an ever more dire climate, the dirty energy cabal added more and more bricks to their wall of deception. In 1998, one year after the Kyoto Protocol was adopted, the American Petroleum Institute (API), including members BP, Chevron, ConocoPhillips, ExxonMobil, and Shell, planned for "victory" over climate-mitigation strategies, using a campaign of uncertainty about the science, according to a leaked API memo.[9] Joining the major coal companies, their game plan included "forged letters to Congress, secret funding of a supposedly independent scientist, [and] the creation of fake grassroots organizations."[9] That the companies knew exactly how dangerous fossil fuels were was a fact confirmed by internal memos that warned that their fossil fuels were causing the climate crisis, and that "the relevant science is well established and cannot be denied."[9]

The API and GCC claimed victory when George W. Bush pulled the US out of the Kyoto Protocol. In fact, a senior State Department official thanked the GCC for providing "input" that helped them make that decision. Exxon and the other members greatly magnified the scientific unknowns into "weapons of mass confusion."[21]

If Exxon and Big Oil have been instrumental in dragging humanity to the brink of extinction, then certainly Charles and David Koch, the princes of fossil fuels, deserve their own special award as "slayers of mankind." Exxon can take a great deal of credit for pulling the veil over the eyes of the public on the scientific veracity of a climate crisis being human caused, but what they really accomplished was laying the ground work for the Koch brothers, who then began assembling the Climate Denial Death Star, opened up their piggy bank, and with the help of an eager and complicit Republican Party destroyed public climate policy at the state and federal level.

The American Legislative Exchange Council (ALEC) is a corporate lobbying juggernaut wedded to conservative lawmakers with a focus on advancing federal and state legislation on an expansive menu of issues for the benefit of corporations. It was formed in 1973 in the wake of the Powell Memorandum mentioned in chapter 15. It has become increasingly conservative, partisan, and controversial, with deep ties to the Koch brothers. Literally thousands of bills have been mass produced by ALEC committees that have appeared in legislatures throughout the country, fueling a hard right, free market dogmatism, and corporate-friendly ideology. At the top of this legislative heap are bills intended to subsidize and protect fossil fuel revenue streams, fend off regulations and clean energy legislation, and institutionalize climate denial. Virtually all of ALEC's dirty energy portfolio has Koch finger prints all over it.[27]

Ditching Renewable Energy with Renewable Stupidity

As late as 2010, both political parties shared at least lip service to a human-caused greenhouse gas phenomenon creating a climate crisis. A case could be made that 2008 Republican nominee John McCain had better climate credentials and rhetoric than Barack Obama. None other than Newt Gingrich, former Republican Speaker of the House, cut advertisements sharing a couch with Nancy Pelosi with their agreeing on the urgent need to address the climate crisis, sponsored by former Vice President Al Gore's Alliance for Climate Protection. But then the Koch brothers, whose business empire is heavily weighted on fossil fuels, sprang into action. Koch Industries and their Americans for Prosperity right wing political action group, launched "an all-fronts offensive with television advertising, social media and cross-country events aimed at electing lawmakers who would ensure that the fossil fuel industry would not have to worry about new pollution regulations."[34] Since 1997, the Koch Brothers have spent over $100 million on 84 different climate denial groups.[35]

In 2010 the infamous "Citizens United" Supreme Court ruling opened the flood gates and corporate money rushed in, all but drowning an already feeble democracy. The Koch brothers gleefully opened up their fat corporate wallets even wider, spending hundreds of millions more on behalf of Republicans who denounced concern for climate legislation. But they pulled from behind the curtain an even uglier step sister—Republicans who didn't march to the Koch Brother Band became targets of an enormous hit job. Their money dried up, they were challenged in the primaries, and many were eliminated even before the general election. Rather than sign off on their own political obituaries, 165 Republicans running for Congress signed off on the planet's obituary, in the form of Americans for Prosperity's "No Climate Tax" pledge,[36] and most of them won. Eighty-three of the 92 new members of the 2010 Congress signed the pledge.

The Koch brothers were assisted by other business groups, like the vast army of lawyers and lobbyists for the US Chamber of Commerce, who met regularly and devised a lawsuit strategy that would undermine Obama's attempts at climate mitigation. The effort culminated in the Supreme Court's deciding in a strictly partisan vote to put a halt to the implementation of Obama's "Clean Power Plan." The leader of that effort, Scott Pruitt, was rewarded with becoming the point man for the Trump Administration's effort to dismantle all of Obama's environmental initiatives as the head of the new EPA.

The intransigence of Congressional Republicans on climate is perpetuated by relative indifference and shocking ignorance by their voting base, a testament to the very successful, cynical, and sinister campaign of Big Oil, Big Coal, and the Koch brothers. An Alabama Congressman, Mo Brooks, has suggested that sea levels are rising because rocks are falling from the White Cliffs of Dover into the English Channel and perhaps other coastlines like California's.[65] A

Republican Congressman from Michigan, Tim Wahlberg, said that "God will take care of climate change if it is real."[67] For people with this level of ignorance and lack of critical thinking to be making public policy obviously imperils us all. But make no mistake, this is also corporate sown, fertilized, and harvested insanity.

Virtually all of the Republican candidates for President in 2012 and 2016 staked out positions of outright climate denial, or joined their colleagues in pathetic verbal dodging of the issue, with excuses like, "I am not a scientist," which of course implied the non sequitur that therefore I can't listen to those who are, and can't do what they insist must be done. Several of them, like Jon Huntsman and Newt Gingrich, flip flopped into positions of climate nonsense, despite having once engaged in public and legislative mitigation efforts. Some like Sen. Ted Cruz called it a conspiracy concocted by liberals.[64] Then there was the 2012 nominee Mitt Romney, who mocked any concern for the climate after having been very supportive of addressing the issue when he was governor of the liberal state of Massachusetts. At the 2012 Republican Convention he smugly declared, "President Obama promised to slow the rise of the oceans and to heal the planet. My promise is to help you and your family." Then in 2015 and 2017 when Romney no longer needed to placate the Koch brothers or other dirty energy donors, he said climate change is real and urged President Trump not to pull out of the Paris Climate Accord.[61,62]

In the 2016 Republican primaries, fossil fuel millionaires and billionaires poured an unprecedented $107 million into various presidential contenders' Super PACs.[64] The Koch network itself spent $250 million on various races during the rest of the 2016 election. American voters have allowed big money from big corporations to become normalized such that they have become inoculated against the outrage that should be provoked. We have become so fatigued about corporate campaign spending that the Koch brothers even bragged publicly in the spring of 2018 that they were going to spend $400 million to control the fall 2018 midterm elections. Truth in advertising requires some candor here. Our supposed democracy is on the auction block to the highest bidder, and corporations inevitably are able to outbid the citizenry.

On the eve of President Trump's decision on the Paris Accord, he received a letter from 22 senators urging him to withdraw. Over the past three election cycles, those senators had received $10,694,284 from the fossil fuel industry.[63] It was, of course, a gesture of complete detachment from the reality of climate science. Furthermore, the US economy will crawl into hibernation in a cramped den of dirty energy while the rest of the world eagerly exploits the enormous economic opportunities of clean energy. And it has quickly and decisively knocked the US off the leadership podium on the most important issue in the history of mankind.

Among all organizations in the entire world, climate denial is now the exclusive calling card of the American Republican Party and is the single largest impediment to the global community staving off climate disaster. The fossil fuel industry has purchased the Republican Party leadership en masse, and the return on investment has been spectacular for them, and a spectacular disaster for you and me. Corporate money not only buys members of Congress, it buys public opinion. Only 31 percent of Republicans believe global warming is happening and human caused, compared to 66 percent of Democrats.[66] In the era of "fake news," corporate-fueled political polarization now amounts to far more than differing opinions. It amounts to those who accept science, empiricism, expertise, reason, and reality itself--and those who don't.

Corporate influence not only dictates who arrives in Congress and the White House, it dictates what they do when they get there. A Princeton study that reviewed data for 20 years examined the relationship between public opinion, assessed from 2,000 public opinion surveys, and compared it to government legislation and regulations. They found that opinions of the bottom 90 percent of income earners in America had virtually no influence on government policy at all.[68] Not surprisingly, the researchers also found that corporations and wealthy individuals had dramatically more influence. "When only the affluent strongly support a proposed policy change, that policy is adopted 46 percent of the time; when only the middle-class strongly support a policy, that policy is adopted only 24 percent of the time."[69] Put another way, the relatively powerless majority only get what they want when their interests align with the economic elites. They even found when the middle class strongly opposes a policy it slightly increases the chance the policy will be adopted.

Large corporations now spend 34 times as much on lobbying as labor unions and public interest groups.[70] The nonprofit Represent.Us found that over five years, the 200 most politically active companies spent $5.8 billion influencing the government and received $4.4 trillion in taxpayer support.[71] In one year alone, 2017, the total amount of money spent on lobbying was $3.37 billion.[72] As they put it, its "a vicious cycle of legalized corruption." While the fossil fuel industry is hardly the only industry that has the government in a head lock, there is no other industry whose interests are as much at odds with those of the public.

Ask Your Doctor if the Paris Climate Accord is Right for You

At its core, the climate crisis is a public health crisis. Although physicians have been somewhat late to the game, they are now getting on board in a big way. Let's take a cue from those relentless, repetitive TV drug ads: "Ask your doctor if the Paris Climate Agreement is right for you." Here's what your doctor says.

In 2009, Utah Physicians for a Healthy Environment conducted a public seminar at the University of Utah Medical School on how damage to the climate threatens public health. The same year, 29 distinguished medical experts wrote in one of the world's most distinguished medical journals that global warming was, "the biggest global health threat of the 21st century, and will put the lives and wellbeing of billions of people at increased risk."[37] They said, "The rich will find their world to be more expensive, inconvenient, uncomfortable, disrupted and colourless, in general, more unpleasant and unpredictable, perhaps greatly so. The poor will die."[55]

In April 2016 the historically conservative American Medical Association filed an amicus brief before the Supreme Court in support of the Clean Power Plan. In March 2017, 11 of the major medical specialty organizations in the US formed The Medical Society Consortium on Climate and Health.[38] The new organization represents more than 400,000 physicians--about half of all the healthcare professionals in the country. Announcing this unprecedented move, physician leaders said, "Climate change [already] threatens the health of every American." And here's why.

To begin with, hotter temperatures themselves aggravate heart disease, increase pregnancy complications, and will increase the number of heat strokes. But that's just the tip of the (melting) iceberg.

Air pollution is intimately related to climate disruption. The chemical reaction that forms atmospheric ozone is catalyzed by heat, and ozone levels are increasing worldwide. A much prolonged, if not year-round forest fire season, and more intense mega-fires send smoke circling the entire globe, increasing everyone's exposure to smoke pollution. The area in the Western US scorched by wildfires has doubled in the last 30 years.[39] More global warming will increase forest fire decimation exponentially.[40] The average burn time has increased from 6 to 52 days, and the burned area has increased 12 times.[57] At this rate, the West will be basically denuded of forests within a few decades. In 2012, it was estimated that the 339,000 people die every year from wildfire pollution,[56] and that number will have only increased since then.

Hotter temperatures and more drought, like what the American West is experiencing, is increasing dust storms already.[41] A single gram of dust can harbor as many as 1 billion microorganisms. Diseases like SARS, valley fever, influenza, and meningitis can be spread by dust. In the Great Basin of the American West, that dust also contains heavy metals like mercury, and radioactive metals like uranium, strontium, cesium, and plutonium.

After Hurricane Florence hit the Carolinas in September, 2018, there were wide spread reports of swarms of mosquitoes that were hyper-aggressive, and triple the size of typical mosquitoes.[88] The reports weren't just urban legend. These mosquitoes are known by the common name of "gallinippers." Because of their size, their bite is very painful. They are the only mosquito species capable of biting through cattle hide.

This species of mosquito is known to emerge after heavy rainfall because the large females preferentially lay their eggs in soil near bodies of water that regularly flood, and then the flood waters cause them to hatch. Add monster mosquitoes to the long list of scourges from the climate crisis.

Vector-borne diseases, spread by mosquitoes and other insects, are rapidly increasing their geographic range and season of distribution. Mosquito populations are very temperature sensitive, increasing tenfold with every 0.1 degree C increase in temperature.[52] We can expect -- and in many cases, are already seeing -- increases in West Nile virus, dengue fever, coccidioidomycosis, Zika, Chagas, Lyme disease, yellow fever, chikungunya, Rocky Mountain spotted fever, Rift Valley fever, Japanese encephalitis, leishmaniasis, hantavirus, and malaria. For example, cases of the debilitating and occasionally fatal coccidioidomycosis, or valley fever are up 400% since 1998, and new research identifies the dust storms from the climate crisis as the likely cause.[71]

Unfortunately, to whatever extent public anxiety is provoked by global warming, it does not often translate into fear of mass starvation. But that is, in fact, the greatest risk. Agriculture will be crippled. The global food supply will be reduced by increased temperatures and the attendant increased water requirements of plants. Every day above 30 degrees C causes a drop in staple crop yields like corn, wheat and soybeans of about 5 percent.[53] By 2100, projected temperature increases worldwide are expected to decrease corn yields 50 percent, soybeans 40 percent and wheat 20 percent.[42] Increased humidity and a longer growing season may help somewhat with corn,[72] but that may be an anomaly. More droughts and floods will further decrease crop production. Weeds and insects thrive under warmer temperatures. Rising levels of CO_2 reduce the concentrations of protein and essential minerals in most plant species, including wheat, soybeans, and rice, according to the EPA. Increased heat threatens livestock. Milk production from dairy cows drops with increased temperatures. Ozone is a powerful oxidizing agent acting on, among other things, human lungs and plants. The US Department of Agriculture says, "Ozone causes more damage to plants than all other pollutants combined,"[43] and it increases water stress, especially for wheat and soybeans.

But among the most important effects of climate disruption are extreme variations in water supply. Protracted, severe droughts, interrupted by occasionally by severe flooding is the likely future of the Western United States, especially the Southwest and California.[23] Essentially those areas that already suffer from too much or too little precipitation will only see those same patterns become more extreme. The Great American Dust Bowl of the 1930s may pale in comparison to what's in store for the American Southwest. A recent study indicates that continuing business-as-usual burning of fossil fuels will result in a 99 percent chance of unleashing a hellish mega drought, one that persists for at least 35 years.[24] That's as much of a virtual certainty as the Kardashians taking more pictures of themselves naked, or as much as Donald Trump continuing to brag about his electoral college victory. One of the study's authors said, "This will be worse than anything seen during the last 2,000 years and would pose unprecedented challenges to water resources in the region."

Furthermore, the study found that even if, in the unlikely case that global warming increases rainfall in the West because a warmer atmosphere increases evaporation from the soil and from reservoirs and makes trees and plants take in more water from the soil, the chance of a disastrous mega drought is still 70 percent. Such a mega drought will feature massive and persistent toxic dust storms, most of the trees in the West will die, and agriculture would be virtually impossible. On our current trajectory of greenhouse gas emissions, southern Europe will see permanent and severe drought by 2080. "The same will be true in Iraq and Syria and much of the rest of the Middle East; some of the most densely populated parts of Australia, Africa, and South America; and the breadbasket regions of China."[87] The American Association for Advancement of Science said the climate crisis will have "massively disruptive consequences to societies and ecosystems, including widespread famines, lethal heat waves, more frequent and destructive natural disasters, and social unrest."[31]

The twin sister of global warming and increased atmospheric CO_2 is ocean acidification. Water bodies absorb about 25-40 percent of the carbon dioxide that human activity emits into the atmosphere, and that extra CO_2 acidifies the oceans. Marine life with calcium carbonate shells (mollusks, crabs, and corals) are unable to adapt to the increased acidity and increased temperatures.[25]

The Great Barrier Reef off the coast of Australia is the most magnificent underwater ecosystem in the world and one of its largest living organisms--or was before global warming. Under normal circumstances, corals can live over 4,000 years. But half of the reef has died in 2016 and 2017. Other research estimates that 90 percent of the world's corals could be dead within the next 30-plus years.[26] Corals are known as "foundation species." So much other marine life depends on them for survival, they are the feeding grounds of about a quarter of the ocean's fish.

As important as coral is, one could consider phytoplankton to be even more important to other marine life, as well as providing half the ocean's oxygen. Phytoplankton populations have been declining about 1 percent a year for several decades due to the combination of warming and more acidic waters.[32, 33] There is already about 40 percent less phytoplankton in global oceans compared to 1950. Phytoplankton are positioned at the bottom of the food chain, and what happens at the bottom affects every organism above, and that includes humans.

Populations and economies dependent on fish could be devastated. Even worse, phytoplankton have an enormous role in moderating the climate because on the ocean's surface they absorb CO_2. Loss of phytoplankton means more CO_2 in the atmosphere and more global warming, setting up a feedback loop of ever worsening ecological and environmental catastrophe.

Military experts are warning that climate disruption, already precipitating refugee crises and regional conflicts, will reach an "unimaginable scale,"[44] adding another layer of global health risk, including igniting security threats from more terrorism. People whose lives have been reduced to desperation from climate-related disasters will become more susceptible to terrorist recruiters, especially if the target is the United States, which has just brazenly given the world Trump's middle finger, openly declaring it couldn't care less about the climate. If Trump's top priority is to protect the US from terrorism, pulling out of the climate agreement will likely increase that risk far more than any travel bans or wall building could decrease it.

A "caravan" of desperate immigrants from Central America were shamelessly exploited as political fodder in the run up to the 2018 fall election. Trump pounded his anti-immigrant chest depicting the immigrants as disease-ridden criminals, terrorists, and a security threat to the US, and cruelly tear gassed them. Meanwhile Democrats compared the Central American immigrants to Jews fleeing Germany because of the threat of physical violence and death in their own country and called on people "uncomfortable with spraying tear gas on children" to join the "coalition of the moral and the sane" to "get our country back." But virtually no one told the most important part of the story.

The American southern border has seen a sharp increase in the number of Guatemalans trying to enter the US starting in 2014.[99] Coincidentally, that was the first year of a severe drought tied to an extreme El Nino that struck Central America's "Dry Corridor," which includes Guatemala, Honduras, and El Salvador.[100] Relief from the drought has only come from occasional, devastating flooding, which only adds to the destruction of crops. One third of all employment in Central America comes from agriculture, and that is now failing across the entire region.[101]

Guatemala is rated as one of the top ten of the world's nations most vulnerable to the climate crisis, which means an agricultural crisis, now evolving into a human crisis. The weather patterns wreaking havoc on Central American agriculture are exactly what climate scientists have predicted.[102] An inter-agency study from the UN interviewing families trying to leave Central America discovered that the driving force for this exodus was not violence per se, but the drought and its downstream consequences--lack of food, no income, and no work, all related to crop failures.[103]

Mendez Lopez is a subsistence farmer in Guatemala and was interviewed by the National Geographic.[104] The multi-year drought completely wiped out his corn fields. He has a rapidly dwindling source of food, no income, and soon will have no way of feeding his six children. He said, "This is the worst drought we've ever had. We've lost absolutely everything. If things don't improve, we'll be forced to migrate somewhere else. We can't go on like this." The reporter described the physical appearance of Lopez and his neighbors as gaunt, skin on bone, after being forced to survive for months on corn tortillas and salt.

It's not just subsistence farming that has withered. Other regional cash crops like coffee have been decimated by another climate related plague, "leaf rust," a fungus that used to die with cool evenings, but no longer does because of warmer night time temperatures.[105]

Robert Albro, researcher at American University says, "The main reason people are moving is because they don't have anything to eat. This has a strong link to climate change – we are seeing tremendous climate instability that is radically changing food security in the region."[106] The UN says that almost half of Guatemalan children under five years old suffer malnutrition.[107] In rural areas of the country it's 90 percent.

Children and teenagers are dropping out of school because their families have no money for supplies. Entire villages are unraveling because there is no money to plant another crop and not enough government help. Even more people would abandon their homes but have no money for transportation. Rural Guatemalans are foraging the landscape in search of wild malanga roots in a desperate attempt to feed themselves.[108] Not surprisingly, starving people resort to extremes to stay alive, and that includes violence, assaults, robbery, and walking thousands of miles seeking a reprieve.

At the current rate of greenhouse gas emissions, by 2100 the world can expect 2 billion of its inhabitants to become climate refugees[109] like the "Central American caravan." President Trump has decried "globalism," extolled "nationalism," and forced the United States to become the only nation to withdraw from the Paris Climate Accord. There is no wall tall enough or border patrol massive enough to stop either the global or the national consequences of the climate crisis.

The President is training guns and tear gas on the trickle of human suffering inching toward our border. But that trickle will become a raging torrent if the United States continues to deny the climate crisis, its role in causing it, and its responsibility to alleviate it.

Finally, the climate crisis is a global psychological and emotional stressor. Extreme heat alone is associated with increased incidence of aggressive behavior, violence,[45] and suicide,[46] and increases in hospital and emergency room admissions for those with mental health or psychiatric conditions. People traumatized economically and/or psychologically by extreme weather events are subsequently plagued by depression, anxiety disorders, substance abuse, and suicide. Lise Van Susteren, a forensic psychiatrist, said, "Expectation of climate-change disasters is causing pre-traumatic stress disorder…the anger, the panic, the obsessive, intrusive thoughts."[47]

Robert Gifford, a University of Victoria psychology professor, said, "I see parallels to the fears we went through in the 1950s about the world ending because of atomic war. There was this general dread among people, and this fear of annihilation."[48] I remember as a child, drills in school to duck under my desk in case of nuclear war (as if that would have helped). Daily dread of the "mushroom cloud" hardly made life in the 1950s a carefree experience for children -- or for anyone else.

Of course every excuse the Trump administration gave for pulling out of the Paris climate agreement, before and after, was deceptive and dishonest. For example, former EPA head Scott Pruitt recently stated that coal industry jobs have increased by 50,000 since Trump took office. However, according to the US Bureau of Labor Statistics, there are only 51,000 total jobs in the coal mining industry.[49] Now add to that studies finding that pollution from coal power plants alone kills 52,000 people a year in the US.[50] Even taking Pruitt's word for it, for every new job in the coal industry, someone else's life will be sacrificed. Go ahead, ask your doctor if the Paris climate agreement is right for you. You know what the answer is.

It's Worse Than You Think

In a tribute to my optimistic personality, another epitaph my wife is inscribing on my tomb stone (she's working on several) is, "He always said, 'No matter how hard life may seem, things are never so bad they can't get worse.'" And with regard to the climate that is now almost guaranteed. The situation is much more dire than most of us are willing to admit. The headline of the NY Times opinion section on Feb. 17, 2019, reads, "Time to Panic." Actually the science has been unequivocal that it's been time to panic for the last 20 years. But instead of panic, we have had paralysis and denial nearly world-wide.

More than half of all industrial emissions of carbon since the beginning of the Industrial Revolution have been released since 1988.[10] Meanwhile, global warming science becomes even more certain and more alarming. After three years in a row of flat greenhouse gas emissions, in 2017 they rose again about 2 percent,[51] driven primarily by increases in carbon emissions in India and China.

Just a few years ago, China announced the cancelling of construction and shuttering of hundreds of coal power plants giving climate scientists hope that the world's largest emitter of CO_2 was going to honor the Paris climate agreement. Satellite imagery, however, shows that hope was misplaced, and in fact, those new plants are still being built, and will trigger a tsunami of deadly, climate-killing coal burning, equal in size to the coal combustion of the existing US fleet[89] and a boost of 25 percent over China's current capacity.

But it's not just China. Brazil has elected as President, Jair Bolsonaro, who is every bit as much an enemy to the environment as Donald Trump. He is more than happy to decimate the Amazon for industrial agriculture and livestock grazing. Britain is embarking on a new era of fracking for natural gas. Norway is pushing ahead for oil drilling in the Arctic, Germany apparently is willing to cut down their Hambach forest to dig for more coal, and India is increasing the carbon emissions of their 1.3 billion people faster than any other country.

Nearly 200 of the world's scientific societies have endorsed the concept of a primarily human-caused climate crisis that is already starting to threaten the health and wellbeing of millions, and soon-to-be billions of people in the next few decades. The total number of scientific organizations who dispute this is zero. Nor is there any dispute that we are nearing our last hour to turn avert disaster. If you were watching a basketball game where the score was 200 to 0 with one minute left in the fourth quarter, and you decided to bet your entire nest egg on that losing team, no one would argue that you were not certifiably insane. Our own Pentagon,[15] the insurance industry,[16] the World Bank,[17] the United Nations,[18] the American Meteorological Society,[19] and virtually every other government in the world accepts the science.

Almost weekly, more studies are published strongly suggesting that the chaos and destruction built into the greenhouse gas phenomenon has been underestimated and that climate-related extreme outcomes are happening even faster than worst-case predictions of even a few years ago.[20]

The premise of the 2004 science fiction movie, *The Day After Tomorrow*, depicts climate catastrophe precipitated by disruption of the North Atlantic Ocean circulation, the end result being rapid global cooling and the abrupt onset of a new Ice Age. In the context of global warming, the plot seemed preposterous. But there is a kernel of scientific evidence to somewhat support the premise. In the late winter of 2018, the North Pole is experiencing is warmest winter on record, with a current temperature above freezing, 35°F, 50 degrees above normal, and 78 degrees warmer than parts of Norway. The Arctic Circle in Alaska just finished averaging 14°F above normal. A warming Arctic triggers more weather extremes throughout North America, like record-breaking cold in the Northeast.[54]

The Atlantic Ocean's circulation pattern has slowed about 15 percent in the last 70 years, and it is due to massive fresh water pouring of the melting Greenland Ice Sheet secondary to warming global temperatures. That circulation pattern acts as a conveyor belt bringing warm water from the Gulf Stream to Northern Europe. If that slowing continues, and there is every reason to think that it will, the end result will be disruption of weather patterns in all countries that are adjacent to the Atlantic Ocean and changing rainfall patterns at the equator. That will result in more weather extremes in Europe, much colder winters and hotter summers.[21] Further slowing will cause warmer waters on the East Coast of North America and likely decimate fisheries.

Similarly increased glacial melting in the Antarctic will further exacerbate the climate instability from global warming. The Indian

Ocean has seen a massive increase in temperature, almost double what the rest of the globe has experienced. This is likely linked already to some of the deadliest heat waves the world has seen so far, and the disruption of the historical monsoon cycle in Asia.[22]

James Hansen, the former NASA scientist and highest profile muse of the climate crisis, and other esteemed climate scientists believe that for the earth to return to a climate similar to the one where human civilization developed we will have to return to a concentration of CO_2 in the atmosphere of no more than 350 parts per million (ppm), which spawned the activist organization, 350.org. In 2018, we passed 411ppm. Because of the lag time between atmospheric CO_2 and the climate consequences, we have already baked into the system much more of a climate crisis than we are currently experiencing, even if we emit no more CO_2 starting tomorrow.

Conventional projections of the earth's warming in relation to CO_2 levels arise from calculations that all the scientists know are wrong;[59] it's called "Charney sensitivity," (named after Steven Colbert's brother, Charney...fake news; it was actually named after Jules Charney who chaired one of the first scientific committees on the issue[60]). Charney's sensitivity is known by all climate scientists to be, out of necessity, an oversimplification, and offers likely an overly optimistic forecast.

Current computational limits do not allow the latest models to address known climate tipping points such as the disappearance of summer Arctic ice cover, the melting of the ice sheets of Greenland and Antarctica, the loss of Amazon and boreal forests, the collapse of the gulf stream, and the release of methane from permafrost and ocean methane hydrates.[58] If these feedback mechanisms kick in, Earth will lose its ability to buffer greenhouse gases, and the climate will be abruptly catapulted past the point of no return into a hell where it disintegrates civilization. There is evidence some of those tipping points have already been passed. Hansen's team believes that if humans continue business as usual, and reach 450 ppm, we will have at least opened the door to that climate hell. [59]

The agreed-upon target limit for global warming set by the Paris Accord was an increase of 3.6° F (2°C). If we reach that threshold major disruptions of ecosystems and agriculture would follow, and we are headed in a direction to far exceed it. That target was picked precisely because there is broad scientific agreement that at that point those feedback mechanisms would overwhelm any mitigation strategies, dramatic warming would become irreversible, and the exit door from climate purgatory will be locked forever.[73]

A report in Nature[29] indicates that the models that are proving the most accurate in recreating current conditions suggests even greater danger is in store. Specifically, that "if emissions follow a commonly used business-as-usual scenario, there is a 93 percent chance that global warming will exceed 4°Celsius (7.2° F) by the end of this century. Previous studies had put this likelihood at 62 percent."[30] After leveling off for three years, global CO_2 emissions are on the rise again, and fossil fuel companies are still committing enormous budgets to exploration and extraction of yet more carbon-based energy.

In October 2018, a new report[91] by the UN Intergovernmental Panel on Climate Change (IPCC) gave more specifics to the long list of dire warnings of their previous reports. The UN had asked the IPCC to estimate what it would take to limit the rise in global temperatures of 1.5° C, the "aspirational goal" of the Paris Accord, and the consequences of failing to do so. To stay below the threshold of 1.5° C will require a massive and immediate shift away from fossil fuels, transformation away from carbon-intensive, industrialized agriculture, and a replanting of the world's forests. The report, drawing on 6,000 studies, also says that we have about 12 more years in which to cut carbon pollution by 45 percent before we reach the point of runaway climate disaster. The report itemizes some of the consequences.

At an increase of 1.5° C, 80 percent of coral reefs will be gone. At 2°C, 100 percent of them will dead. The disappearance of Arctic ice in the summer is ten times more likely at 2° C versus 1.5° C. At 2°C, there will be billions of people without adequate water, agriculture yields will be severely reduced, and climate refugees will likely be over 1.5 billion by 2060.[92] Hansen said both 1.5°C and 2°C increases would "take humanity into uncharted and dangerous territory," but that, "1.5° C gives young people and the next generation a fighting chance."[93]

Let's go back to Vernal, Utah, mentioned in Chapter 4. Eastern Utah could be considered ground zero for the battle to keep the world's fossil fuels in the ground. In addition to the fracking frenzy for oil and gas in the area, Utah is also "blessed/cursed" with the largest unconventional fossil fuel reservoir in the United States and perhaps the world—oil shale and tar sands deposits that exceed those in Alberta Canada and Saudi Arabia. Using geology-based assessment methodology, the U.S. Geological Survey estimated a total of 4.285

trillion barrels of oil rest in the oil shale of the three principal basins of the Eocene Green River Formation, near Vernal, Utah.[80]

If those deposits are extracted and burned (and the process would be much more carbon intensive than conventional oil and gas drilling), Utah would become home to the largest known carbon "bomb" on the planet, more "game over" for the planet than the much hyped Keystone pipeline.

Apparently that is just fine with Utah's governor and the majority of our legislature. It is certainly not only fine with, but enthusiastically promoted by Uinta County commissioners and the fossil fuel industry. It is also fraught with irony because numerous projections on global warming predict that Utah will become North America's greatest warming "victim" outside the Arctic. Projections as early as 2008 suggested that temperatures may rise by 9 °F in Utah by 2100. Global warming calculations have only become more alarming since.[81]

To keep the climate stabilized enough to maintain civilization as we have come to know it or even avoid mass starvation and global chaos, we will have to stay within a carbon budget. The countries of the world must only allow somewhere between one third and one tenth of the known, economically recoverable reserves of coal, oil, and gas to be extracted and burned in order to avoid the all-important threshold of rise of 3.6 °F.[74] Even the International Energy Agency (IEA) (not exactly an environmental extremist organization) calculates that two-thirds of all proven fossil fuels reserves will have to be left unburned. Addressing participants in the 2012 UN climate talks in Bonn, Fatih Birol, Executive Director of the IEA said, "We cannot afford to burn all the fossil fuels we have. If we did that, it [average global surface temperature] would go higher than four degrees. Globally, the direction we are on is not the right one. If it continues, the increase would be as high as 5.3 degrees [F] — and that would have devastating effects on all of us."

There is no evidence whatsoever that any of these corporations are entertaining any thoughts of self-restraint. On March 30, 2014, the very same day the IPCC released a new report warning that global warming will raise the risks of a host of costly and deadly consequences, from sea level rise to food supply disruptions, ExxonMobil released its own set of climate reports.[75] Exxon's vice president of corporate strategic planning stated to the media:

> "Our analysis and those of independent agencies confirms our long-standing view that all viable energy sources will be essential to meet increasing demand growth that accompanies expanding economies and rising living standards...ExxonMobil's Outlook for Energy and all credible forecasts, including that of the International Energy Agency, predict that carbon-based fuels will continue to meet about three-quarters of global energy needs through 2040. All of ExxonMobil's current hydrocarbon reserves will be needed, along with substantial future industry investments, to address global energy needs."[75]

In other words, to the entire world, Exxon said, "Screw you, we're going to burn everything we've got, and we'll keep plundering every corner of the earth's surface hoping to find more."

The five largest oil companies in the world BP, Chevron, ConocoPhillips, ExxonMobil, and Shell, make a combined profit of $175,000 per minute.[76] As mindless, amoral, and unbridled as a malignancy destroying its host, ExxonMobil, TransCanada, Peabody Energy and the like, employ hundreds of thousands of people working like tumor cells for the relentless destruction of the environment and climate that they themselves depend upon for their very lives. And the rest of us stand by and watch it happen. In fact, if we work for a bank, we may not only be watching it happen, we may be lending them the money to make sure it happens.

Financing Our Self Destruction

As a group, the world's big banks were some of the most outspoken supporters of the Paris climate accord. Despite publicly beating their chests on their "green credentials" in fighting the climate crisis and financing clean energy, some of the largest banks in the world, especially in the US and Canada, are still loaning money to the fossil fuel cabal for even the most "extreme," i.e., carbon-intensive dirty energy projects. The Royal Bank of Canada (RBC), Toronto-Dominion Bank (TD), and JPMorgan Chase, Goldman Sachs, and Deutsche Bank were some of the worst in terms of increasing their loans for extreme fossil fuel projects, like oil shale, ultra-deep water drilling, and tar sands.[90]

Several environmental groups have collaborated on an annual report about the banks that finance the fossil fuel industry. The 2018 edition of this report, "Banking on Climate Change,"[95] revealed how, despite the freshly signed Paris Climate Agreement, in 2017 the banking sector actually increased their financing of extreme fossil fuel extraction by 11 percent. As a group, Chinese and Japanese banks get the lowest rankings in this report for their lending policies' impact on the climate.[95] In fact, "no banks have yet truly aligned their business plans with the Paris Climate Agreement."[95] The international banks that received "F" grades included Mitsubishi UFJ, Agricultural Bank of China, China Construction Bank, Industrial and Commercial Bank of China (ICBC). European banks ranked somewhat better. The big U.S. banks all received poor grades. Bank of America received a D, Citi Group a C-, Goldman Sachs a D+, JP Morgan Chase a D+, Morgan Stanley a C-, and Wells Fargo a D+. Overall, U.S. and Canadian banks increased their financing of dirty energy in 2017. The bulk of that increase was for Alberta tar sands expansion, while European and Asian banks decreased their financing.

In 2015 the world's 20 largest national economies, the G20, rallied the rest of the world to embrace the Paris Climate Agreement. However, since then, those same G20 countries have sabotaged their own agreement by essentially "colluding" (to use Donald Trump's favorite term) with the fossil fuel industry in providing four times more public financing for dirty fossil fuel projects than for clean energy development,[19] via sweetheart loans, guarantees, and various types of financial incentives. Virtually every country has its own set of government policies that subsidize fossil fuels. In the United States that includes a dizzying array of corporate production and operation subsidies, like for "intangible drilling," depletion allowances, artificially regulated and unregulated markets, Limited Partnership exemptions, "last-in, first-out" accounting, failure to pay royalties, and far below-market leasing charges for coal extraction on public lands. There are further subsidies on the consumer side of the ledger that specifically keep the price of fossil fuel consumption artificially low, like the Low Income Home Energy Assistance Programs (LIHEAP).

Many of these subsidies have been around so long that they seem almost like part of the Biblical Ten Commandments, which obscures the accurate perception that they are an accumulation of anachronistic corporate welfare. They are clearly distortions of reasonable public policy, especially given what we now know about the climate and public health consequences of dirty energy. All told, these subsidies for dirty energy are far greater than those that exist for clean energy. In a landmark report, "Hidden Price Tags: How ending fossil fuel subsidies would benefit our health,"[96] the international health advocacy group HEAL asks this question: If fossil fuels need to stay in the ground, why are we handing out public money to support them? And the answer to that question is exactly what you would expect. Because of the political clout of fossil fuel corporations, they are able to cajole, arm twist, campaign contribute, and muscle their way to maintaining public policy that gives them an unfair advantage over clean energy alternatives. For example, during the 2016 election cycle, dirty energy corporations spent a combined $354 million in campaign contributions and lobbying for candidates nationwide, and reaped an astounding $29.4 billion in federal subsidies during the same time. That's an 8,200 percent return on investment, or more bluntly, an 8,200 percent return on their "bribery."[97] A recent analysis[98] found that at oil prices of $50 a barrel, half of oil field drilling would not have occurred without government subsidies.

All these subsidies seem like letting fossil fuel corporations win the lottery year after year, but they have been the mother of all disasters as an investment of public money for everyone else. The G20 spent $440 billion in subsidies for dirty energy in 2014, but a very conservative estimate of the externalities, or public health costs of burning fossil fuels that year, was at least six times that amount, or $2.76 trillion,[96] and that likely doesn't scratch the surface of the costs of long-term climate consequences.

In listening to news reports a couple of years ago about the controversial construction of the Keystone pipeline, I always found it fascinating that the only recognized means of defusing what was described at the time as the most important "carbon bomb" on the planet, the Alberta Tar Sands, is a back-door, after-the-fact, almost sleight-of-hand approach, i.e. obstruct the free flow of the product. I heard numerous commentators state that oil from the Alberta Tar Sands was going to get to market one way or the other. It's as if we are all standing idly by, watching the planet being lit on fire by arsonists, and it is beyond our ability or will power to stop it, by any means other than trickery.

For the moment, leaving aside the issue of whether the lack of a pipeline would slow down or stall the mining of Canada's tar sands, I found it puzzling that neither proponents or opponents felt that there was any means of telling the multi-national corporations that they could not, in fact, finance, build, and light that carbon bomb. It seemed that if our current laws allowed, or even guaranteed our self-destruction, then we were simply powerless to do otherwise.

Despite the steady advancement of the science and the scientifically irrefutable evidence on the climate crisis, API's current position "rejects any federal mandates to reduce greenhouse gas emissions…and condemned President Obama's Clean Power Plan to cut emissions from the country's power plants, the cornerstone of the administration's climate agenda, as destructive 'government interference' in free markets."[77]

Even in 2015, at Exxon's annual meeting, then CEO Rex Tillerson said, "it would be best to wait for more solid science before acting on climate change. What if [after] everything we do, it turns out our models are lousy, and we don't get the effects we predict?"[78] Tillerson's meandering in the fog of self-interest, conveniently blinds him to the much greater likelihood that our models are indeed "lousy," but error in vastly underestimating the danger, not overestimating the danger. Even if the climate crisis were a complete hoax created by the Chinese as Donald Trump has claimed, with the cost of wind, solar, geothermal, and battery storage now on par with fossil fuels, the worst that could happen with a commitment to clean energy is that we would merely avoid all of the health consequences of air pollution, water pollution, oil spills, and fracking caused earth quakes.

I would ask Tillerson and anyone else mired in such absurd rationalization, "If your two-year-old granddaughter wandered out into the street, would you wait until you had absolute proof that a car was barreling down upon her to run out and pluck her from danger? If you did that, and it turned out that no car was about to run over her, would you lament that you had wasted time and effort in removing her from the risk?"

Stephen Hawking was widely regarded as the greatest scientific mind since Einstein. This is what he had to say about global warming. "Trump's action could push the Earth over the brink, to become like Venus, with a temperature of two hundred and fifty degrees, and raining sulfuric acid." He believes continued burning of fossil fuels will make Earth a "sizzling ball of fire" and the end of humanity.[28] While most climate scientists consider Hawking's depiction hyperbolic, Earth doesn't need to reach anywhere near that extreme state in order for human life to be pushed over the brink.

That is why climate deniers in positions of power now have the fate of the world in their hands, in a way no one ever has before. And climate deniers are in positions of immense power thanks to the dirty energy corporations.

The fossil fuel industry has never had friends like Donald Trump, Ryan Zinke, and the Republicans in Congress, and the rest of mankind has never had greater enemies. From whittling down national monuments for the express purpose of more drilling, to abandoning common-sense pollution mitigation and methane capture strategies in fracking, from drilling in the Arctic and in both U.S. coastal waters to ditching the Clean Power Plan and stricter vehicle fuel efficiency standards, whatever modest shackles previous administrations have attached to dirty energy corporations, they have been unleashed by the Trump Administration beyond their wildest dreams.

Noam Chomsky, the liberal MIT professor, linguist, dissident, and famous author said, "(We are) led by leaders who understand very well what they're doing, but are so dedicated to enriching themselves and their friends in the near future that it simply doesn't matter what happens to the human species. There's nothing like this in all of human history. There have been plenty of monsters in the past....But you can't find one who was dedicated, with passion, to destroying the prospects for organized human life. Hitler was horrible enough, but not that."[94]

Lighting the Planet on Fire: Post Script

In 1987, Ronald Reagan, hardly America's foremost environmental president, signed the Montreal Protocol that brought the countries of the world together in an effort to stop the release of chlorofluorocarbon chemicals (CFCs) that were destroying the ozone layer in the earth's upper atmosphere. The Protocol was perhaps the world's most unqualified cooperative environmental success. Hydrofluorocarbons (HFCs) were designed to replace CFCs, but they came with a different dangerous calling card, as bad as destroying the ozone layer. HFCs are intense greenhouse gases.

During President Obama's Administration most of the world's countries agreed to the Kigali Amendment to the Montreal Protocol that would phase out HFCs and replace them with alternatives for refrigeration, HFOs, hydrofluroolefins that are far less potent greenhouse gases. Industries seeking to capitalize on that opportunity formed the Alliance for Responsible Atmospheric Policy, which includes companies like Dow and Carrier, to promote the market for refrigerant alternatives created by the Kigali Amendment.

Enter the Competitive Enterprise Institute (CEI). Their director, Myron Ebell, was tabbed to lead Trump's EPA transition team. Ebell brought a hefty global warming denial to the new administration and the stroke of genius that if HFCs cause global warming, but global warming is a hoax, then what is the point of HFC alternatives. Under pressure from the companies comprising the CEI and the pretense that those alternatives were unnecessary expenses, the Trump Administration decided to withdraw from the Kigali Amendment as of Sept. 24, 2018. Fossil fuel corporations (you may have heard of them), like Exxon and the Koch Brothers, are the largest funders of CEI.[84]

But if it seems like dirty energy corporations are monsters, bent on our destruction, unbelievably they may actually not be the most sinister or the most audacious. According to the Environmental Investigation Agency,[82] many Chinese and Indian companies that make an HFC refrigerant, HCFC-22, are demanding big money to dispose of a by-product of that process, HFC-23, which is a greenhouse gas 14,800 times more potent than CO_2. Despite having cheap destruction technology readily available, they intend to hold their stores of HFC-23 hostage until the rest of the world pays up. There's a word for that: it's called blackmail. Actually, we need another word, because blackmail doesn't do justice to this level of malevolence. This has caught the attention of other manufacturers of HCFC-22 in developing countries who are poised to join in on holding the rest of the world hostage.

Under a UN program, incinerators for HFC-23 are installed at 19 refrigerant facilities, mostly in China and India but also in South Korea, Argentina, and Mexico, to help control the super greenhouse gas. Destruction of HFC-23 is extremely cheap. But refrigerant companies made billions in windfall profits from the sale of carbon credits, by deliberating manufacturing more HCFC-22 than was needed so they could capture and sell credits for destroying HFC-23. This prompted the European Emissions Trading Scheme to ban the trade of HFC-23 credits as of May 1, 2013. When these credits were banned, manufacturers flooded the market in the rush to sell them before they became worthless. Businesses banked their other allowances, lowering demand further. Other carbon markets have followed suit, resulting in the collapse of the HFC-23 credit market. What is at stake here is the greenhouse gas equivalent of one fourth of China's annual CO_2 emissions.

The IPCC report of Oct. 2018 demonstrates that "within the laws of physics and chemistry," with current resources and technology, and feasible financial commitments, we can avert climate disaster, and not destroy human civilization. The missing ingredient, of course, is the political will. Until and unless that arises from the ashes of our present governmental dysfunction, corporations are unrelenting in continuing their destruction, leaving humanity with the nightmare of choking on their waste, hurtling us off the climate cliff and writing the obituary for a livable planet.

Nancy Pelosi and Newt Gingrich in 2008 TV commercial affirming humans cause climate change. Shortly after, the entire Republican Party was essentially "purchased" by the radically anti-science Koch Brothers.

Charles and David Koch—Koch Industries
Their obstruction of addressing the climate crisis will have contributed to incalculable, global-wide, suffering and death.

Nearly one third of the world's population could become climate refugees if greenhouse gases are not drastically curbed.

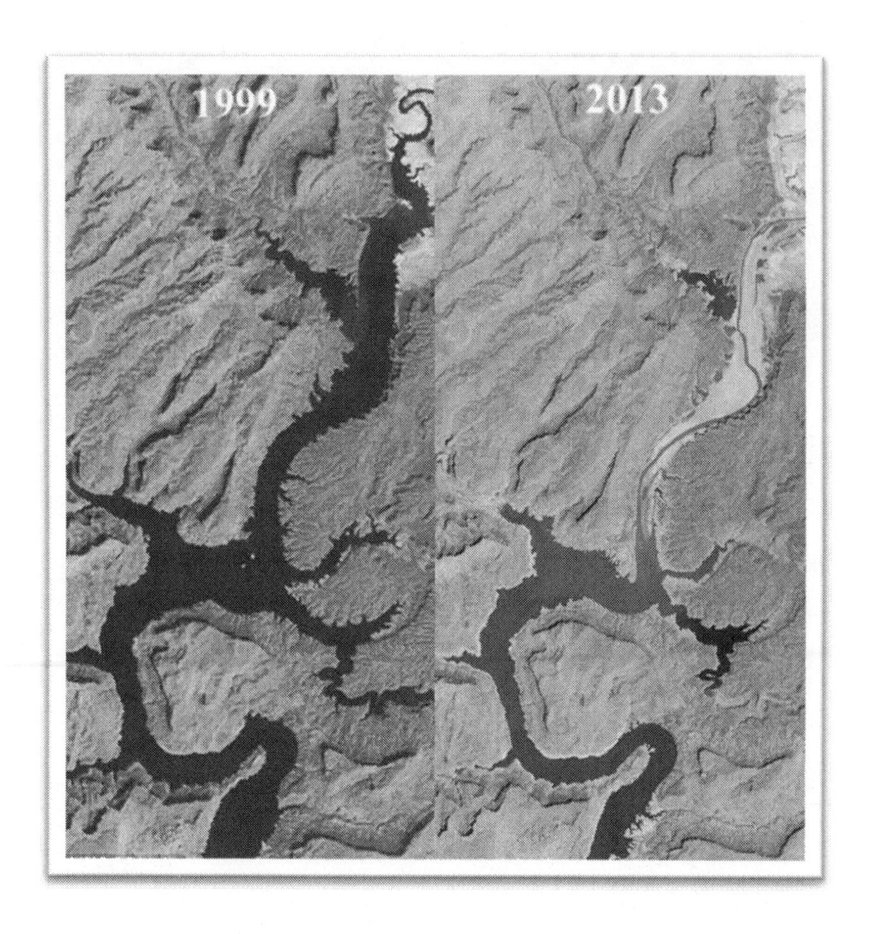

Lake Powell lost 4.4 trillion gallons of water from 1999 to 2103. These satellite photos from NASA show how much the lake has contracted. Continuing business-as-usual burning of fossil fuels will result in a 99 percent chance of unleashing a hellish mega drought in the Western US, one that persists for at least 35 years.

7

The Tobacco Industry:
New Smoke and Mirrors

"It is more profitable for your congressman to support the tobacco industry than your life."

—Jackie Mason

I'm hardly the first person to wonder how Big Tobacco executives sleep at night. Stating that the tobacco industry sets the gold standard for corporations in knowingly, willingly, and unapologetically serving up death on a mass scale may not seem like "news." But this chapter is not just redundant rummaging through the history books. As late as 2017, the tobacco industry was still killing 1,200 Americans every day. And later in this chapter I'll address the tobacco industry's involvement in the exploding, dangerous fad among teenagers and young adults of "vaping" with e-cigarettes, which is poised to undermine all the recent progress in reducing the plague of smoking.

Humans have been tightly connected to tobacco long before the Marlboro Man. Smoking cannabis and tobacco appeared as early as 5,000 B.C. In the Americas, carved stone drawings by the Mayans from 600 to 900 AD depict pipe smoking of tobacco during religious ceremonies. Tobacco was not just the first commercial agricultural crop grown in North America, it was the main tradeable, economic asset for the first settlers of Jamestown. The American Revolution may not have been successful without the profits from selling tobacco. The first commercial cigarettes were made in 1865 by Washington Duke on his 300-acre farm in Raleigh, North Carolina.

Widespread selling of cigarettes became possible following James Bonsack's invention of the cigarette-making machine in 1881. Bonsack's cigarette machine could make 120,000 cigarettes a day. He went into business with Washington Duke's son, James "Buck" Duke. They built a factory and made 10 million cigarettes their first year and one billion cigarettes five years later. The first brand of cigarettes was packaged in a box with baseball cards and were called Duke of Durham. Buck Duke and his father started the first tobacco company in the US, the American Tobacco Company. Philip Morris entered the market with its Marlboro brand in 1902.

Starting during World War I, a long and cozy relationship between the tobacco industry and the federal government began when soldiers were given free cigarettes daily, a practice that continued during World War II. By 1944, servicemen received 75 percent of the 300 billion cigarettes produced annually. American women joined the ranks of proud smokers largely during WWII.

Nicotine is one of the most addictive chemicals known, more so even than heroin or cocaine. Modern cigarettes have been extensively engineered and optimized as nicotine delivery devices developed through major national and international research and development programs. In 1994 the five major tobacco companies submitted to the United States Department of Health and Human Services, application to mix 599 chemical additives to tobacco. More than 100 of those 599 additives have pharmacological actions that camouflage the odor of environmental tobacco smoke, enhance or maintain nicotine delivery, increase the addictiveness of cigarettes, and mask symptoms and illnesses associated with smoking behaviors. Tobacco companies have even engaged in developing genetically engineered tobacco to enhance nicotine delivery.[1, 2, 3, 4]

They have also used reconstituted tobacco and nicotine extracts, to manipulate cigarette nicotine levels and influence people's smoking behaviors. Reconstituted tobacco, known as "sheet," is a major ingredient in modern cigarettes. Sheet is manufactured from recycled stems, stalks, scraps, collected dust, and floor sweepings. Those materials are ground up, nicotine is extracted from them, and chemicals, fillers, glue, and other agents are added to the slurry. The sheet is then pressed out and puffed, with the previously extracted nicotine sprayed onto it, and ground into tiny curls before being incorporated into cigarettes at the desired level.[5]

Tobacco companies extensively studied nicotine extracts as a method to augment nicotine levels in cigarettes, which of course, makes them more addictive.[6, 7, 8, 9, 10, 11, 12] Most of the current additives were added to cigarettes after 1970. Animal research conducted by Philip Morris demonstrated a synergistic interaction between nicotine and acetaldehyde: rats pressed a bar more for the combination than for either substance alone.[13,14] If these findings would be applicable to humans, then smokers would puff more with the combination of nicotine and acetaldehyde. In other words, the tobacco industry deliberately made cigarettes more effective killing machines.

Making cigarettes more deadly was extremely profitable. In the early 1990s, Philip Morris alone paid $4.5 billion in taxes, making it the largest tax payer in the U.S. at the time. Cigarettes became the most heavily advertised product in the United States.[15]

The science on the broad consequences of cigarette smoking was well established in Germany by the early 1940s. The Nazis had funded research on the health consequences of smoking, and German scientists led the world in research on tobacco and health. Third Reich physicians were the first to understand the relationship between smoking and heart disease. Franz H. Müller's 1939 medical dissertation was the world's first controlled epidemiological study of the relationship between tobacco and lung cancer. Public anti-smoking campaigns, restrictions on sales and advertising, and numerous smoking prohibition laws were passed in cities throughout Germany. Taxes on tobacco were raised sharply. Women, especially pregnant women, were often particular targets of this official Nazi anti-smoking effort. Hitler was the first national leader in modern times to publicly denounce smoking and the tobacco industry. By the end of the war, tobacco consumption per capita was significantly less in Germany than in the United States.

Taking an approach to smoking opposite to that of Nazi Germany was not a difficult sales pitch for tobacco manufacturers in the US. Smoking was depicted as glamorous, healthy and a social asset. Movie stars like Claudett Colbert were plastered on smoking ads. In 1946, Brown and Williamson used baseball legend Babe Ruth to pitch Raleigh cigarettes, with the claim that "Medical science offers proof positive … No other leading cigarette is safer to smoke!" Ironically, Babe Ruth later died of throat cancer. Later on, Mickey Mantle pitched Viceroys and Camels.

Senior scientists and tobacco executives in the United States were not oblivious to the research that had taken place in Germany.

Evidence suggests they knew full well about cigarettes' cancer risks by the 1940s.[17]

Because there was also a growing public perception that cigarettes might have health consequences, the tobacco companies began specifically courting the endorsement of physicians in response to that perception. In the 1930s cigarette advertisements appeared regularly in top tier medical journals like The New England Journal of Medicine and The Journal of the American Medical Association. Ads for Lucky Strikes and Camels told customers, "20,679 physicians say 'Luckies are less irritating,'" and "More doctors smoke Camels than any other cigarette!"

R.J. Reynolds took the marriage of tobacco and the medical community a step further, creating a Medical Relations Division. This division was based out of an R.J. Reynolds advertising firm and sought researchers who could substantiate the medicals claims that the company was making in advertisements. This allowed the company to reference research findings in their advertisements, both to consumers and to physicians.

In the United States, by the early 1950s, cohort studies had verified the association between smoking and lung cancer, that lung cancer death rates were substantially higher in cigarette smokers than in cigar or pipe smokers, that the strength of the association between smoking and lung cancer increased with the amount of smoking (although we have since discovered that the relationship is not linear), and that people who quit smoking experienced a lower death rate than their counterparts that continued to smoke.[18] Many of the industry's own scientists privately acknowledged shortly thereafter that smoking was a serious health threat.

In 1953, a young chemist at R.J. Reynolds, Dr. Claude Teague, conducted a comprehensive literature search on smoking and cancer in which he referenced 78 scientific papers on the topic of smoking and cancer. Based on his own review, Teague concluded: "Studies of clinical data tend to confirm the relationship between heavy and prolonged tobacco smoking and incidence of cancer of the lung. Extensive, though inconclusive testing of tobacco substances on animals indicates the probable presence of carcinogenic agents in those substances."[19] Teague was employed at RJ Reynolds for 35 years (1952-1987) and held various executive-level positions at the company, including that of director of research and development.

In 1953, sensing growing public concerns about health risks from smoking, JAMA banned tobacco ads from its pages and the AMA banned smoking ads from its conventions. The number of physicians who smoked started dropping. During my residency in Boston from 1978 to 1980, smoking in the doctors' lounge was still commonplace. One of the few anti-smoking zealots at our medical Mecca, Massachusetts General Hospital, was Dr. Hardy Hendren, a world renowned pediatric urology surgeon. Because of smoking's strong connection to bladder cancer, Hendren was an in-house anti-smoking bull dog, who was known to literally snatch cigarettes out of the mouths of his colleagues as they sat in the doctor's lounge. This led, on more than one occasion, to lively, spirited, entertaining, and often hostile exchanges between the snatcher and the snatchee.

By 1954, the tobacco manufactures responded by collaborating in launching an advertisement entitled, "A Frank Statement to Cigarette Smokers" which appeared in 448 newspapers in 258 cities reaching an estimated 43,245,000 Americans.[20] The advertisement questioned research findings implicating smoking as a cause of cancer, promised consumers that their cigarettes were safe, and pledged to support impartial research to investigate allegations that smoking was harmful to human health. In two 1954 speeches made by Philip Morris vice president George Weissman, he promised: "[I]f we had any thought or knowledge that in any way we were selling a product harmful to consumers, we would stop business tomorrow."

But the advertising campaigns began to change. Cigarette companies began promising consumers that their brands were better for them than their competitor's brands because the smoke was less irritating, smoother, and milder. In the 1950s and 1960s, in response to information linking cigarette smoking with cancer, the tobacco industry propagated massive amounts of advertising that promoted filters and lower tar cigarettes as technological cures for whatever ills could possibly be attributed to smoking. Filtered cigarettes, and low tar and nicotine brands began to capture an increase in market share.

In 1964, the Surgeon General Luther Terry released the report, "Smoking and Health: Report of the Advisory Committee to the Surgeon General of the United States," confirming the link between lung cancer and chronic bronchitis and cigarette smoking. He said that the nicotine and tar in cigarettes cause lung cancer. In 1965, the United Stated Congress passed the landmark Cigarette Labeling and

Advertising Act. It said that every cigarette pack must have a warning label on its side stating "Cigarettes may be hazardous to your health." Nonetheless, tobacco corporations successfully fought tooth and nail against every substantive regulation for the next three decades while tens of millions of people died early deaths in the name of tobacco profits.

In 1970, Richard Nixon signed the Public Health Cigarette Smoking Act banning cigarette ads on radio and television. Nevertheless, pictures of the Marlboro Man became the most widely recognized advertising image in the world. Because Philip Morris was spending $2 billion a year in media advertising, many publishers and editors were willing to print health articles that omitted or minimized the dangers of smoking. In 2005, tobacco companies made a settlement with the National Association of Attorneys General that included an agreement to remove tobacco ads from the school library editions of *Time, People, Sports Illustrated* and *Newsweek*.

In the 1990s, the tobacco industry realized that they were losing customers. Each day, 1,200 smokers died and 3,500 of them quit. If corporate profits were to be maintained, they would have to invest in attracting new customers--kids. The deeply cynical and morbid marketing strategy of luring children into cigarette addiction became an integral part of the tobacco industry's business model that emerged after the widespread publicity over cigarettes' health consequences. The importance of the youth market was illustrated in a 1974 presentation by RJR's Vice-President of Marketing who explained that the "young adult market . . . represent[s] tomorrow's cigarette business. As this 14-24 age group matures, they will account for a key share of the total cigarette volume - for at least the next 25 years."[21, 22]

Not about to be deterred by restrictions on television ads, the industry began to sponsor youth activities, rock concerts, sporting events and distribution of specialty items bearing logos and issuing coupons and premiums. Expanding into alternative, unregulated media outlets, Philip Morris developed a record label that specifically targeted young females. Available only with the purchase of cigarettes, the CDs were packaged with two packs of Virginia Slims.

In Oct. 2008 in New York City, an exhibit displayed hundreds of print ads and television commercials, titled "Not a Cough in a Carload: Images Used by Tobacco Companies to Hide the Hazards of Smoking." The advertising world's best artists and copyrighters had helped sell cigarettes to the world for decades. In addition to physicians, Santa Claus was used to sell Pall Malls, cartoon characters, like the Flintstones and penguins, were cheerleaders for Winston and Kool, children appeared as accessories for their smoking parents, and babies were added to pictures for brands like Marlboro. But none of these were as successful as R.J. Reynolds' "Joe Camel."

Joe Camel was actually born in Europe. The caricatured camel was created in 1974 by a British artist, Billy Coulton, for a French advertising campaign that subsequently ran in other countries in the 1970s. Joe Camel first appeared in the U.S in 1988, in materials created for the 75th anniversary of the Camel brand.

In 1991, the Journal of the American Medical Association published a study showing that by age six, nearly as many children could correctly respond that "Joe Camel" was associated with cigarettes as could respond that the Disney Channel logo was associated with Mickey Mouse, and alleged that the "Joe Camel" campaign was targeting children, despite R. J. Reynolds' contention that the campaign had been researched only among adults and was directed only at the smokers of other brands. At that time, it was also estimated that 32 percent of all cigarettes sold illegally to underage buyers were Camels, up from less than one percent. Subsequently, the American Medical Association asked R. J. Reynolds Nabisco to pull the campaign. R. J. Reynolds refused, and the Joe Camel Campaign continued unabated.

Typical Joe Camel ad of the early 1990s

Joe Camel catapulted a dramatic leap in cigarette sales to teenagers from $6 million to $476 million. The average age of a smoker's first use of cigarettes was 14.5 years. Collegiate smoking jumped 28 percent between 1993 and 1997, due largely to the rise in adolescent smoking that had begun in the early 1990s. I don't know what happens to camels when they have outlived their usefulness--sent out to pasture, to stud, or start a pyramid scheme (sorry), but an agreement with the Federal Trade Commission forced Joe into retirement in 1997. But Joe had certainly done his job, luring millions of children and teenagers to an early death.

Not satisfied with just clever advertising to children and manipulation of cigarettes to increase their addictive potential, Big Tobacco executives found other ways to create a "smoke" screen (pun intended) that hid the danger of their product. Testifying before Congress on May 17, 1994, Joseph A. Califano, Jr., former U.S. Secretary of Health, Education, and Welfare, stated:

"The evasions, lies, and transfer of documents overseas by the tobacco industry to prevent any Government agency or cigarette-injured patient from finding them has distorted U.S. Government policy for 30 years…the President of the United States, the Secretary of Health, Education, and Welfare, and the Surgeon General were all victims of the concealment and disinformation campaign of the tobacco companies."

When smoking rates began to drop in the United States, and regulation and public pressure began to take their toll on company profits, the tobacco corporations moved into overseas markets. Wherever US cigarettes go, smoking rates start to rise. Smoking rates in Japan, South Korea, Thailand and Taiwan rose 10 percent following the massive inflow of American cigarettes after the US forced these countries to open their markets to domestic tobacco exports, threatening them with trade sanctions if they did not comply. How Russia got hooked on cigarettes, and who helped the tobacco companies accomplish that dastardly deed is particularly galling (see Chapter 18).

Personifying the "success" of American tobacco companies is Aldi Rizal, a two pack a day chain smoking two year old in Indonesia. Over 22 million people viewed a You Tube video of him smoking.[23] Although now at age four, Aldi supposedly has quit, there has been a rush of imitator smoking toddlers in Indonesia. Indonesia is a smoking "free for all."

After Phillip Morris (PMI) targeted Indonesia for its growth campaign, youth smoking rates in the country doubled. Tobacco ads targeting youth are ubiquitous, vendors often ply cigarettes next door to schools, and cigarettes are sold with impunity to children. PMI resurrected the Marlboro Man, they run sexy TV ads, send squads of pretty girls into crowded market places to hawk their cigarettes, sponsor popular rock concerts and sporting events like badminton tournaments, and are actively exploiting the explosion in the use of social media. The World Health Organization estimates that 1 billion people will die from smoking cigarettes during the 21st century,[49] and 50-70 percent will be in third-world countries.

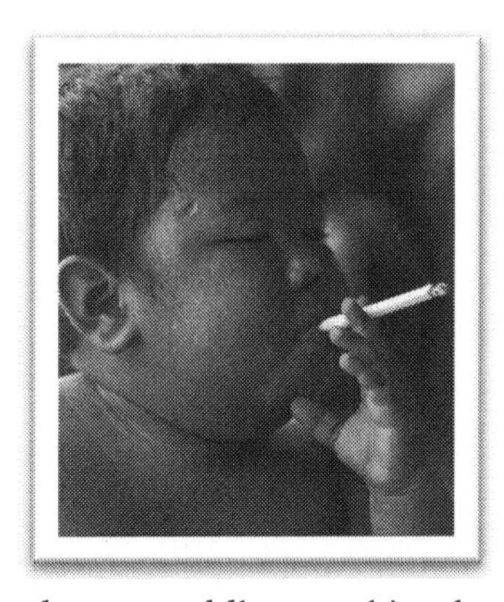

Children and even toddler smoking has become an epidemic in Indonesia.

Neither the tobacco companies or their CEOs have ever shown any remorse about what they did or continue to do. Recall the testimony in 1994 from the CEOs of the seven largest tobacco corporations before Congress,[24] unanimously declaring that nicotine is not addictive, but also admitting that they were capable of controlling the amount of nicotine in cigarettes.

The companies maintained the stance that smoking had not been proven to be injurious to health through 1999. In October of that year, Philip Morris announced to the public on its web site that "[t]here is an overwhelming medical and scientific consensus that cigarette smoking causes lung cancer, heart disease, emphysema and other serious disease in smokers."[25] In fact the carefully worded message only acknowledged that there was a medical and scientific consensus that smoking caused disease, not that Philip Morris accepted this consensus.

For those of you that think the tobacco industry's inhumane business model is a relic of the past, the latest Surgeon General's report found that today's cigarettes have had deliberate design and ingredient changes that have increased their toxicity and addictiveness, making them even more likely to cause lung cancer and COPD compared to cigarettes manufactured in the 1970s.[26, 27] In addition to increasing the levels of nicotine, the industry discovered that adding ammonia compounds increase the speed at which nicotine is delivered to the brain, and adding sugars makes nicotine even more addictive and the smoke easier to inhale. Other chemicals have been added specifically to attract new tobacco users. Levulinic acid has been added to reduce the harshness of nicotine and make the smoke less irritating. Chocolate and licorice have been added for sweetness and to make the cigarettes more appealing to young people. Menthol cools and numbs the throat to reduce irritation. If you have asthma, normally you'd need a doctor's prescription to get a bronchodilator. Or you can just smoke a cigarette. The tobacco company villains have even gone to the extent of adding bronchodilators to their cigarette ingredients to dilate small airways, accelerating the delivery of smoke to the lungs.

The Surgeon General's report focused on two specific changes as the most likely reason for the increased risk of developing lung cancer from today's cigarettes: an increase in the levels of highly carcinogenic tobacco-specific nitrosamines (TSNAs) in American cigarettes (the report linked this increase to tobacco blends used in US cigarettes compared to cigarettes sold in Australia and Canada), as well as the curing process now being used. As a result, exposure to tobacco-specific nitrosamines is much higher among US smokers than among their counterparts in Australia and Canada.[28]

Manufacturers have also introduced ventilation holes in cigarette filters that caused smokers to inhale more frequently and vigorously, thereby drawing in more nicotine and carcinogens. These design changes lower machine-measured tar and nicotine, allowing the manufacturers a better tar and nicotine rating, but provide absolutely no health benefit.

The tobacco industry is a wicked enterprise indeed. The American Cancer Society (ACS) estimates that tobacco killed 100 million people in the 20th century. About 42 million adults and nearly 3 million children still smoke in the US. Approximately half of continuing smokers will die prematurely as a result of their addiction, losing an average of at least a decade of life. The ACS estimates that tobacco will kill 1 billion people in the 21st century.[29,30]

The term "corrupt tobacco company" already seems redundant, but in addition to all the other business practices that make a mockery of every ethical standard one could possibly construct, secret documents exposed by the World Health Organization reveal "a 50-year conspiracy to resist smoking restrictions, restore smoker confidence and preserve product liability defense." WHO documents show an "industry-wide collusion," on "business practices of buying scientific and other expertise to create controversy about established facts, funding political parties, hiring lobbyists to influence policy, using front groups and allied industries to oppose tobacco control measures, pre-empting strong legislation by pressing for the adoption of voluntary codes or weaker laws, and corrupting public officials."[31]

A 1999 US Justice Department lawsuit against Philip Morris and R.J. Reynolds ended with a 2006 judicial ruling requiring that four cigarette companies broadcast "corrective statements" to compensate for decades of lies and propaganda campaigns. After years of stalling, the ads finally appeared in 2017 stating, "Smoking causes heart disease, emphysema, acute myeloid leukemia and cancer of the mouth, esophagus, larynx, lung, stomach, kidney, bladder and pancreas." Then the ads throw in this admission, "Cigarette companies intentionally designed cigarettes with enough nicotine to create and sustain addiction."[32]

The final wording of the ads was the result of hundreds of hours of negotiation between the Justice Department, public health groups and the tobacco industry. But the ads were deliberately designed to have minimal appeal and efficacy, involved no graphics or imagery, and a computer generated rather than a human voice.

The companies will have to spend $30 million on these ads, but that is the proverbial drop in the bucket of the $8 billion a year they still spend on marketing, persuading more people to pick up the deadly habit.

It would seem that with all the negative publicity that Big Tobacco has endured in the last two decades, many people believe that they are rapidly approaching their own demise. Nothing could be further from the truth. "Big Tobacco is thriving, profitable, and increasing its sales."[33]

In 2016, their sales amounted to 300 cigarettes for every man, woman and child on the planet. Nearly every major investment company in the world invests in tobacco because they are "defensive" stocks, in other words, stable, predictable, and very low risk. They rarely decline in value and reliably offer nice dividends. Selling death and disease can be very good business.

Vaping: Big Tobacco Finds Another Way to Kill People

"Vaping" with nicotine-laden electronic cigarettes is a rapidly expanding new fad with America's teenagers. A recent study showed, "about 13 percent of 8th graders, 24 percent of 10th graders, and nearly 28 percent of 12th graders at US schools reported using a vaping device in the past year, according to data from the 2017 Monitoring the Future (MTF) survey."[34] A study by the National Institute on Drug Abuse released in December 2018 showed that the number of high school seniors who had vaped in the previous 30 days had doubled in the last year, the fastest growth of any used substance in the last 44 years.[54]

The e-cigarette market is expected to see continued exploding growth and be worth $34 billion by 2021,[37] and "Big Tobacco" has elbowed their way in on the market in a "Big" way. Corporations like Reynolds, American, and Altria are buying up smaller companies in a move to dominate the emerging market. Altria just spent $13 billion for a 35 percent stake in the largest vape company Juul.[53] With a market valuation now of $38 billion, Juul is worth more than Ford Motor Co.

"Big Vape" is copying the tried and true playbook of Big Tobacco. Several big tobacco companies recently paid $1.2 million to lobbyists to fight a bill before the California legislature that would have prohibited the sale of e-cigarette vending machines and establish other regulations comparable to those for tobacco products. In California, they've spent $10 million in lobbying over last nine years. British American Tobacco (BAT), becoming another powerhouse in the vaping industry, claims that e-cigarettes produce 95 percent fewer toxins than smoked cigarettes. But this is another example of "follow the money." This claim comes from research funded by BAT. Big Vape is "enlisting veteran tobacco industry law firms to contest federal oversight in court and partnering with tobacco-funded libertarian groups to fight regulation in states and cities, and they're rallying armies of ex-smokers who believe vaping saved their lives."[37]

When Big Tobacco was facing stiff regulation by the FDA in the late 1990s they turned to, among other groups, the Competitive Enterprise Institute (CEI), the Heartland Institute, and Grover Norquist's Americans for Tax Reform. All of these groups receive substantial funding by Big Tobacco and other corporations, who then chimed in publicly to condemn the FDA's attempt to reign in tobacco products and indoor smoking, claiming that nicotine was no more addictive or harmful than caffeine. The CEI is now painting the same disingenuous argument for the vaping industry.[40]

The Consumer Advocates for Smoke-Free Alternatives Association (CASAA) calls itself a grass roots organization whose mission is to protect the rights of vapers. In 2016, the American's for Tax Reform sent Paul Blair to team up with CASAA and the American Vaping Association on a tour of several states to promote the "Right to Vape," a campaign of opposition to looming regulation from the FDA. CASAA and CEI even teamed up on a lawsuit against the Federal Department of Transportation, trying to force them to allow e-cigarettes on airplanes. Their argument could have fit right in with the Mad Hatter's Tea Party of Alice and Wonderland. "Banning e-cigarettes would actually increase transportation-related deaths by driving nicotine-dependent passengers to drive rather than fly, and would undermine rather than promote passenger comfort by subjecting passengers to nicotine withdrawal symptoms that are a common cause of 'air rage.'"[37] Sounds like they are threatening some kind of terrorism if people are not allowed to use e-cigarettes on airplanes.

Aaron Biebert directed an absurd movie tribute to the vaping industry called *A Billion Lives,* the premise of which is that cigarettes will kill one billion people over the next century, and those lives could be saved by vaping instead of smoking. Biebert blames a conspiracy of "corrupt federal health officials," drug companies (for trying to squash their competition), and anti-smoking puritanicals for lying about the risks of nicotine and vaping and therefore are responsible for the deaths of one billion people. Biebert calls vaping a virtual "public health miracle."[37] Parents, psychologists, public health officials and addiction specialists are calling it something much different—a public health disaster.

Big Tobacco has a long history of trying to spread the fog of "reduced harm" for various alternative products to traditional smoking. For example, R.J. Reynolds released a reduced risk cigarette called "Eclipse," and Philip Morris's iQOS,[41] both of which claimed to reduce the harm of tobacco by heating tobacco rather than burning it. To further their "reduced harm" narrative, Big Tobacco turned to an organization billed as an independent scientific body, the American Council on Science and Health (ACSH), but which is, in reality, a front group for the tobacco, agrichemical, fossil fuel, and pharmaceutical industries.[44] The longtime Medical/Executive Director of ACSH, Dr. Gilbert Ross, seems to be an appropriate choice to head such an organization. He was convicted of defrauding Medicaid, sentenced to almost four years in prison, and forced to make restitution of over $600,000 to the state of New York.[45] Researchers from the Mayo Clinic have proven that "reduced harm" cigarettes, are just as toxic as the original.[42]

The vaping delivery device that is exploding in popularity is called a JUUL. Most cartridges or "pods" of vaping liquid contain nicotine, and the pods for JUUL, contain extremely high concentrations of it. One pod inserted in a JUUL contains as many "puffs" as an entire pack of cigarettes. One thing is already obvious about vaping: it is a much more potent delivery system for nicotine than cigarettes, and the addiction potential is much greater as well.

Anecdotes of teenagers trying to quit vaping tell a very disturbing story of even greater difficulty than quitting cigarettes. The method of tapering protocols doesn't appear to work with vaping.[55] Teenagers trying to quit are experiencing all the symptoms of substance addiction—insomnia, anxiety, paranoia, volatile emotions, hyperactivity, and depression—which, when added to a baseline of a typical teenager's volcanic disposition, can create nightmarish behavior affecting the whole family.

One of the more cynical marketing schemes of e-cigarette corporations is the recent offering of "college scholarships" for winning essay contests on the benefits of vaping as opposed to smoking.[43] It is an obvious scheme to make vaping appealing to young students. It is also a scheme to get their brands attached to an official university website and sends a very deliberate and misleading message that the university endorses this. The shady strategy has been successful in getting vaping scholarships listed in financial-aid directories from such well known universities as Harvard, UC Berkeley, and the University of Pittsburg, to name a few, and even some that have taken public positions against vaping. It is worth noting that Harvard and California State University at Long Beach removed the vaping scholarships' listing when asked about it by the Associated Press.[43]

Research on the health repercussions of vaping is still early and incomplete. Although initial results indicate fewer consequences than traditional tobacco, vaping is emerging from the newest research as a serious health hazard, made even more so by the perception that it's essentially benign. Teens who use e-cigarettes regularly had about three times the amount of toxic chemicals in their urine, like acrolein (which has a strong association with multiple sclerosis), acrylamide, acrylonitrile, crotonaldehyde, propylene oxide (probable carcinogens), and diacetyl (which causes "popcorn lung"), compared to teens who never did, a finding that was particularly true if they chose fruit flavored products.[36] Even though many of these chemicals are approved food additives, when they are heated they can be transformed into toxic byproducts.

Propylene glycol, a typical ingredient in a pod, is theoretically benign. It's used in theatre fog machines and is an FDA-approved ingredient in things like cake mixes. But when exposed to a vaping heating element, it releases the potent carcinogen formaldehyde.

In addition, a wide variety of heavy metals contaminate the aerosol particles in vaping--cadmium, nickel, arsenic, chromium, manganese, zinc, and lead,[35] and the proven carcinogen common to cigarette smoke and fossil fuel combustion--benzene.[37] Diacetyl is a compound that is released by some of flavorings in e-cigarette fluids, and it has been linked to a debilitating lung disease, bronchitis obliterates, i.e. popcorn lung.[50] "Vapers" often comment that their lung capacity and exercise stamina have been plummeting since taking up the practice.

Vaped nicotine accelerates heart rate, constricts blood vessels, increases blood pressure, increases vulnerability to blood clots, and harms the lungs and GI tract. Researchers have stated, "There is decreased immune response and it also poses ill impacts on the reproductive health. It affects the cell proliferation, oxidative stress, apoptosis, DNA mutation by various mechanisms which leads to cancer. It also affects the tumor proliferation and metastasis and causes resistance to chemo and radio therapeutic agents."[52]

In trying to make the case that vaping is a positive for society, Big Vape claims that e-cigarettes are safer than smoking tobacco, and that access to e-cigarettes helps smokers quit tobacco. They say it's the equivalent of providing "methadone or clean needles to IV drug users."[37] The research is conflicting at best as to whether vaping actually leads to quitting tobacco. In fact, a 2016 Surgeon General report indicates that teenagers that had tried vaping were actually twice as likely to take up smoking.[39] There is also significant evidence that most adult e-cigarette users continue to smoke cigarettes,[46] and there is even stronger evidence that adolescents that try vaping are more likely to become eventual smokers.[47, 48]

The most recent study published just before this book went to press showed that 90 percent of smokers who "vaped" were still smoking cigarettes a year later, and that patients who tried to quit cigarettes by vaping were 50 percent less likely to be successful than those who did not.[51]

A tried and true political tactic is to accuse your opponents of exactly what you are guilty of. In Dec. 2108, the Competitive Enterprise Institute (CEI) posted commentary on their website that claimed opponents of vaping were fear mongering, that "e-cigarettes are vastly safer than smoking, help smokers quit, and are a net positive for public health." It went on that opponents were just in it for the money, "driven by a need to defend and expand the financial resources they need. This campaign to restrict or ban e-cigarettes does a huge disservice to public health."[56] Sounds like Big Tobacco has a friend. They do actually—themselves. That's who funds CEI.

Just like Joe Camel, vaping is deliberately designed and marketed to attract adolescents, with a wide variety of sleek packaging and flavored scents like chocolate, cherry, vanilla, and crème brûlée.

. Users report chocolate flavored vape tastes just like a candy bar. In fact, surveys show that teenagers believe they are just vaping flavoring. To that end, the Food and Drug Administration and the Massachusetts Attorney General are currently involved in an investigation of Juul Labs as to whether its marketing practices were intended to deliberately entice teenagers. That investigation should take about 30 seconds.

Even more, there are now a growing number of pro-vaping blogs, online forums, "meet ups," expos, and conventions like VapeCon, where vaping is promoted as a cultural movement itself, an "entire way of life," far more and far better than just another deadly addiction that kills people. These conventions take place throughout the US, in countries like the Philippines, South Africa, Denmark, Spain, Canada, and Paraguay. Girls and young women can put on their bikinis and vie for such honors as being Miss VapeCon.

"Con" is certainly an appropriate moniker, because an entirely new generation of teenagers are, once again, being conned into surrendering their lives and their futures to the clutches of Big Tobacco.

Executives from the seven major tobacco companies testify under oath before Congress in 1994 that nicotine is "not addictive."

Scenes from recent vaping events, celebrating dense clouds of likely highly addictive nicotine

8

Monsanto:

Controlling Everything You Eat

"Genetic Power's the most awesome force the planet's ever seen, but you wield it like a kid who's found his dad's gun.... and before you even knew what you had you patented it and packaged it and slapped it on a plastic lunch box, and now you're selling it."

—Ian Malcolm from Jurassic Park

The biotech industry has been selling to the world the idea that without their genetically modified crops (GMOs) a worldwide apocalypse will doom us all. You, your children and all your loved ones will starve to death, probably by next week. Think of Mel Gibson in a sequel to the "Road Warrior"—"Food Warrior," where instead of gasoline, it's food, like *Cap'n Crunch*, that grotesque, degenerate barbarians kill each other over. Well, who wouldn't want to take a pass on that kind of nightmare?

According to their script, their GMO food is necessary to increase or even maintain food production and avert mass starvation in an increasingly dangerous world. Even though almost all of these claims relate to unproven, future, or merely theoretical GM crops, anyone who doesn't buy the biotech sales pitch is disdainfully portrayed as being dismissive of the plight of the world's poor, naive, scientifically ignorant, or all of the above. UK environment secretary Owen Paterson has questioned the morality of delaying the production of genetically modified crops. He recently called opponents of GMOs "wicked."[1] Count me as one of the "wicked," and likely you as well after reading this chapter.

George Monbiot wrote over a decade ago, "The biotech companies are not interested in whether or not science is flourishing or people are starving. They simply want to make money. The best way to make money is to control the market. But before you can control the market, you must first convince the people that there's something else at stake."[2]

The purpose of GMOs

To that end, stripping away the biotech rhetoric, there are currently only two objectives behind creating genetically modified organisms for food—to either allow a plant to survive being drenched with pesticides, or to create a plant that produces its own endogenous pesticide. Undeniably the end result is plants that are contaminated either on the surface or throughout the entire plant with pesticides.

Pesticides (a term that includes insecticides and herbicides) are biologic poisons. Their progenitors were chemical warfare agents developed in World War I, like the organophosphate nerve gases. They work by electrochemical disruption of nerve cells. Beneficial insects, animals, and humans are composed of the same cells, or biologic infrastructure, as are weeds and harmful insects. It takes little imagination to see that more than pests can be the victims of these poisonous chemicals.

Sitting in the driver's seat of the GMO industry, and the corporation most responsible for a world awash in pesticides, is Monsanto, a chemical company founded in 1901 in St. Louis, MO. An exposé on Monsanto in the German magazine der Spiegel in Oct. 2017 begins, "Some companies' reputations are so poor that the public already has low expectations when it comes to their ethics and business practices."[3]

Until twenty years ago, Monsanto was only on the public's radar as the manufacturer of the notorious Agent Orange, a toxic chemical used by the US military in the Vietnam War to destroy vegetation in the places where the Viet Cong liked to hide. The scheme was an environmental disaster. That the US military is slow to acknowledge culpability for anything they do might be the greatest understatement of all time. So when the Dept. of Veteran Affairs officially lists 14 serious diseases—including cancer, diabetes, and Parkinson's disease—as related to exposure to Monsanto's Agent Orange, we can be well assured that represents just the tip of the iceberg.

About 1.5 million US servicemen were exposed to Agent Orange to varying degrees, and about 40,000 have filed disability claims related to Agent Orange exposure. According to William Sanjour, who led the Toxic Waste Division of the U.S. Environmental Protection Agency (EPA), thousands of veterans were disallowed benefits because "Monsanto studies showed that dioxin [as found in Agent Orange] was not a human carcinogen." An EPA colleague discovered that Monsanto had apparently falsified the data in their studies. Sanjour says, "If [the studies] were done correctly, they would have reached just the opposite result." The Vietnamese government claims that 400,000 Vietnamese people were killed or disabled by Agent Orange, and 500,000 children were born with birth defects as a result of their exposure. The plague upon the health of Agent Orange's victims will continue for decades, and in some cases continue to afflict subsequent generations. Monsanto has paid nothing to any Vietnamese victims, despite the fact that this was essentially chemical warfare upon their civilian population. [4,5]

Agent Orange was just one of Monsanto's many toxic products that were spread around the globe for decades. It may not have even been their most dangerous. Ninety-nine percent of all the highly dangerous PCBs (polychlorinated biphenols) manufactured in the US were manufactured at a single Monsanto plant in Sauget, Illinois. Ninety-three of Monsanto's manufacturing plants ended up as Superfund sites. That alone might make you pause and ask yourself, "Do I want a company that leaves a trail of toxic waste wherever they go to be in charge of the food I eat?" Whether you like it or not, they are.

In the last two decades, Monsanto has transformed itself into the world's largest seed company, the largest producer of GMOs and is still one of the world's largest pesticide companies. Monsanto has spent millions cultivating an environmentally friendly, almost heroic public image of itself, a vital part of the effort to feed a world busting at the seams with over 7 billion people. Specifically, Monsanto claims or implies that because of its products, farmers are reducing chemical use, increasing yield, protecting their crops from drought, saving you personally from starvation, and in their spare time they helped write the New Testament (Ok, I made that last one up). *Forbes Magazine* gave Monsanto its "Company of the Year Award" in 2009. In 2013, the Chief Technology Officer of Monsanto, Robert Fraley, received the World Food Prize, agriculture's version of the Nobel Prize in recognition of people who improve the "quality, quantity or availability" of food in the world. I'm surprised they didn't claim it was actually Monsanto, not Jesus, that fed the masses with five loaves of bread (gluten free of course) and two genetically modified fish.

Less well known, however, is that the foundation that administers the World Food Prize, accepts large contributions from industrial agriculture, including a $5 million pledge from Monsanto in 2008. It's like Fox News giving Sean Hannity an award for having the best looking hair on cable TV.

Perhaps it is no surprise that readers of *Natural News* overwhelming awarded Monsanto a slightly different award: "World's Most Evil Corporation," an award Monsanto wins virtually every year. For 46 years my wife's basic life tenet has been that everything I'm doing is wrong. After researching Monsanto, it is clear Monsanto is even out-doing me.

On May 25, 2013, two million people worldwide took part in a "March Against Monsanto." That event was planned by an activist in Salt Lake City, Utah, my hometown, capital city of the most conservative state in the country. Nobody ever protests against any business in Utah, where the unfettered free market is virtually sanctified by the conservative culture of the Mormon Church, where business leaders make up the overwhelming majority of ecclesiastical leaders. Nonetheless, hundreds of protestors covered Utah's capitol grounds listening to speakers berate "the world's most evil corporation," shouting, "We've basically been sentenced to death, to illness, to infertility."

Numerous anti-Monsanto documentaries have been done; even *Fortune* magazine once labeled Monsanto America's most feared corporation. The Geneva-based Covalence group placed Monsanto dead last on a list of 581 global companies ranked by their reputation for ethics. What has Monsanto done to achieve this lofty perch? Nothing less than seek to monopolize the world's food supply with an expensive, risky, and dangerous, industrialized crop system. And that is exactly the strategy that was adopted in January 1999, where a representative from Arthur Anderson Consulting Group (ACG) explained how his company had helped Monsanto create that plan.

As detailed by Jeffery Smith, in his book, *Seeds of Deception,* ACG asked Monsanto what their ideal future looked like in 15 to 20 years. Monsanto executives described a world with 100 percent of all commercial seeds genetically modified and patented. ACG then worked backwards from that goal, and developed the strategy and tactics to achieve it.

Monsanto's mercenary attitude was revealed by Phil Angell, Director of Corporate Communications for Monsanto, in a 1998 *New York Times* piece summing up how the company thinks about the safety and health of consumers. He said, "Monsanto should not have to vouch for the safety of biotech food. Our interest is in selling as much of it as possible. Assuring its safety is FDA's job."[6] He might as well have said, "We've found a great way to profit from what everyone on earth puts in their mouths, and we don't really care what the consequences are." Can you imagine a pharmaceutical company that states publicly and without embarrassment, they shouldn't have to vouch for the safety of their drugs, that their only objective was sell as much of their snake oil as possible? How about General Motors admitting that kind of attitude about the safety of their cars? How about gun manufacturers? Oh wait, that is exactly the attitude of gun manufacturers (see Chapter 12).

In the last 20 years, by virtue of their dominant role in processed foods, GMO soy and corn ingredients are consumed by almost every member of the general public. If you eat any food from a box or a can, you are eating GMOs. By 2012, in the U.S., 90 percent of sugar beets (representing half of overall sugar production), 85 percent of soybeans (which are found in 70 percent of all supermarket food products), 85 percent of corn, and most of the canola oil and alfalfa were GMO. Nine percent of the entire terrestrial land mass of the United States is planted in GMO crops.

Genetic Engineering: Playing Russian Roulette

Thanks largely to Monsanto, for the first time, scientists from the biotech industry can bring characteristics from anywhere in nature into the DNA of a plant, not just from other similar plant species, but all the way from organisms on the opposite end of the animal and plant kingdom. The essential identity of the plant can be breached. A plant can be made to behave like a pig, a cow can be made to behave like a mouse, and a Fox News commentator can be made to behave like a journalist (only in the laboratory so far).

Genetically engineered plants and animals are created using horizontal gene transfer, or "horizontal inheritance." Horizontal gene transfer is achieved by inserting a gene from one species into a completely different species, which can create entirely unpredictable, if not chaotic results. This is in contrast to vertical gene transfer, which follows from natural reproduction and genetic evolution. Vertical gene transfer is the transmission of genes from one generation to the next through sexual or asexual reproduction.

Genetic engineering takes artificial combinations of genes that have never existed together, forcibly inserts them into random locations in the host genome, and then clones the results, thus the name "Frankenfood." And just like in the story of Frankenstein, the consequences and ultimate end result may be far reaching, far different, and far more destructive than what was intended, all while being denied by the creators. Taking microscopic genes out of one species and artificially inserting them into another can be expected to work no more smoothly than transplanting a monkey brain into a horse, or dog's brain into a cat. A cat that's happy to see when you come home just doesn't seem right.

GMO cheerleaders, including many biotech scientists, believe they can apply the principles of vertical inheritance to horizontal inheritance. But according to Dr. David Suzuki, world renowned geneticist, this entire concept is "just lousy science," riddled with flaws. Individual genes are not solo players; they never function in an independently in a void. Rather they are musicians in an entire genome "orchestra" and their conduct is directed by "epigenetics," which are essentially the chemical bath that surrounds the gene and has a profound influence on how those genes are expressed (see Chapter 5). Whole sets of genes are turned on and off by that epigenetic bath, affecting the expression and functioning of any particular organism, and we call that entire orchestration an "activated genome." Much like the proverbial "fish out of water," it's naive and a dangerous mistake to assume a gene's traits are expressed predictably and properly, regardless of where they're inserted. One study showed 5 percent of an organism's genes changed their expression, i.e., the way they synthesized proteins, after one single gene was added to their genome.[7]

The very nature of what we eat as food is being dictated by, and manipulated for profit by, a handful of corporations. The very best that can be said about GMOs is that we are all being herded involuntarily into an experimental pen, forced to play Russian roulette with our health. The long-term consequences are not very clear yet, thanks largely to the biotech industry owning the playing field, but there are at least strong hints of serious adverse outcomes. And make no mistake, the end game is not feeding the world or improving public health; it is profit.

Higher organisms carry their DNA around in mini-biologic packages inside cell nuclei. They release genes into the world only under relatively controlled acts of reproduction. On the other hand, much like drunken drivers (or members of Congress), bacteria and viruses are far more "reckless" in throwing genes around in a chaotic fashion. Using microorganisms as a means of delivery for foreign genetic material is like firing loose cannons of DNA with your eyes closed, never knowing where the cannon balls will end up. Yet that is often the delivery system used by the genetic engineering process.

The kind of horizontal gene transfer that is currently used to create new GMO crop seeds tends to make the human body respond as if to a foreign invasion. Many chronic diseases have a common denominator in the biologic process of inflammation. If foods with horizontally transferred, "invader" genes cause inflammation, then it stands to reason that GMOs could precipitate multiple chronic diseases. While it does not prove causation, it is worth noting that inflammatory-based chronic diseases, like Alzheimer's, diabetes, inflammatory bowel disease, and atherosclerotic heart disease have increased right alongside the proliferation of GMO foods in the US.

The health and well-being of every human being is closely tied to their microbiota, the 100 trillion microorganisms that inhabit their intestines. GMOs and pesticides alter the intestinal microbiota, which is known to play a role in inflammatory bowel disease and in colon and rectal cancer.

The American Cancer Society has found that young and middle-age adults are being stricken with rising rates of colon and rectal cancer. Colorectal cancer rates dropped steadily between 1890 and 1950, but they have been increasing in every generation born since. People born in 1990 now face a life time of double the risk of colon cancer and quadruple the risk of rectal cancer compared to those born 40 years earlier.[8] The rise of those cancers almost certainly has to be of environmental origin because human genes do not change in that short a time frame. The timing of this rise in GI cancers corresponds roughly, but not exactly, with the introduction of GMOs into our food supply.

Monsanto's signature genetic triumph is the insertion of a gene for the production of an insecticide, Bt toxin, into the genetic material of plant seeds. Bt toxin (Bacillus thuringiensis) is a gram-positive, soil-dwelling, natural bacterium, commonly manufactured as a "natural" pesticide. When eaten by an insect the toxins attack the gut cells of the insect, punching holes in the lining, the Bt spores spill out of the gut and germinate in the insect causing death within a couple of days. Farmers have used Bt toxin from soil bacteria as a natural pesticide for years. They spray it on plants, where it washes off and biodegrades in sunlight.

However, Monsanto's Bt-toxin is genetically built into every cell of the plant. The toxin cannot be washed off in your kitchen sink. You cannot eat any part of that plant, without eating the toxin. Furthermore, the plant DNA-produced version in Monsanto's GMOs is thousands of times more concentrated than the spray. It is designed to be much more toxic. One of organic farmers' concerns is that Bt crops will add so much of the toxin to the environment that insects will develop resistance to it, a trend that is already being observed.

Monsanto claimed that Bt toxin is broken down in the human digestive system, so there would be no risk to humans. A new study shows that to be Monsanto propaganda, or "fake news" to use current vernacular.

The only human study on GMO consumption showed that even if you stop ingesting GMOs, harmful GM proteins may be produced continuously inside your gastrointestinal tract. Those GMO genes inserted into the soy end up contaminating the normal bacterial flora inside the intestines, where they continue to function and can transform your intestinal bacteria into a mini-pesticide factory.[9]

Blood samples from 93 percent of pregnant mothers and 80 percent of fetuses show the presence of active Bt toxin, i.e., poison.[10] To expect us all to believe that having the human body act as a constant, walking poison factory with no consequences is asking a bit much.

Is There Clinical Evidence of Trouble?

If indeed you are what you eat, at the microscopic level and in the most intimate way possible, your personal health is currently at the mercy of Monsanto and their biotech cohorts.

The Union of Concerned Scientists feels strongly enough about Monsanto misrepresenting the safety of their products that they created a series of advertisements challenging Monsanto's claims. In December 2013, hundreds of scientists from the European Network of Scientists for Social and Environmental Responsibility, signed a joint statement saying that GM foods have not been proven safe and that existing research raises concerns. The number of signees on to that letter is now almost 300.[11]

Studies in humans are limited, not only much to Monsanto's liking, but due at least in part to Monsanto and other biotech companies' refusal to provide GMOs to independent scientists. Monsanto requires that a contract be signed stipulating that the seeds must not be used for research without their approval. But despite Monsanto stonewalling, independent researchers have completed enough studies to conclude that all is not well with GMOs.

Bt toxin has been shown to be harmful to mammalian red blood cells and bone marrow.[12] From studies on mice and in farm workers, the EPA's Scientific Advisory Panel concluded that Bt proteins could act as "antigenic and allergenic sources."[13, 14] Numerous animal studies have linked Bt toxin and GMOs to infertility, immune dysfunction, gastrointestinal disease, accelerated aging, and diminished kidney function. [15, 16, 17, 18, 19, 20, 21] The question of how safe are GMOs cannot be answered without simultaneously addressing the question of how safe are the pesticides that are GMOs' always present ugly step sisters. And herein lies a likely even larger threat.

Glyphosate, the primary active ingredient in Roundup, is the most common herbicide used with GMOs. It was invented in Switzerland in 1950, but first discovered to act as an herbicide by Monsanto scientist John Franz. The compound acted by destroying the metabolism of plants by preventing them from forming essential amino acids. Glyphosate is systemic-acting, affects the entire anatomy of green, leafy plants, killing within a matter of days.

The majority of food staples in the United States now originate on industrial farms planting monocultures of crops genetically engineered to withstand Roundup, "Roundup Ready," which are repeatedly sprayed to exterminate weeds in crop fields. Not surprisingly, glyphosate residue remains on the crops, which then enters the food chain of livestock and humans. The overwhelming majority of the world's high volume grains are now "Roundup Ready." Glyphosate residues cannot be removed by washing, nor can the food be decontaminated by cooking. The chemical remains stable, in food, for a year or more, even if the foods are frozen, dried or processed. Eating strictly organic doesn't protect you. Several studies have shown that it is even present in the air we breathe, water we drink, and even in rainfall over large areas of the country.[22]

A recent study of non-farmworker, urban dwellers in Germany found glyphosate in the urine of every person tested, and at levels 5 to 20 times the legal limit for drinking water.[23] It is no surprise that 93 percent of Americans have detectable glyphosate in their urine.[24]

A January 2014 study[143] published by a German research team found glyphosate was significantly higher in the urine of chronically-ill people compared to healthy people. German researchers leading the University of Leipzig study concluded "the presence of glyphosate residues in both humans and animals could haul the entire population towards numerous health hazards."

In the United States, the EPA has set a legally enforceable maximum contaminant level (MCL) for glyphosate of 700 micrograms per liter in drinking water, which is 7,000 times higher than the limit set in Europe. Not surprisingly, urine testing of US residents showed levels 10 times higher than those of European residents. This reflects the pervasive use of Roundup in the United States compared to Europe.

One of the reasons why glyphosate is presumed benign to humans is that supposedly it does not accumulate in fat. Senior Monsanto scientist (forgive me if "Monsanto scientist" seems like a contradiction in terms) Dan Goldstein recently stated, "If ingested, glyphosate is excreted rapidly, does not accumulate in body fat or tissues, and does not undergo metabolism in humans. Rather, it is excreted unchanged in the urine."[139] Regulatory bodies worldwide seem to have taken Monsanto's word for it that glyphosate is not bioaccumulative, and is therefore benign. But there is much more to the story.

Since the introduction of GMO seeds in 1996, the amount of glyphosate used on crops in the US has increased almost 10 fold. Ramon Selder, a professor of microbiology and a retired senior scientist and team leader for the EPA's biosafety program, wrote in an April 2014 Op Ed that the concentration of Roundup typically found as a residue in commercially grown GMO fields is "sufficient in laboratory assays to: induce hormone disruptions during frog development (mixed-sex frogs); kill young trout and tadpoles; stop the growth of earthworms in soil; inhibit activities of beneficial soil and human gut bacteria; and stimulate the growth of human breast-cancer cells assayed under laboratory conditions."[25]

For years, Monsanto has publicly claimed that Roundup is benign to humans because glyphosate inhibits an enzyme supposedly found only in plants and microbes, but not in people, EPSP synthase. The problem, or more accurately, the deception with that contention is three fold. One, glyphosate is not the only potentially toxic ingredient in Roundup. Two, just because glyphosate targets that enzyme doesn't mean that other adverse effects are not triggered. Three, EPSP synthase is certainly found in the 100 trillion bacteria that constitute the bacterial flora of the human GI tract. A healthy intestinal bacterial flora is essential to good health for any human. All of these deceptions will be addressed in later parts of this chapter.

Glyphosate has been shown to cause birth defects in animals by interfering with the molecular mechanisms that regulate early development in frogs and chickens, creating deformities of embryos at very low concentrations.[26, 27] Glyphosate behaves as an "endocrine disruptor" or hormone mimicker [28] (see Chapter 4). As such, it has the potential to precipitate a wide variety of maladies, including birth defects, hormone and reproductive disorders,[29, 30, 31] cancer,[32, 33, 34] fatty liver disease even at very low doses of exposure,[35] and impaired brain development through inhibition of the thyroid gland.

Exposure to pesticides, including glyphosate, is associated with neurodegenerative diseases like Parkinsonism. [36, 37, 38] One mechanism may be that pesticides can prevent the brain from disposing of its own "toxic waste." In vitro studies have revealed toxicity to genes from the primary metabolite of glyphosate.[39] Workers occupationally exposed to pesticides show increased DNA damage.[40, 41]

Glyphosate enhances the adverse effects of other food-borne chemical residues and environmental toxins, increasing levels of systemic inflammation, and as previously mentioned, the common denominator of many chronic diseases like gastrointestinal disorders, obesity, diabetes, heart disease, depression, autism, infertility, cancer, and Alzheimer's disease.[42]

Research by Russian scientists found that feeding hamsters GMO soy resulted in complete sterility after two or three generations.[43,44] Many older studies that Monsanto claimed exonerated Roundup from causing any health affects made the mistake of only testing the main active ingredient, glyphosate. But Roundup contains other chemicals as adjuvants and surfactants that facilitate binding to the plant. These chemicals were billed by the manufacturers as benign and, by and large, regulators bought into the pretense. Those adjuvant chemicals can be kept secret by the manufacturers. But a 2014 study found that the toxicity of eight out of nine pesticide formulas was greater than the toxicity of their active ingredient alone.[45] Roundup was the most toxic of the pesticides studied, and the commercial formulas of the pesticides were as much as 1,000 times more toxic than the "active ingredient" by itself.

I'm a big fan of functioning kidneys, as are most physicians. Personally, whenever I'm not passing a kidney stone, I consider them two of my top ten organs. Agricultural workers in India, Sri Lanka, Vietnam, Africa, and Central America probably feel the same way, especially now that their kidneys may be one of those things that Monsanto is willing to sacrifice for quarterly profits.

In Central America, they call it the "malady of the sugar cane," an epidemic of a fatal kidney disease beginning 15-20 years ago that has been the demise of tens of thousands of field workers with hundreds of thousands losing some degree of kidney function. The WHO has determined the common denominators of the disease—exposure to arsenic or cadmium, pesticides, consumption of "hard water," low water intake, and high temperatures and dehydration.

In many of these countries, there is culture of ghostly silence about the disease. Workers don't want to know if they have it, and they are largely asymptomatic until they reach the late stages of the disease. They all fear dialysis because of the ironic and false perception that it's the dialysis that ultimately causes death rather than prolonging life. The agricultural industry claims they couldn't be responsible, and industry certainly doesn't encourage a diagnosis. In Nicaragua, workers are reportedly fired if they test positive.[46]

Of course Monsanto is equally quick to absolve themselves, despite the fact that appearance of the disease corresponds to the beginning of widespread use of their Roundup in the fields.

The best hypothesis on the cause of the disease seems to be the ability of Monsanto's glyphosate, the primary, active ingredient in Roundup, to bind to the heavy metals found either in naturally occurring hard water, soil, or present in fertilizers. Hard water contains metals such as calcium, magnesium, strontium, and iron, along with carbonate, bicarbonate, sulfate, and chlorides. Glyphosate was originally patented as a chelating agent, a compound that binds strongly to metals. Its ability to bind to these heavy metals and carry them to the kidney makes it a prime suspect. The pesticide's "half-life," i.e., resistance to breakdown, can increase from several weeks in normal water to many years in hard water. The onset of the disease corresponds chronologically with the widespread use of Roundup in these stricken countries. In Sri Lanka alone, the disease now afflicts 15 percent or 400,000 working-age people in the northern part of the country, with a death toll of about 20,000, according to the Chronic Kidney Disease of Unknown etiology (CKDu) study.[47]

Thousands of Indian farm workers have experienced allergic reactions after picking Bt cotton. In 2006, over 1,800 sheep, buffalo and goats died within days after grazing on post-harvest Bt cotton plants.[48] When given the choice, pigs, chickens, buffalo, geese, elk, deer, raccoons, squirrels, mice, and rats in South Africa, India, Illinois have all been seen avoiding genetically modified organisms. Animals aren't exactly telling anyone why they are rejecting GMOs, but just the fact that they are should give us pause. In search for the causes of serious diseases of entire herds of animals in northern Germany, especially cattle, glyphosate has repeatedly been detected in the urine, feces, milk, and feed of the animals.

Pigs share more with humans than you might think. Male pigs do pretty much nothing but drink beer, watch football on TV, and leave the toilet seat up and their dirty socks lying around constantly. But both sexes of pigs actually share similar digestive systems with humans. However, pigs fed a diet of only genetically modified grain showed markedly higher rates of severe stomach inflammation than pigs who dined on conventional feed.

Glyphosate was originally patented as an antibiotic, (U.S. Patent number US7771736 B2), so it is not surprising that it might affect the bacteria or microbiota in the human colon, the same site of concern as Bt toxin. There is evidence that glyphosate destroys beneficial bacteria and allows harmful ones, such as salmonella, and clostridium, to flourish in the intestines of farm animals.[49, 50, 51, 52] After rats were fed GMO potatoes for only 10 days, their stomach lining cells and intestines were significantly altered.[53]

I recently provided anesthesia for a 38-year-old patient who had to have emergency surgery to remove his colon due a "toxic megacolon" secondary to inflammatory bowel disease (IBD). IBD is a chronic disorder that afflicts 1.4 million persons in the U.S., and the incidence has been steadily rising worldwide in the last two decades. In some countries the rate is five times greater than it was in the 1950s. Developing countries such as India, China, and South Korea, where IBD was almost unheard of 30 years ago, have seen a dramatic increase that is now paralleling the incidence in the Western world.

IBD results from a loss of homeostasis, or balance, between the immune system and the microbes that inhabit the intestine. Not only is IBD a devastating and debilitating chronic illness, it is also one of the three highest risk factors for the development of colorectal cancer.

Fifty years ago, IBD was almost exclusively diagnosed in adults. "These days, treating children with IBD is business as usual in our clinics," says Maria Oliva-Hemker, M.D., chief of the Gastroenterology & Nutrition division at Johns Hopkins Children's Hospital. The change in this disease pattern must be either genetic or environmental. But significant genetic changes do not occur in the time frame of one or two generations, so the real culprit must be primarily environmental, or a combination of genetic susceptibility and environmental triggers. The chronological rise and geographic spread in the incidence of IBD corresponds almost exactly to the onset of the GMO/intensive pesticide model of agriculture. That in itself doesn't prove the connection but does add yet another strong piece of circumstantial evidence. Meanwhile, virtually all the other possible environmental factors have not shown a connection—including smoking, appendectomy, oral contraceptives, diet, breastfeeding, infections, vaccinations, antibiotics, and childhood hygiene.[54]

There is a consistent pattern of bowel inflammation in a high proportion of children with the autism spectrum disorder (ASD).[55, 56, 57] ASD is increasing in frequency in the United States at an alarming rate. Even if some of the increase is an expanded definition of the disorder and better awareness and identification of afflicted patients, consider this trend. The autism rate in 1981 was one in 10,000. In 2007, it was 1:150. In 2009, it was 1:100. In 2012, it was 1:88. In 2014, it is 1:68. If this rate of increase continues, by 2025 it will be 1:2, i.e. 50 percent. The CDC says that since the 2012 estimate of 1 in 88 children identified with ASD, the criteria used to diagnose, treat, and provide services have not changed, but the rate has increased another 30 percent.

Meanwhile, autism rates in Europe have remained virtually flat for the last decade. Recent estimates in European countries range from 1 in 5,000 in Germany to 1 in 700 in Portugal. So what are Americans doing to harm themselves and their children's brains that Europeans aren't, besides watching *Keeping Up with the Kardashians*? No one knows for sure, but one thing to consider is the massive increase in GMOs and the concomitant upsurge in pesticide and herbicide use.

In more than 60 developed countries worldwide, there are significant restrictions or outright bans on the production and sale of GMOs. In the US, federal agencies have approved the GMO/pesticide industrial agriculture system with safety assumed solely on the basis of studies conducted by the same corporations that created them and profit directly from their sale.

On a June 12, 2013, autism conference in Edinburgh, Scotland, Dr. Martha Herbert, a pediatric neurologist, an expert on autism from Harvard Medical School and author of *Autism Revolution*, explained her conclusion that the culprit is an environmental toxin in autistic children that interferes with nutrient absorption from the intestines. Humans only absorb nutrients courtesy of the bacteria in our gut, which is disrupted by glyphosate. Because it is a chelating agent, binding essential minerals such as cobalt, zinc, manganese, calcium, molybdenum, and sulfate, glyphosate obstructs the absorption of essential nutrients.

Recently, the connection between the GI tract and the brain has been solidified by evidence that the "trillions of bacteria, viruses, archaea, and eukaryotes that make up the gut microbiome" are altered in a wide range of neurologic disorders--like autism, depression, and Parkinson's.[131] The neonatal period is, simultaneously, a critical window of development for microbes to inhabit the infant gut as well as rapid brain growth and development. In the early 2000s animal studies showed that both beneficial and adverse manipulation of the bacterial content of the GI tract was associated with changes in neurological development of the animals.[132] Autism is well known to occur four to five times more often in boys than girls. Researchers found that manipulating the intestinal microbiome resulted in altered levels of the neurotransmitter serotonin in male mice but not in their female counterparts.[133] Several studies show that organophosphate chemicals (nerve agents and pesticides) cause greater neurotoxicity in male animals than females.[136, 137, 138] Other studies show that altering the gut microbiome impaired a person's ability to excrete other known neurotoxicants like mercury, arsenic, and nicotine.

Several studies show that the gut microbiome is abnormal in children with poor neurological development, and others that show a wide range of abnormal GI symptoms among people with Autism Spectrum Disorder, and that changes in the gut bacterial flora are associated with changes in the symptoms of autism.[134, 135]

Strong circumstantial evidence suggests that by altering the microbiome of the gut, Bt-toxins from GMOs, and the pesticides used concomitantly with them, could be causing microscopic leaking in the intestinal walls of newborns, facilitating the absorption of incompletely digested foods and toxins into the blood. That, in turn, triggers systemic inflammation, precipitating autoimmune diseases and food allergies. Furthermore, since the blood-brain barrier is not developed in newborns, toxins in the blood may migrate into brain tissue itself, interfering with normal brain development. This is a very plausible biological pathway for autism. The brains and cerebrospinal fluid of autism patients consistently show a marked inflammatory process.[58]

In Benton, Yakima, and Franklin Counties in Washington beginning in 2010, a mysterious mini-epidemic of a fatal brain birth defect, anencephaly (essentially a missing brain) and severe spinal cord birth defects emerged at rates 4 to 8 times the normal rate. Toxic environmental exposures almost certainly have to be the cause, and pesticides have to be at the top of the list of suspected environmental exposures.

Coincidently, also beginning in 2010, Benton County started a noxious weed eradication program, spraying glyphosate, along the Yakima River, which all three counties used for irrigation water. No one bothered monitoring glyphosate concentrations in the water. There is the distinct possibility of a direct causal link between the ability of glyphosate to inhibit folic acid [folate] production, and a deficiency of folic acid [folate] in pregnant women, which is a known contributor to the risk of anencephaly.

The Washington State Health Department investigated the epidemic and has come up with no answers. But Sarah Barron, the nurse who first made the observation and alerted authorities, wonders if perhaps they didn't find anything because they were half-hearted in looking for the culprit. The state hasn't spoken to any of the families who had the babies with birth defects. Not a single one. They never asked key questions like what they ate, or if they'd been exposed to pesticides sprayed in this agricultural area. Andrea Jackman, whose daughter, Olivia, was born with spina bifida, another type of neural tube defect, is outraged. "Nobody's asked me anything," said Jackman.[59] It's almost like the state doesn't want to know.

Dr. Medardo Vasquez is a neonatal specialist at the Children's Hospital in Cordoba, Argentina, where GMOs and the attendant explosion in pesticide use has taken over agriculture. He makes this statement, "I see new-born infants, many of whom are malformed. I have to tell parents that their children are dying because of these agricultural methods. In some areas in Argentina, the primary cause of death for children less than one year old are malformations."[60]

"Freedom," "fair and balanced," "non-partisan," and "independent" are all comforting sounding words, but they are often churned out by propaganda machines. The name "Independent Women's Forum" sounds benign enough, but its agenda is as conservative as Ann Coulter. They support right wing politicians, and their organization and personnel have strong ties with the secretive foundation network of the Koch Brothers. The American Chemistry Council is among their many corporate funders, so it is perhaps of little surprise that IWF and Monsanto sponsored a lecture series whose theme could have been titled, "Don't Worry, Be Happy" about all the pesticides in your family's food.[61] The event was titled "Food and Fear, How to Find Facts in Today's Culture of Alarmism," and the setting was the campus of Washington University in St. Louis, giving it a mirage of academic legitimacy.

Typical of their panel of "experts" was Julie Gunlock, the director of IWF's "Culture of Alarmism" project, and author of *"From Cupcakes to Chemicals: How the Culture of Alarmism Makes Us Afraid of Everything and How to Fight Back."* Apparently we should all just trust Monsanto to keep our food safe.

The American Academy of Environmental Medicine (AAEM) believes GMO food is a serious health risk. The AAEM called for a moratorium on the production of GM foods. They stated, "Several animal studies indicate serious health risks associated with GM food, including infertility, immune problems, accelerated aging, insulin regulation, and changes in major organs and the gastrointestinal system...There is more than a casual association between GM foods and adverse health effects. There is causation..."[62] The AAEM didn't mince words. They called on all physicians to prescribe diets without GMO foods to all patients, for a moratorium on growing GMOs, long-term independent studies to investigate the health effects, and labeling of all GMO foods so that consumers could exercise the choice to avoid them.

"Death Star" Harvesting

One of the reasons the health consequences of glyphosate is increasingly becoming a grave public health concern is the recent adoption of industrial agriculture's technique of "Spraying crops to death," meaning that Roundup is being sprayed directly on the crops shortly before harvest to facilitate a so called "clean harvest" (love the irony) by uniformly killing off all plants (including the crops, mind you) on the field. If maturation of the crops has been compromised by excessive moisture, as is sometimes the case, herbicides are used to bring the crops to maturity by means of a "death-spray"—Big Ag's "Death Star." The technique accelerates the drying out of the crops and kills the weeds at the same time.

This Death Star strategy is now common for staples like potatoes, grains, canola, and legumes, and even for crops that are not GMOs. For example, spraying herbicides on a field of potatoes just prior to harvest, makes the skin firm and reduces the potato's vulnerability to late harvest blight and germination, improving their shelf life. The toxin is absorbed by the whole potato when sprayed on the leaves, so when eating McDonald's fries, tater tots, and hash browns,[63] you are swallowing Big Ag's parting gift of poison. On the plus side, when you find potato chips in the cushions of your couch from your 1994 Super Bowl party, you can still eat them because they will taste the same.

An advertisement from our friends at Syngenta explains just how safe is pre-harvest desiccation: "For professional producers chemical desiccation now counts among the standard measures to assure high quality production [...]. In this context one also speaks of the 'economic maturity' of crops, as the usage of herbicides allows for a safe termination of the harvesting procedure."[63]

To facilitate the Death Star technique, the EU raised the legal limit of glyphosate in wheat and bakery items to 100 times the legal limit for vegetables. For livestock feed grains, it was raised even more. Never mind that enforcement of these limits is essentially non-existent.

Yet, in tacit acknowledgement of glyphosate's affects, and in a revealing gesture of their priorities, the EU does not allow Roundup to mess around with their brewing of beer. They prohibit its use on malting barley or during seed propagation, because that would impair germination capacity. Beer does not brew with grains plastered with Roundup. For bread and livestock grain however, the impact on germination is not a problem. Theoretically, European authorities do not allow the desiccated grain to be used for fodder in the same year, but again, actual enforcement is pretty much a fantasy.

Pause for a moment and let sink into the deepest part of your cerebral cortex just what is happening with Death Star harvesting. As late as seven days before crops are harvested and sold to bakeries, Big Ag soaks the wheat in Roundup to dress the kernels up in the appearance of uniform maturity. In today's world, "looking good" is priority number one. As for you, the consumer, it is exactly the same as adding glyphosate—again, think "poison"— as a flavoring ingredient right into the bread dough, as if it were maple syrup. And the same is true for livestock feed. Hey, who wouldn't want mayo, pickles and Roundup on their burger?

Desiccating crops with Roundup just before harvest is undoubtedly a primary reason Americans are showing more Roundup in their bodies than ever before. In the last twenty years, the number of people that tested positive for the ingredients in Roundup shot up 500 percent, and the amount in their bodies shot up more than ten times, 100 times more than what was found in rats.[64] Even more worrisome is that children are generally showing higher levels than adults.

Roundup's supposed safety is also tied to the insistence that glyphosate breaks down quickly in the soil. Despite surprisingly few studies on the issue, there is emerging evidence that is also not the case.

Furthermore, many of the studies used to exonerate glyphosate did not study the supposedly inert ingredients that are added to glyphosate to make Roundup and other herbicide formulas. A recent study on human cells showed that these other "safe" chemicals are many times more toxic than the active ingredient and that pesticide formulas are frequently contaminated with toxic heavy metals, like arsenic, cobalt, lead, and nickel, making the end products act as powerful endocrine disruptors (see Chapter 4).[147]

A friend of mine sent me an email after being alarmed by observing a neighbor's son, as a city employee, spraying the herbicide Roundup along the Jordan River Parkway in the center of Salt Lake City. She asked him what he was spraying and he said, "nothing dangerous, just Roundup." He wasn't wearing protection of any kind.

The very same day, the World Health Organization (WHO) announced that they had concluded that glyphosate was a "probable carcinogen." The World Health Organization's cancer agency, International Agency for Research on Cancer (IARC), published the declaration in one of the world's most prestigious medical journals.[65] More about that later. So Big Ag's Death Star harvesting technique involves giving our staple crops a nice frosting of cancer-causing chemicals just before harvest, and they will still be there when your kids open their bag of Cheetos.

Monsanto: A Legal Bully

Monsanto's former motto, "Without chemicals, life itself would be impossible," has been replaced by "Imagine." Ok, let me enjoy "imagining" that "impossible." Their web site home page claims it "help[s] farmers around the world produce more while conserving more. We help farmers grow yield sustainably so they can be successful, produce healthier foods . . . while also reducing agriculture's impact on our environment." The company holds about 650 seed patents, primarily cotton, corn, and soy seeds, and about one third of all biotech product research and development.

Monsanto came to own such a vast supply by buying major seed companies to stifle competition, patenting genetic modifications to plant varieties, and suing small farmers. Patenting seeds is the real reason the biotech industry promotes GMOs. Monsanto has cross-licensing arrangements with BASF, Bayer, Dupont, Syngenta and Dow. They have agreements to share patented genetically engineered seed traits. So the giant seed corporations are not competing with each other, they are competing with peasants and small farmers over control of the seed supply.

Monsanto has come under harsh criticism for aggressive litigation against individual farmers for patent violation claims on GMO seeds. Documentaries such as *Food, Inc.* and *The Future of Food* told some of these farmers' stories, forcing Monsanto to respond by constructing a special section on its website to rationalize these lawsuits. As explained in a 2013 detailed report by Food and Water Watch,[66] Monsanto ensures its right to sue farmers through the company's technology licensing agreement on every bag of GMO seed. "Any farmer who buys Monsanto's seed is bound to it, either by signing a contract or simply opening the bag. Monsanto prohibits farmers from saving any seed."

Monsanto's contracts allow them to investigate farmers' fields at any time and to access farmers' records filed with the USDA Farm Service Agency. These records tell Monsanto how many bags a farmer bought and exactly how many acres he planted the seed on, making property investigations and prosecution very easy. Just to make sure Monsanto left no stone unturned, they set up a toll-free "snitch line" where neighbors and community members are encouraged to anonymously spy on and report farmers that may be using Monsanto's seeds without a license. Monsanto has hired private investigators to "videotape farmers, sneak into community meetings, and interview informants about local farming activities."[66]

The corporation has gone so far as to take farmers to court demanding thousands of dollars in damages and legal fees even from farmers who never planted the company's seeds in the first place. Defying all common sense, even farmers who do not grow their products must be held accountable for their neighbors' GMO crops, and Monsanto effectively eliminates the company's responsibility for contaminating the fields of farmers who do not want to raise GMOs. In Australia, GMO genes were found up to 3 km from where they were intentionally planted. In Hawaii, 30-50 percent of papaya was found to be contaminated with GMO genes.

It is nothing less than legal terrorism for Monsanto to go after those who inadvertently and against their wishes end up with Monsanto's traits in their fields. GMOs contaminating non-GE crops is inevitable, since cross-pollination is inevitable within the same species or with close relatives. Monsanto's business model of suing farmers as "intellectual property thieves" whose only crime was to have farms in the vicinity of GMO cultivation is like a rapist suing his victims for allowing themselves to be assaulted. That Monsanto usually wins these cases speaks volumes about our legal system.

Canadian scientists found that nearly half of the 70 "certified seed" samples tested were contaminated with the Roundup Ready gene.

Another study in the US found that virtually all samples of non-GE corn, soy beans, and canola seed were contaminated by GE varieties.[67] In August 2006, trace amounts of Bayer's experimental genetically engineered Liberty Link rice was found to have contaminated 30 percent of cultivated rice land in Texas, Louisiana, Missouri, Arkansas and Mississippi.

"Farmers have been sued after their field was contaminated by pollen or seed from someone else's genetically engineered crop [or] when genetically engineered seed from a previous year's crop has sprouted, or 'volunteered,' in fields planted with non-genetically engineered varieties the following year," according to the Center for Food Safety.[140]

Monsanto has sued 410 farmers and 53 small businesses in more than 27 states, netting themselves over $21.5 million in awards from the courts. Monsanto keeps staff on hand solely for the purpose of investigating and prosecuting farmers. Even for the farmers who win their cases the process takes years of legal battle, causing stress and a significant financial burden. Many farmers settle out of court rather than try to defend themselves. The company investigates roughly 500 farmers each year.[68]

Environmental contamination by GMOs via cross-pollination also poses a serious threat to biodiversity. Contamination of the very building blocks of nature should be of grave concern. Plant traits introduced by genetic engineering are more likely to escape into the wild than the same traits introduced conventionally. Andrew Kimbrell, director of the Center for Technology Assessment, believes such escapes are inevitable. He said, "Biological pollution will be the environmental nightmare of the 21st century."[144]

Biotech Industry Infiltrates Academia

The land-grant university system includes some of the largest state universities—including the University of California system, Pennsylvania State University, and Texas Agricultural & Mechanical University—with 109 locations and a presence in every state. Farmers, consumers, policymakers and federal regulators depend on land-grant universities as a source of credible, independent research.

In the last three decades, federal policies, like the Bayh-Dole Act of 1980 (see Chapter 3) began encouraging land-grant universities to partner with the private business on agricultural research. That sounds benign enough, but the end result is that many public universities have essentially prostituted themselves to corporations by accepting donations and funding for research, allowing the goals of legitimate academic investigations to be subsumed into goals of industry and discouraging independent research that might be critical of industrial agriculture. Industry funds now are the dominant source of research money for these agricultural academic centers. Big Ag and Big Biotech have managed to thoroughly stack the scientific and academic deck heavily in their favor. Most food conferences and fellowships are also funded by Big Food companies, including Monsanto, which has a chilling effect on their independence and integrity.

A report from Food and Water Watch[69] details this disturbing trend, and Monsanto has certainly stepped up to the plate to take full advantage. For example, grants from Monsanto, Syngenta, and SmithBucklin, totaling $18.7 million between 2006 and 2010, represented almost half of the research grant money received by the University of Illinois Crop Sciences Department during that time.[70]

This is a "can't lose" investment opportunity for corporations. Their funding corrupts the public research mission of land-grant universities, warps the science that is supposed to guide farmers in improving their yields and livelihoods, and it does all this under the guise of academic respectability and impartiality. It should come to no one's surprise that industry-funded academic research routinely produces favorable results for industry sponsors.

Some research programs make no pretense about how this works. For example, the University of California Department of Nutrition conducts research into the benefits of eating chocolate, funded by the candy manufacturer Mars. Gee, I wonder what the results will show. A study supported by the National Soft Drink Association found that soda consumption by school children was not linked to obesity. An Egg Nutrition Center-sponsored study found that frequent egg consumption did not increase blood cholesterol levels. Deans and college administrators might as well be forced to wear jackets with the logos of all their corporate sponsors.

More than 15 percent of university scientists acknowledge having "changed the design, methodology or results of a study in response to pressure from the source of funding. A peer-reviewed analysis of dozens of nutrition articles on commonly consumed beverages found that industry-funded studies were four to eight times more likely to reach favorable conclusions to the sponsors' interests.[71]

About half of the authors of peer-reviewed journal articles on the safety of GMO foods had an affiliation with industry. All of these produced favorable results to industry sponsors.[72] Monsanto and the rest of Big Ag, like Chiquita, Dole, Sysco, Kraft, Cargill, Coca-Cola, Tyson, MacDonald's, and Walmart, have even shared board members with several universities. The University of Georgia's Center for Food Safety offers seats on its board of advisors for $20,000 to industry sponsors, where they can help direct the center's research efforts. South Dakota State University (SDSU) president David Chicoine joined Monsanto's board of directors in 2009, receiving $390,000 in annual salary from Monsanto, more than his salary as university president.

Shortly before Chicoine was placed on Monsanto's payroll, the company sponsored a $1 million plant breeding fellowship program at SDSU. Chicoine's Monsanto appointment also coincided with a new SDSU effort to enforce university seed patents by suing farmers for sharing and selling saved seed. One farmer sued by SDSU claimed SDSU took a page out of Monsanto's play book, running a fake "seeds wanted" ad in a newspaper, entrapping him in a deliberate set up.

Donations for "naming rights" is part of the unholy union between corporations and academia. Walmart donated $50 million to the U. of Arkansas to build the Sam Walton Business School, Cargill donated $10 million to build the Cargill Plant Genomics Building. Iowa State University now has a Monsanto Student Services wing in the main agriculture building, so named because of their $1 million gift. The University of Missouri sports a Monsanto Auditorium. Monsanto donated $200,000 to the University of Illinois's College of Agriculture for the Monsanto Multi-Media Executive Studio. Where better to hold rigorously unbiased, academic seminars on agriculture and crop science?

In addition to industry's funding their research, professors often augment their academic salaries with high-priced consulting for industry. A 2005 survey found that over 30 percent of land-grant academic scientists acknowledged consulting for agricultural corporations. One Cornell professor was a paid Monsanto consultant while also publishing journal articles promoting the benefits of Monsanto's rBGH for dairy farms.

As intended, these donations/investments purchase for the company, influence, credibility, and brand power at university ivory towers. Monsanto is not only gaining access to research that is publicly accepted as legitimate and independent, but it is profiting highly from it. The company's signature products, rBGH, the artificial growth hormone for cows and Roundup Ready seed technology, were made possible through research conducted by publicly funded universities.

A microcosm of the incestuous relationship between industry and academia emerged in 2015 over the controversy of whether glyphosate was contaminating human breast milk. That year, the first-ever testing for glyphosate in the breast milk of U.S. women was done by a small non-profit, Moms Across America.[73]

Their small survey found high levels in three out of the 10 samples tested, 760 to 1,600 times higher than the European Drinking Water Directive allows for individual herbicides. The shocking results contradicted industry pronouncements that glyphosate does not bioaccumulate. Certainly, the last thing a newborn should be ingesting is Roundup from its mother's milk. These results are even more shocking given that nine of the ten nursing mothers tested had made a conscious effort to avoid pesticides and GMOs.

When this information was widely covered by the national media, Monsanto did what they always do when their profits are threatened. They unleashed their attack dogs against the message, the messengers, and the media. The first attack came from a seemingly "independent" source, a press release from Washington State University, with one of their biology professors, Michelle McGuire, stating, "The Moms Across America study flat out got it wrong." Ouch! Those white coats in the academic ivory tower put those uppity moms in their place but good. Well, not so fast. It turns out that the research McGuire was standing on to tear into those moms had not yet been published, probably was not even underway at the time, and was not so independent after all.[74]

The journal where McGuire's "set the record straight" study was eventually published, claiming no glyphosate exists in human breast milk, was published in the American Journal of Clinical Nutrition, which is copyrighted to the American Society of Nutrition (ASN). Monsanto and several other Big Ag corporations are "sustaining partners" of ASN. Michelle McGuire is a spokesperson for the ASN.

McGuire's breast milk samples were, stop me if you're surprised, tested at Monsanto's laboratories and one of their subcontractors. When the study was published in 2016, one third of the listed authors were Monsanto employees. Monsanto devised the study and the analytical method. Despite the authors' claiming no conflict of interest, a research gift of $10,000 from Monsanto to McGuire was ultimately disclosed, in addition to paying for the study's expenses.

But there is an even more sinister flip side to corporate infiltration of academic agriculture. The amount of grant and contract money that professors generate plays a huge role in the granting of tenure and salaries. This overtly and covertly discourages academic research on environmental, public health, and food safety risks related to industrial agriculture, and explains the sparse research on alternatives to the dominant agriculture model. Academics unwilling to wear corporate badges are placed at a huge career disadvantage.

Scientists that swim against the tide in pursuing potential critical inquiries into biotech seeds often face reprisals. One university investigator anonymously told the prestigious journal *Nature* that a Dow AgroScience employee threatened that the company could sue him if he published certain data that cast the company in a bad light.[75]

Further hampering independent research on GMOs are common seed licensing agreements that can specifically bar research on seeds without the approval of the corporate patent holder. This can effectively bar independent research on important characteristics of GMOs like yields and health consequences.

In 2009, 26 scientists sent a letter to the EPA calling their attention to this observation.

> "These agreements inhibit public scientists from pursuing their mandated role on behalf of the public good unless the research is approved by industry. As a result of restricted access, no truly independent research can be legally conducted on many critical questions regarding the technology, its performance, its management implications, [insecticide resistance management], and its interactions with insect biology."[76]

Professor Seidler amplified those same concerns writing, "Patent enforcement by American agrichemical seed companies have prevented US scientists from researching what some exclaim are 'problems' associated with GMO crops. We will not know the facts as long as the seeds and plants that we, our children, pets and livestock consume are not made available for conducting long-term, controlled experiments."[77]

In a pathetic commentary on how corrupt and dysfunctional the relationship between industry and academia has become, the scientists who wrote this letter withheld their names for fear of being blacklisted and losing private-sector research funding.

Ironically, media coverage of GMO news and opinion pieces are often covered with a veneer of condescension toward those who question the safety of GMOs, frequently putting doubters in the same uninformed, unsophisticated, anti-science corner as vaccination opponents, global warming deniers, and conspiracy theorists. This is exactly what Monsanto had hoped for. In fact, the media usually completely misses the fact that GMO doubters are asking for independent science, not corporate science, to be the standard by which the issue is analyzed public policy enacted. Outside the US, scientists that have done independent research on GMOs and have found alarming health consequences in animals have been viciously attacked by the biotech industry, in some cases literally, as detailed later in this chapter.

When the media reported the study showing intestinal inflammation in pigs fed GMOs, predictably, on cue, the GMO hit men jumped all over the study like flies on molasses: "Scientists Shred Study That Says Genetically Modified Food Makes Pigs Sick,"[78] "You Can Put Lipstick On A Pig (Study), But It Still Stinks"[79]

This latter article was written by notorious *Forbes* "science" writer Dr. Henry I. Miller, founding director of the FDA Office of Biotechnology from 1989 to 1993. Adorned with an impressive appearing CV and a condescending attitude, Miller's name pops up often in major periodicals denigrating those who criticize GMOs, Big Ag, or pesticides. Miller declares himself in his *Forbes* columns as a "debunker of hypocritical, dishonest, junk science." He is an avowed Ayn Rand admirer, a lover of free markets, and best friend for life (BFL) of the nuclear, pesticide, chemical, and GMO industries. Typical of GMO cheerleaders, he arrogantly declares anyone who sees problems with these industries and products as "fear mongering Luddites."[80, 81, 82]

Miller's attitude reveals a lot about how and why the FDA became the lap dog of the biotech industry. Miller trumpets, "Genetically engineered plants, which have achieved monumental economic and humanitarian successes, receive disproportionately intense scrutiny from no fewer than three government agencies, and none has been shown to be harmful in any way."

Documents released in the massive lawsuit against Monsanto in the fall of 2017 revealed e-mails showing that Monsanto asked Miller to write an article defending glyphosate after the IARC had declared it a probable carcinogen. Miller agreed to after he asked Monsanto to write the first draft. The article[127] was eventually published in *Forbes* magazine under Miller's name without any acknowledgement of the origin and ghost writing from Monsanto.[128]

This blatant fraud prompted *Forbes* to end their relationship with Miller, but *Newsweek* and the *Wall Street Journal* apparently aren't that concerned about the ethical scandal, publishing another one of his GMO cheerleading pieces even after he was exposed.[129, 130]

In the *Newsweek* piece, Miller had the audacity to claim that organic farming is more harmful to the environment than industrial agriculture, and cites only sources connected to Big Ag. But the frosting on his multi-layered fraud cake is that he claims the money spent on political lobbying by the organic food industry is "Goliath," spending "billions" more than Big Ag, i.e. "David."

The Empire Strikes Back

Jeffery Smith highlights many scientists that have been threatened, denied tenure, or even fired for publishing studies critical of Roundup and GM crops. The biotech industry has succeeded in muzzling their scientific critics throughout the world.

Dr. Arpad Pusztai was one of the world's leading nutrition scientists for thirty years, a world-renowned expert on food safety, who worked at the UK's leading food safety research lab, the Rowett Institute. He became the target of world-wide media attention in August 1998, when he said on a BBC TV "World In Action broadcast" that he would not eat GMOs because they had been inadequately tested, and that it was very, very unfair to use our fellow citizens as guinea pigs." He cited fresh results from his own research.

This was truly a scientific bombshell. Pusztai's statement obviously threatened to derail the multimillion dollar PR campaign of the biotech industry to create public confidence in GMO foods. Reminiscent of the days of Galileo, a few days after his public appearance, and reportedly after a call from the office of British Minister Tony Blair himself, Pusztai was suspended until his date of retirement, his research team was disbanded, and its documents confiscated. What's more, Arpad Pusztai was hit with a gag order, banning him from speaking to the media under threat of legal action, so that he could not respond to any of the attacks that were being directed at him. Several years later, a British newspaper reported that Blair's interference was prompted by a call from Monsanto to President Bill Clinton, who then called Blair telling him to stop Pusztai. (Bill Clinton had once praised Monsanto in a State of the Union address and was spending billions in tax payer money to back the GMO industry).[83]

The Rowett Institute was apparently ordered to shoot the messenger by the heads of the US and British governments. The Institute changed its story several times about Pusztai, first claiming that he had used potatoes not intended to be used as food, then that he had mixed up the results with other studies, then that Pusztai was a confused old man (a whopping 68 years old at the time), that his potatoes weren't GMOs at all, and finally that his research had not been peer reviewed.

Pusztai then sent his results to 24 independent scientists throughout the world. They supported his conclusions and vouched for his intellectual prowess and lack of cognitive decline. Then a second review committee was appointed by the UK Royal Society. It again concluded that Pusztai's results were inconclusive and even flawed (many Royal Society members worked as consultants for biotech firms). A world leading British medical journal, the Lancet, found the judgement of the Royal Society "a gesture of breathtaking impertinence"[84]

But the damage to Pusztai's credibility had already been done. The news about his "mistakes" was effectively distributed all over the world. People were led to believe that there was no scientific basis for his warning about GMOs. It is important to note that the respected Rowett Institute had been starved of funding by the Thatcher government and they had become increasingly dependent on industry funding. It was later revealed that Monsanto had given the Rowett

Research Services a $224,000 grant prior to Pusztai's now notorious interview at BBC. Ironically, one year after the Pusztai interview, Andrew Chesson, the top scientist at Rowett Institute at the time, the same scientist who presided over the firing and gagging of Pusztai, said, "Potentially disastrous effects may come from undetected harmful substances in Genetically Modified Foods."[85]

Other examples of intimidation of scientists, from just the last few years, include Ignacio Chapela, an untenured Assistant Professor at Berkeley, whose paper revealing GM contamination of legacy maize in Mexico (Quist and Chapela, 2001) was damaging PR for the biotech industry that had just given Berkeley $25 million in research grants. An intensive internet-based campaign to discredit Chapela was launched, which turned out to be the work of the Bivings Group, a public relations firm specializing in viral marketing--and frequently hired by Monsanto (Delborne, 2008). But the worst was yet to come. Despite a favorable vote of 32 to 1 for Chapela to receive tenure, Berkeley's Chancellor, Robert Berdahl, denied it. Chapela made this response to the campaign and his tenure denial.

"I am living proof of what happens when biotech buys a university. The first thing that goes is independent research. The university is a delicate organism. When its mission and orientation are compromised, it dies. Corporate biotechnology is killing this university."[86]

Vicki Vance, a professor at the University of South Carolina, doesn't see genetically modifying plants as inherently dangerous, but after Monsanto attempted to muzzle her research, she now has a problem with mega corporations "allegedly" using their money, and power to cover up the risks of genetic technology. Vance has studied "small interfering ribonucleic acid molecules," siRNA, in plants for most of her career.[87]

Medical researchers are hoping to perfect a process for humans to assimilate tumor suppressor RNAs missing in cancer patients. If they can be replaced—an experimental treatment known as microRNA replacement therapy—then proliferation of cancer cells could be suspended. But leave it up to Monsanto to take pretty much the opposite approach. They want to use RNAs to manufacture pesticides, and insert them into crops to wipe out major pests like the corn root worm when they chew on those crops. In order to gain public acceptance of the idea, Monsanto is busy trying to prove that humans can't absorb siRNA from crops that they might eat.

Alas, a research team from China demonstrated that mice and humans can ingest small plant RNA, which then makes it to their blood via the intestines, and ultimately reach their cells, and that the RNA is capable of directing and interfering with mammalian genes, a phenomenon referred to as "trans-kingdom gene regulation."[88] They found that one particular molecule of RNA from rice could inhibit a protein that promotes the elimination of low-density lipoprotein, or "bad" cholesterol from blood. This would have earth-shaking implications for humans, indicating that eating foods containing modified RNA could aggravate heart disease and other health issues tied to cholesterol.

It appears that the muscle of the biotech industry prevented that research from being published in the major science journals, like *Science, Cell* and *Molecular Cell.* The principal author, Chen-Yu Zhang, was told their findings were just "too extraordinary." Apparently "extraordinary" is the new term for throwing a hand grenade on Monsanto's business model and causing their executives to run around with their hair on fire. With the help of Monsanto, studies began to be published that disputed Zhang's experiments.

Professor Vance ultimately got involved when she basically replicated Zhang's research but could find no takers when she tried to get her findings published, nor when she tried to get funding for more research. She was harassed by Monsanto lawyers and "uninvited to conferences" to speak, even after she was on the printed program.

Dr. Gilles Eric Seralini, a molecular biologist at the University of Caen, is perhaps the most well-known researcher to become the victim of coordinated campaigns of professional and personal harassment.[89] Seralini published a study in the *Journal of Food and Chemical Toxicology* showing severe organ damage, particularly to the liver, kidneys, and pituitary gland, in rats fed either GM corn and/or low levels of Roundup in their diet. Additional unexpected observations were higher rates of large palpable tumors and a 600% increase in mortality in most treatment groups compared to a control group.[90]

Following a familiar pattern of virtually all publications that demonstrate the potential health risk of GMO plants, the biotech industry went ballistic, and Seralini became an immediate target. It even led to an unprecedented retraction by the journal editors, citing the results as "inconclusive," and erasing this important study from the scientific literature. Furthermore, the retraction is being used to promote GM foods by throwing into doubt a study that raises legitimate concerns about GM food safety. Seralini was widely branded as a junk scientist, by such publications as *Popular Mechanics*, *Reason,* and *Forbes Magazine.*

After the retraction, and completely ignored by the media, over 180 scientists signed a letter to the journal condemning the retraction, calling it an "attack on scientific integrity,"[91] and demanding a reinstatement of the study and a full apology to Dr. Seralini. The letter stated that the retraction was "antithetical to scientific progress. It also sends an unmistakable message to other researchers exploring the potential risks of GM crops and their associated pesticides: to the effect that their scientific research may not be funded or published, that their professional and personal reputations could be attacked, and that their careers could be ruined. Moreover, erasing the evidence provided by the Séralini study could potentially put at risk the health of people, livestock, and the environment."

Ignored by the media as well was all the incriminating evidence that clearly taints the criticism. Before the retraction of the Seralini paper, a former Monsanto scientist was brought into the journal as biotechnology editor. The criticisms lobbed at the Seralini paper was that the proper strain of rats was not used and their numbers were too small. But the strain of rat was the one required by the FDA for drug toxicology, and the toxic effects were unambiguously significant.

Dr. Gilles Eric Seralini, is perhaps the most well-known researcher targeted by a coordinated campaign of professional and personal harassment by Monsanto

In fact, Monsanto published a similar study in the same journal eight years before using the same number and strain of rats. But their study was only for 90 days, a strategy that stacks the deck against finding any long-term harm. Industry-funded studies typically last only 90 days, which is too short a time period for most laboratory animals to manifest chronic or life threatening illnesses. Industry has absolutely no motivation to do longer studies, because if the results showed consequences, the entire industry's viability and fortune would evaporate almost overnight. That regulatory authorities allow 90 days to be the protocol for such studies is more evidence that the biotech industry has co-opted government agencies.

To absolutely no one's surprise, no harm was found. In contrast, the Seralini study continued for two years, and importantly, did not see any tumors until after nine months. Finally, critics threw in for good measure a personal attack that Seralini was biased against GMOs and couldn't be trusted. The loudest critics came from the Science Media Centre of the British Royal Institution. The Science Media Centre has a long history of quelling GMO controversies. Its funders include companies that are some of the largest producers of GMOs and pesticides, like BASF, Bayer, and Syngenta, a fact virtually ignored by the media that was more than happy to pile on Seralini for his "bias." The underlying message is that GMO critics are inherently biased, and GMO cheerleaders are saintly in their objectivity. Michael Hansen, senior scientist at Consumer's Union, explained in an interview with Steve Curwood on his program *Living on Earth*:[92]

> "Well, basically what Dr. Séralini did was he did the same feeding study that Monsanto did and published in the same journal eight years prior, and in that study, they [Monsanto] used the same number of rats, and the same strain of rats, and came to a conclusion there was no [tumor] problem. So all of a sudden, eight years later, when somebody [Seralini] does that same experiment, only runs it for two years rather than just 90 days, and their data suggests there are problems, [then] all of a sudden the number of rats is too small? Well, if it's too small to show that there's a [tumor] problem, wouldn't it be too small to show there's no problem?"

The illicit relationship between Monsanto and other Big Ag corporations and the FDA gives birth to corrupt corporate welfare. USDA spends around $2 billion every year on agricultural research, funding its own scientists and those at land-grant universities. Unfortunately, USDA prioritizes research dollars for commodity crops like corn and soybeans, which are the building blocks of processed foods and the key ingredients in factory-farmed livestock feed. This parallels corporate-funded research, prioritizing industrial agriculture, and specifically ignoring health and environmental consequences. Many USDA research projects become essentially a subsidy for Big Ag. Between 1994 and 2002, USDA funded more than 3,000 plant biotechnology studies: none investigated possible unintended toxins and only two examined potential allergens in GMOs.

The International Agency for Research on Cancer (IARC) located in Lyon, France, is an arm of the World Health Organization and the UN. The IARC is responsible for creating a public inventory of cancer-causing substances in the interest of public health protection, and whose scientific conclusions have been regarded above reproach for 50 years. Their working groups are typically about 20 scientists selected from a world-wide pool, specifically for their expertise and lack of conflicts of interest on a specific issue. IARC conclusions are based on studies published in independent scientific journals and specifically exclude confidential industry-funded research, in sharp contrast to the methods used by the EPA and other regulatory bodies that routinely use industry research.

After working on an evaluation of pesticides for over a year, on March 20, 2105, in a "shot heard round the world" the IARC working group announced their finding that several pesticides were probably human carcinogens. One of that those so classified was glyphosate, the bedrock of Monsanto's empire.

Upon hearing the news, Monsanto apologized to the world for causing so much misery and death, promised to clean up the global mess they had created, and announced compensation to everyone who has been harmed by Roundup in the last 40 years. Just kidding, of course. If your name is the title of document followed by word "Papers," that usually means you've done something really bad. The famed French journal, *Le Monde*, won a prestigious international award, for their "Monsanto Papers" where they exposed the company's real response, which was to release the Monsanto "Kracken."[93]

First, they issued this defiant public statement: "All labeled uses of glyphosate are safe for human health." Without even a pause for introspection, hesitation or the stirring of a conscience regarding the damage they are doing to the entire world's population, Monsanto demanded a retraction from the WHO.[94] Christopher Wild, IARC director, recently said, "We have been attacked in the past, we have faced smear campaigns, but this time we are the target of an orchestrated campaign of an unseen scale and duration." He was talking about Monsanto's response to the IARC's declaring glyphosate genotoxic (it causes DNA damage), carcinogenic to animals, and a "probable carcinogen" to humans. Using a "no holds barred" arsenal of public berating, false accusations, legal harassment and intimidation, intrusive demands, calls for Congressional investigations, spies posing as fake journalists, paid off academics and social media trolls, and arm twisting media outlets like Reuters, Monsanto essentially waged an all-out campaign to literally destroy all their scientific critics—the IARC, its scientists, and even loosely affiliated public institutions like the highly respected National Institute of Environmental Health Sciences (NIEHS) in the US, and the 180 scientists that are members of the Ramazzini Institute of Italy.

Common sense and reality was turned on its head. Monsanto was demanding that these international institutions, established for the protection of the public, must now be accountable to Monsanto. Monsanto shoved people onto the international stage that actually called for the IARC to be abolished. The closest Monsanto came to actual scientific exoneration of glyphosate came in the form of six articles that appeared in the journal, *Critical Reviews in Toxicology*. But that journal is openly funded by Monsanto, and the sixteen authors had been recruited by a crisis management consulting group that specializes in ginning up scientific reports that defend companies whose profits are threatened by regulatory or legal action.[95]

After the election of Donald Trump, the notorious lobbying organization, the American Chemistry Council, overtly joined Monsanto's war in launching the "Campaign for Accuracy in Public Health Research," a vicious PR and legislative initiative to silence glyphosate's scientific critiques and attack their funding.

The IARC, NIEHS, and Ramazzini Institute are three pillars of independent science whose work is critical to public health protection from environmental contaminants. It is a David vs. Goliath battle because these three institutions have no staff or money to defend themselves, only their integrity and their publicly funded research. Monsanto and the ACC freely spend millions on their weapons of war.

GMO's Broken Promises

Monsanto's GM seeds have to be purchased each year and require expensive and toxic pesticides, which the same company just happens to produce. It doesn't take the geniuses at *Forbes* magazine to figure out that if you own the rights to all the food grown everywhere, you literally rule the world. Monsanto is the Incredible Hulk of corporate aggression.

In pursuing this business model, Monsanto has managed to do more damage to the world's food supply and public health than any other single entity. Monsanto's products are grown on about 40 percent of the total crop acreage of US agriculture. You simply cannot avoid Monsanto's genetically modified food no matter how hard you may try.

Once Monsanto's products are introduced, it is virtually impossible to revert back to traditionally grown crops. Farmers must invest in the matching herbicide and pesticides, sign licensing agreements, and become liable for any natural spreading of the genetically modified traits.

Broad-based genetic and chemical contamination of all agriculture by GMOs and pesticide drift are eroding farmers' ability to grow truly organic food and consumers' ability to purchase it, regardless of price. Monsanto's ruthless business practices, high seed prices and vicious legal attacks have played a key role in the disappearance of small and medium size farms, bankrupting small farmers, and driving world agriculture further towards huge monocultures and into the controlling web of gigantic agribusinesses and food processing corporations.

Dr. Oscar Zamora, vice chancellor of the University of the Philippines at Los Banos, says: "For every application of genetic engineering in agriculture in developing countries, there are a number of less hazardous and more sustainable approaches and practices with hundreds, if not thousands, of years of safety record behind them. None of the GE [genetically engineered] applications in agriculture today are valuable enough to farmers in developing countries to make it reasonable to expose the environment, farmers and the consumers to even the slightest risk."[96] Exactly none of the supposed benefits of GMO crops—increased yields, more food production, controlled pests and weeds, reductions in chemical use in agriculture, or drought-tolerant seeds—have actually materialized. The Global Citizen's Report on the State of GMOs[97]points out that in fact the opposite has occurred. "Failure to Yield," a report by the Union of Concerned Scientists, has established that genetic engineering has not contributed to yield increases in any crop. Based on 293 test cases, research found that in developing countries, organic methods produced 80 percent higher yields than industrial farms.[98]

The International Assessment of Agricultural Knowledge, Science and Technology for Development (IAASTD) report, authored by more than 400 scientists and backed by 58 governments, stated that GM crop yields were "highly variable" and in some cases, "yields declined." The report noted, "Assessment of the technology lags behind its development, information is anecdotal and contradictory, and uncertainty about possible benefits and damage is unavoidable." They determined that the current GMOs "have nothing to offer the goals of reducing hunger and poverty, improving nutrition, health and rural livelihoods, and facilitating social and environmental sustainability."

Meanwhile, a recent study by the United Nations Special Rapporteur on the Right to Food reported that alternative, agroecological systems doubled crop yields over a period of three to 10 years in field tests conducted in 20 African countries.[99]

Even a 2014 USDA report declared, "There are no significant differences in yields of GMO and non-GMO crops." When asked about the USDA report, the chief technology officer for Monsanto declared, "American farmers are smart and wouldn't adapt a technology that didn't have tangible benefits."[100] Declaring the value and superiority of GMOs on the premise that American farmers are "smarter" than the science is like saying global warming isn't happening because humans are too smart to usher in their own self-destruction (see chapter 6).

Herbicide-tolerant (Roundup Ready) crops were supposed to control weeds and Bt crops were intended to control pest insects. But everything about GMOs is an aberration, if not a defiance of biological science. Non-Monsanto scientists warned it was biologically inevitable that genes for herbicide resistance would be transferred to weeds, leaving this herbicide strategy useless after a few years. And that is exactly what has happened. Instead of controlling weeds and pests, GMO crops have led to the emergence of super weeds and super pests.

Palmer's amaranth, or pig weed, is just such a super weed. It's the Godzilla of plant pests. It can reach eight feet in height, its stem can be sturdy enough to damage farm machinery and must be removed by hand. We could build Donald Trump's wall with it.

One hundred and thirty types of weeds in 40 states are now herbicide resistant, increasing costs, cutting yields, and leading to the use of more powerful and increasingly toxic chemical herbicides. Half of all surveyed farms, 60 million acres, are now infested with Roundup-resistant weeds. Andrew Wargo III, the President of the Arkansas Association of Conservation Districts said, "It is the single largest threat to production agriculture that we have ever seen."[101]

The USDA regulates GMO crops, but the EPA regulates herbicides, so there is no coordinated regulation of Monsanto's Roundup Ready system, and it is this system that is responsible for the recent explosion of glyphosate resistant weeds in the last few years.

Like a drunken gambler oblivious to the reality of the odds stacked against him, Monsanto is doubling down on this failed technology and now recommends to farmers even more toxic herbicides, leaving in their wake even more crop damage and widespread environmental destruction. Monsanto is even paying farmers now to start using dicamba and 2,4-D, even more toxic chemicals than Roundup. The latter is one of the original ingredients in Agent Orange. And, predictably, Monsanto has now developed GMOs that are resistant to dicamba and 2,4-D, hoping to perpetuate the entire failed, but very profitable strategy. Already one fourth of the planted US soybean crop in 2017 came from dicamba-resistant GMO seeds.

Again, predictably, the use of these new toxins is a looming disaster. In 2107, dicamba drift ruined 4 percent of the US soybean crop, 3.6 million acres, according to the EPA. In addition, 2,708 complaints of destructive dicamba drift have been filed in 25 states, from Mississippi to Minnesota, over damage to tomatoes, watermelon, cantaloupe, vineyards, pumpkins, organic vegetables, residential gardens, fruit trees, shade trees, wild flowers and shrubs. Most incidents, however, went unreported, and the problem was likely even more wide spread.[102]

Not surprisingly, pollinator insects have been affected by the new pesticides, and honey production is down sharply in areas where dicamba was sprayed (see Chapter 9). All this has led to heated, bitter arguments among farmers, and at least one killing: a farmer now charged with murdering his neighbor.

This is exactly the chemical arms race against pests that was widely predicted by critics of biotechnology. The Union of Concerned Scientists points out that the use of multiple herbicides will speed up even further the continued evolution of weeds resistant to chemicals. It is another stroke of irony and hypocrisy that Monsanto denigrates GMO opponents as "anti-science" simpletons, but persists in a business model that is itself, pure anti-science ideology.

The same path blazed by chemical-resistant weeds is now being followed by chemical-resistant insects. By 2008, GM crops in the US were requiring 26 percent more pounds of insecticides per acre than acres planted with conventional varieties. [103, 104, 105]

In Brazil, China, and Argentina, the introduction of Bt toxin and Roundup Ready crops have resulted in the increased use of insecticides, in some cases as much as 300 percent.[106]

Monsanto claims that only pest insects are affected. But studies have shown that bees, butterflies, and ladybugs are adversely affected, and that, not surprisingly, soil micro-organisms, necessary for overall soil health and crop production, decline in the presence of Bt toxin. If those soil micro-organisms decline when smothered by lethal chemicals, then it may not be much of a surprise to find that the quality of the crops might be reduced. Professor Seidler points out that soybeans harvested from organic farms had higher concentrations of protein and essential amino acids, and higher concentrations of two minerals, and no Roundup residues (Food Chem. 2014).[107]

Furthermore, already eight species of pest insects have developed resistance to Bt toxins, either in the field or laboratory, including the diamond back moth, Indian meal moth, tobacco budworm, Colorado potato beetle, and two species of mosquitoes. The genetically engineered Bt crops continuously express the Bt toxin throughout their growing season. Predictably, long-term exposure to Bt toxins promotes eventual resistance in insect populations, and by incorporation of the toxin into every part of the plant, that degree of exposure will accelerate that resistance.

Monsanto's Behavior Overseas

Arguably Monsanto's business practice overseas has rivaled their biologic transgressions. In 2012, Argentina's tax agency, AFIP, raided a Monsanto contractor and found what they described as "slave-like" conditions among workers in its cornfields.

The agency says the contractor, Rural Power SA, hired its farmhands illegally, would not allow them to leave the fields, and withheld their salaries. These "slave" workers were forced to work in the cornfields 14 hours per day and buy their own food at the company store at inflated prices. The AFIP says it holds Monsanto responsible for its contractor's inhumane labor conditions.[108]

Monsanto's aggressive presence in India has been particularly devastating. The world bank forced India to open the country's seed and agricultural sectors to global corporations, and in turn, it increased demand for Indian agricultural products. The end result was that India began focusing on export-oriented cash crops, especially cotton. Seizing on this opportunity, Monsanto introduced its GMO insect-repelling Bt cotton seeds in 2002, and within 8 years it controlled most of the Indian cotton production. More specifically, Monsanto gained control of 95 percent of the cotton seed market, and farmers saw their seed prices jump 8,000 percent.

Indians farmers have never seen the much-hyped benefits of Bt cotton. Monsanto claimed yields would be 1500 kg per acre, but the actual yield averaged only a third of that. Moreover, the seeds cost twice as much as regular seeds and require much more water, all at a time when the climate crisis is making precipitation more undependable. Bt cotton was engineered to be resistant to bollworms, but the worms have developed resistance, and pesticide use has increased dramatically, in some cases 13 fold.[109]

Now Monsanto is selling Bollgard II with two additional toxic genes in it. New pests have emerged, and farmers are using more pesticides. And the cycle repeats itself. Furthermore, the farmers have to buy the seeds every year. Increased debt, soaring seed prices, and global competition have combined to create a crisis for India's small farmers.

Monsanto has played a key role in a growing epidemic of suicides among small farmers in many countries but especially India. In the past 16 years, well over 270,000 small farmers have committed suicide,[110] most of them small cotton farmers where Monsanto controls

95 percent of the cotton seed and profits off of suing farmers and trapping them in debt.[111]

"We are ruined now," said the 38-year-old wife of one farmer suicide victim. "We bought 100 grams of BT Cotton. Our crop failed twice. My husband had become depressed. He went out to his field, lay down in the cotton and swallowed insecticide."[112] The suicide epidemic has been labeled "GM Genocide" in India.

Presently 5 million Brazilian farmers are locked in a lawsuit with Monsanto, suing the biotech giant for $7.7 billion for charging the farmers royalties if they plant Monsanto seeds from the previous year's harvest, a practice dating back to the dawn of human agriculture. Monsanto is trying to charge the farmers an ongoing 2 percent per year royalty for reusing Monsanto's patented seeds.

The US war on drugs has been widely panned from just about every angle. But the environmental devastation it has wrought upon other countries has received little attention. In Colombia, Monsanto has received $25 million from the US government for providing Roundup Ultra in the anti-drug fumigation efforts of "Plan Colombia." Roundup Ultra is a highly concentrated version of Monsanto's glyphosate, with additional ingredients to increase its lethality. Colombian communities and human rights organizations have charged that the herbicide has destroyed food crops, water sources, and protected areas, and has led to increased incidents of birth defects and cancers.

Dr. Andres Carrasco, a well-respected embryologist in Buenos Aires, worked for nearly thirty years in embryonic development, and was President and Assistant Secretary of Conicet (Brazil's National Commission for Scientific Research). He published a study in 2009 showing that Roundup, at much lower doses than those commonly found on crops, caused deformities in chicken embryos that resembled the kind of birth defects being reported in areas like La Leonesa, Argentina, where Argentina's "Big Ag" heavily uses glyphosate to treat GMO crops. Dr. Carrasco has been visited at his laboratory by groups of men refusing to identify themselves, physically threatening him and demanding to see details of his research.

On August 7, 2010 Carrasco was scheduled to give a lecture about his study, but a mob of about 100 people attacked Carrasco and his delegation preventing him from ever reaching the school where the lecture was to take place. Dr. Carrasco and a colleague locked themselves in a car as the mob yelled threats and beat on the vehicle for two hours. One delegate was hit in the spine and became a paraplegic. Another person received blows to the head requiring treatment. A former human rights official was knocked unconscious. Witnesses said the angry crowd was the work of local officials and agribusiness bosses, and local police deliberately refused to intervene, according to Amnesty International.[113] The Argentine government had published a report[114] earlier in 2010 showing that over the last ten years of intense agricultural expansion and increased herbicide use in La Leonesa, the very place Carrasco was attacked, there had been an increase of birth defects of nearly 400 percent and childhood cancers had increased 300 percent.

After the 2010 earthquake in Haiti, Monsanto appeared to seize on what they thought was an opportunity to infiltrate Haitian agriculture. In May 2010, Monsanto announced it was making a $4 million gift of seeds "to support the reconstruction effort" in Haiti and provide seeds to Haitian farmers who were in need of seeds. This smelled a little like altruism, something that many observers would not have thought possible at Monsanto. But, Haitian farmers already had enough seed to plant that season, according to several reports. The Catholic Relief Service in Haiti concluded that the seeds were not needed and advised against accepting them.

The first batch of Monsanto seeds did arrive in Haiti, but they were preceded long before by Monsanto's reputation. The seeds had been treated with fungicides and herbicides considered so dangerous that even a lax EPA had advised that workers should wear special protective clothing when handling them. Some bags had labels warning that consumption of the seeds was dangerous for humans and animals. Monsanto's passing mention of these toxins to Haiti's Ministry of Agriculture officials in an email contained no explanation of the dangers or any offer of special clothing or training for those who would be farming with the toxic seeds.

The Papaye Peasant Movement, one of Haiti's farmer associations, condemned the "gift" and even held a demonstration on June 4, 2010, where thousands marched and symbolically burned the Monsanto seeds. The leader of the Peasant Movement of Papay, Chavannes Jean-Baptiste, and Father Jean-Yves Urfié, a former chemistry professor, saw the donation as a plan to get peasant farmers to continue buying more expensive hybrid seeds and institute large-scale agribusiness in Haiti. Urfié said he would like Haiti to send a "clear sign to farmers all over the world" by being the first country to "stand up to Monsanto and say: We don't want Monsanto here, because it has already proved that it is dangerous."[141]

Monsanto Infiltrates Every Branch of Government

Corporate capture of the very federal regulatory agencies designed to contain corporate malfeasance has been written about numerous times. In a piece for *In These Times*, Elizabeth Grossman and Valerie Brown give just a taste of the details of how "glyphosate is a clear case of 'regulatory capture' by a corporation acting in its own financial interest while regulatory agencies dismiss disturbing evidence, or leave serious questions about public health unanswered or ignored."[115]

Monsanto has also enjoyed an especially disturbing, politically cozy relationship with the EPA, successfully silencing whistleblowers within the EPA dating back to the early 1990s and continuing to the present day. E.G. Vallianatos, an EPA employee for twenty-five years, through five different administrations—Republican and Democratic, has written a book about how badly industry has co-opted the regulatory functioning of the EPA, *Poison Spring: The Secret History of Pollution and the EPA*. He writes:

> "With highly placed industry appointees in both the White House and Congress, chemical and agricultural giants essentially control the actions of the EPA. Here's how it works. Corporate lobbyists meet almost daily with EPA scientists and managers, muscling their science and pressuring them to see the world through industry eyes. This task is made easier when, as often happens, industry's chief lobbyists are former EPA political appointees, and senior EPA officials are former industry heavyweights . . . With industry constantly browbeating Congress to shrink the size of the EPA, it becomes logical for EPA managers to encourage agency scientists to think of their own well-being first, trumpeting the economic benefits of new chemicals and downplaying worries that might prevent a new product from reaching the market . . . EPA scientists quickly learn that challenging the corporate agenda can bring career-ending payback—their decisions would be questioned, their promotions and careers would be put at risk . . . The EPA's

> industry bosses handpick the scientists to collect data or— inside the agency—to adopt industry information and rubber-stamp it as government policy. A product labeled 'EPA Approved' thus loses any real integrity."[116]

For the purposes of federal regulation, a Bt toxin crop is not a food but classified as a pesticide and therefore falls under the jurisdiction of the EPA. The scientific, if you can call it that, justification for allowing this is the concept of "substantial equivalence," i.e., GMOs are substantially equivalent to their natural counterparts. My wife says that I am the "substantial equivalent" of a real husband. Substantial equivalence is an absolutely meaningless statement meant to sound scientific. It's as much gibberish as Hans Solo telling Chewbacca to "turn on the frivoliter bolsters and activate the force field." Moreover, it isn't true in any sense. Organic soybeans have been shown to harbor a healthier nutritional profile than GMOs—specifically more antioxidant phytochemicals, more glucose, fructose, sucrose, maltose, total protein, and zinc, and less total saturated fat. The only knock was they also contained less fiber.[117]

The FDA operates with at least as much, if not more dysfunction that the EPA. The supposed safety of genetically modified food is based only on a claim made by the FDA that they were not aware of any information that GMOs were different in any substantive way from their natural counterparts. That scornful claim and its back story deserve to be stripped naked and laid bare in the public square for all to see. The supposed "safety" of GMOs is purely a triumph of dark political maneuvering, not an expression of anything that can be called science.

In May 1998, the Alliance for Bio-Integrity led an unprecedented coalition of public interest groups, scientists, and religious leaders in filing a landmark lawsuit against the FDA to obtain mandatory safety testing and labeling of all GMO foods. From 44,000 pages of evidence,[118] it is clear the FDA's own records reveal it declared GMOs to be safe despite broad disagreement from its own experts -- all the while claiming an overwhelming scientific consensus supported its stance. Moreover, FDA scientists had urged their superiors to first require long-term studies because of real concerns about GMO safety, but those never happened.

The FDA's public posture was essentially the end result of a political shakedown started during the Reagan Administration and continuing through Clinton's, where the FDA was told to cooperate with the agenda of the biotech industry. Henry Miller worked at the FDA from 1979 to 1994, and was in charge of biotechnology during that time. According to Miller, "US government agencies have done exactly what big agribusiness has asked them to do and told them to do." How did Monsanto and the biotech industry emerge as master puppeteers over federal government agencies?

In late 1986, four Monsanto executives paid a visit to Vice President George H. W. Bush at the White House, making a cunning pitch—"We (the biotech industry) need to be regulated." Government guidelines, the executives reasoned, would reassure a public that was growing uneasy, if not skeptical, about the safety of the radical new science of genetic engineering.[119]

In the months that followed, the Reagan Administration did exactly what was asked. The FDA, EPA, and the USDA were all used as props for the biotech industry through the next three White House Administrations. The FDA's motivation was so strong that it not only disregarded the warnings of its own scientists about the unique risks of GMOs, it covered them up and claimed that no such input had been received. The FDA then created a special position, Deputy Commissioner, to oversee GMOs and appointed Michael Taylor, Monsanto's former attorney to the position. That's like having a tobacco executive crafting the government's regulations on cigarettes. Rather than clean house on this shameful industry/regulator incest, the Obama Administration institutionalized it by making the same Michael Taylor the US Food Safety Czar.

Under Taylor's influence, warnings from FDA scientists were persistently overridden, drafts of FDA policy increasingly contradicted the scientists' assertions, and the FDA took the position that no safety testing was necessary. The "substantial equivalence" concept was attacked by numerous agency experts asserting that it ignored the recognized potential for bioengineering to produce unexpected and unpredictable toxins, carcinogens, and allergens, hazards not ordinarily involved with conventional breeding.

I would love to use the Trojan horse metaphor here, but Monsanto operatives infiltrating government agencies didn't happen by stealth, under cover of darkness; it happened in broad day light. When dairy producers who did not use Monsanto's products began to label their products as "Organic" and "Hormone Free," Taylor quickly went to work and they were greeted by a two-pronged Monsanto attack. Taylor tried to prevent them from doing so from within the walls of the FDA, and Monsanto slapped them with a lawsuit, claiming that those labels amounted to negative advertising against hormone-produced milk.[120]

In 1994 Taylor wrote a guidance document from within the FDA requiring that any food label describing the product as bovine-growth-hormone-free must also include these words: "The FDA has determined . . . no significant difference has been shown between milk derived from [BGH] and non-[BGH] supplemented cows."

The numerous FDA in-house critiques are summed up by Dr. Linda Kahl, who lambasted the agency as "trying to fit a square peg into a round hole . . . [by] trying to force an ultimate conclusion that there is no difference between foods modified by genetic engineering and foods modified by traditional breeding practices." It is little wonder that GMO critics cynically refer to the FDA as the "Fraud and Death Administration." Dr. Samuel S. Epstein (see Chapter 2), professor emeritus of environmental and occupational medicine at the University of Illinois at Chicago School of Public Health and Chairman of the Cancer Prevention Coalition said, "I've seen first-hand how Monsanto and the FDA resorted to scientific deceit of the highest order to market genetically engineered milk."

See no evil, hear no evil, speak no evil, is certainly how the FDA has treated GMOs. FDA spokespersons often claim that no GMO food has ever harmed a consumer. Never mind that they could not possibly know that. But the government routinely distorts the fact that in 1989, a GMO food likely killed dozens of Americans and permanently crippled over 1,500 others. That product (a supplement with the amino acid L-Tryptophan) contained at least one unusual toxin never found in conventionally produced batches. Many experts think the bioengineering process was the most likely cause of the toxicity.[121]

The end result of the FDA taking Monsanto under its wing is that the FDA essentially relegated responsibility for assessing the safety of GMO foods to the industry itself. The biotech companies literally get to decide when or whether to consult with federal agencies and also what scientific data to submit. In essence, people who make GMOs are the same people who test GMOs for safety, says Bruce Blumberg, PhD, a developmental and cell biologist at the University of California, Irvine. "Americans think the FDA and EPA are testing GMOs and making them safe, but that's simply not true."[142]

Typical of the EPA's capitulation to Monsanto is their ruling of July, 2013, to permit an increase in the allowable concentrations of glyphosate in the food you eat.[122] No miraculous new science had emerged showing glyphosate was safer than they thought. All the science showing the opposite was still being ignored. The only thing that changed was that Monsanto needed higher concentrations so that Big Ag would have no problem using even more Roundup to fight the emergence of resistant weeds and to utilize the previously mentioned Death Star harvesting method. And we're not talking a small increase. The amount of allowable glyphosate in oilseed crops such as flax, soybeans, and canola will be increased from 20 parts per million (ppm) to 40 ppm, which is over 100,000 times the amount shown to induce the proliferation of breast cancer cells in vitro. The EPA is increasing limits on allowable glyphosate in other food crops from 200 ppm to 6,000 ppm. A study published in June, 2014 showed that just prior to being harvested, GM soy raised in Iowa had very high levels of glyphosate, almost double the levels that Monsanto itself said in 1999 would be considered extreme.[123]

Whenever Monsanto and Big Ag wanted to sell and use more Roundup, the EPA magically found a way to accommodate that by increasing the supposed "safe level" for human consumption. This kind of incest between Monsanto and the EPA did ensure "safety" of Monsanto's bottom line, but certainly not your health.

Virtually every branch of the US government, including the Supreme Court and the World Bank, has acted as Monsanto's handmaiden, often times using taxpayer money to do so.[124] The Department of State has hosted meetings to discuss the merits of GE technology in target countries in recent years. Cables from embassies in China, Hungary, Ukraine, France and even the Vatican show a relentless drive to convince those countries of the benefits of GE crops, whether or not the countries themselves want or need it.

Michael Taylor is hardly the only key government official to also wear a Monsanto hat. The Monsanto "revolving door" is spinning out of control, as many former employees of the biotechnology industry are now working in government posts or have become official advisors to governments. The regulated have become the regulators, with predictable results. William Ruckelshaus, former head of the EPA under Nixon and Reagan, once said, "At EPA you work for a cause that is beyond self-interest...You're not there for the money, you are there for something beyond yourself." But on leaving the EPA, he himself became a Monsanto director.

Marcia Hale, assistant to President Clinton, Michael Friedman, Commissioner of the FDA under Pres. Obama, Mickey Kantor, Secretary of Commerce and US Trade Representative under President Clinton, Linda Fisher, assistant administrator at the EPA, Suzanne Sechen, a primary reviewer for bovine growth hormone in the FDA—all have enjoyed revolving door careers at Monsanto. The head of the main research arm for US government agricultural research formerly worked for Danforth Plant Science Center (funded by Monsanto), and a former Monsanto employee is on the government committee tasked with legalizing GM salmon.

Supreme Court Justice Clarence Thomas is a former Monsanto attorney. Thomas wrote the majority opinion in the 2001 Supreme Court decision J.E.M.Ag Supply, Inc. v. Pioneer Hi-Bred International, Inc., which found that "newly developed plant breeds are patentable under the general utility patent laws of the United States." This case benefitted all companies which profit from GMOs, of which Monsanto is the largest.

In the US, it is now standard practice for biotechnology firms to employ former members of Congress and Congressional and White House staff to give the industry an inside track. In Argentina, representatives from biotechnology corporations Monsanto, Syngenta, Bayer, Dow, and Pioneer sit on a prominent national panel that directly advises the government agency that decides about the applications that these same companies submit.

One of the most blatant examples of government and Monsanto incest came in the form of the "The Monsanto Protection Act," of 2017, so labeled by Food Democracy Now. The Republican-controlled House passed an amendment to a spending resolution that instructed the Secretary of Agriculture to allow GMO crops to be cultivated and sold even when the courts had found they posed a potential risk to farmers of nearby crops, the environment, and human health.

Destroying the biosphere

When I was a boy, I remember road trips with my father were punctuated by a steady splattering of insects on our windshield. You literally had to stop at least every two hours to clean off your windshield. There were bugs splattered all over the front grill of the car. Those days are gone, but why should you care? Isn't that a good thing?

"Three-quarters of flying insects in nature reserves across Germany have vanished in 25 years, with serious implications for all life on Earth, scientists say," was the lead to a story in *The Guardian* newspaper in October, 2017.[125]

The decline in pollinators and butterflies has received a fair amount of media attention. But this study shows the staggering magnitude of loss of all insects. And the study[126] was not done over agricultural fields, but in nature reserves. While some aspects of climate change could contribute to the decline, global warming should be more favorable to insects, making the decline even more ominous. Pesticides are overwhelmingly the primary suspects. Another study from Germany published one day later showed a 15 percent decline of bird populations in just the last 12 years.

On February 11, 2019, the lead story on Huffington Post carried this headline: "Insects Are Dying En Masse, Risking Catastrophic Collapse of Earth's Ecosystems." The most extensive review[149] of insect populations ever done, found a "shocking" decline in global populations. Over 40% of insect species are threatened with extinction. The total mass of insect populations is declining 2.5% annually, causing the authors to warn that, "If insects species losses cannot be halted, this will have catastrophic consequences for both the planet's ecosystems and for the survival of mankind."[148] The main contributors to the decline are habitat loss from intensive agriculture, climate change, and global chemical contamination."

This may come as a surprise to devoted viewers of reality TV, but the food we eat actually originates in soil, not on grocery store shelves. Perhaps even more basic to the ecosystems that support human life are the microorganisms in soil, critical to soil health and therefore agricultural production. The Soil Association is drawing attention to emerging science that glyphosate may be altering the normal balance of soil microorganisms, like bacteria and fungi, favoring those that are detrimental to plant health, and increasing crop vulnerability to pathogens.[145] Research is even stronger showing that Roundup is toxic to earthworms, a harbinger of and major contributor to soil health.

Monsanto's entire business platform was built in total disregard for its effect on virtually all life forms--from microorganisms, to insects, to birds, to wildlife, to livestock, and to humans. Echoing Rachael Carson, Prof Dave Goulson of Sussex University, UK, and a contributor to the above-mentioned bird population study said, "We appear to be making vast tracts of land inhospitable to most forms of life, and are currently on course for ecological Armageddon. If we lose the insects then everything is going to collapse."[146] And Monsanto couldn't care less.

Post Script

At the time of this writing, Monsanto has just lost the first of thousands of lawsuits filed for causing cancer. In this case, the victim was a grounds keeper, Dewayne Johnson, who contracted non-Hodgkin's lymphoma after using Roundup on a daily basis for years. The jury awarded Johnson $39 million for compensatory damages, and $250 million in punitive damages. A judge reduced the award to $78.5 million, a sum accepted by Mr. Johnson. Monsanto has also just been bought by chemical giant Bayer (see Chapter 9), who will drop the Monsanto name in hopes of cleansing the giant public relations stain that has come to define that name. Don't expect that to transform their behavior into a beacon of humanitarianism. Nor can you expect all the problems endemic to GMOs, industrial agriculture, and the chemical contamination of our food supply, will disappear with the stroke of a pen.

On May 25, 2013, two million people worldwide took part in a "March Against Monsanto." "Millions Against Monsanto" is now an ongoing campaign of opposition to the biotech giant.

On June 4, 2010, 10,000 Haitian farmers marched in protest of Monsanto's "charitable" donation of GMO seeds.

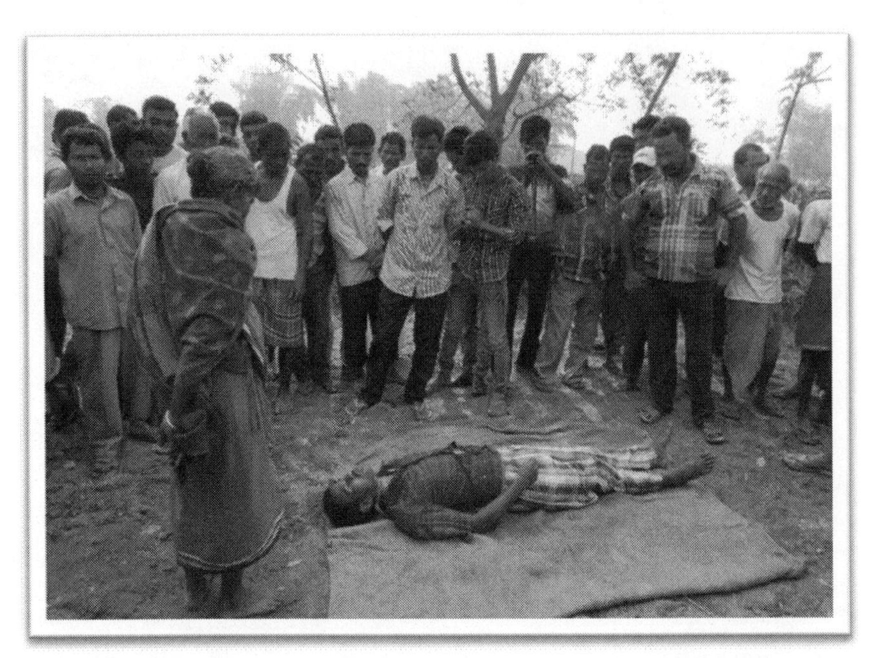

Suicides among farmers in India have become an epidemic. Despite denying any responsibility, Monsanto's business practices and GMO products have likely contributed to the tragedy..

Monsanto's notorious Agent Orange herbicide was sprayed all over Vietnam in the Vietnam War. Now Monsanto is producing GMO crops resistant to its toxic ingredient 2,4-D, because Roundup Ready GMOs have created such widespread weed resistance.

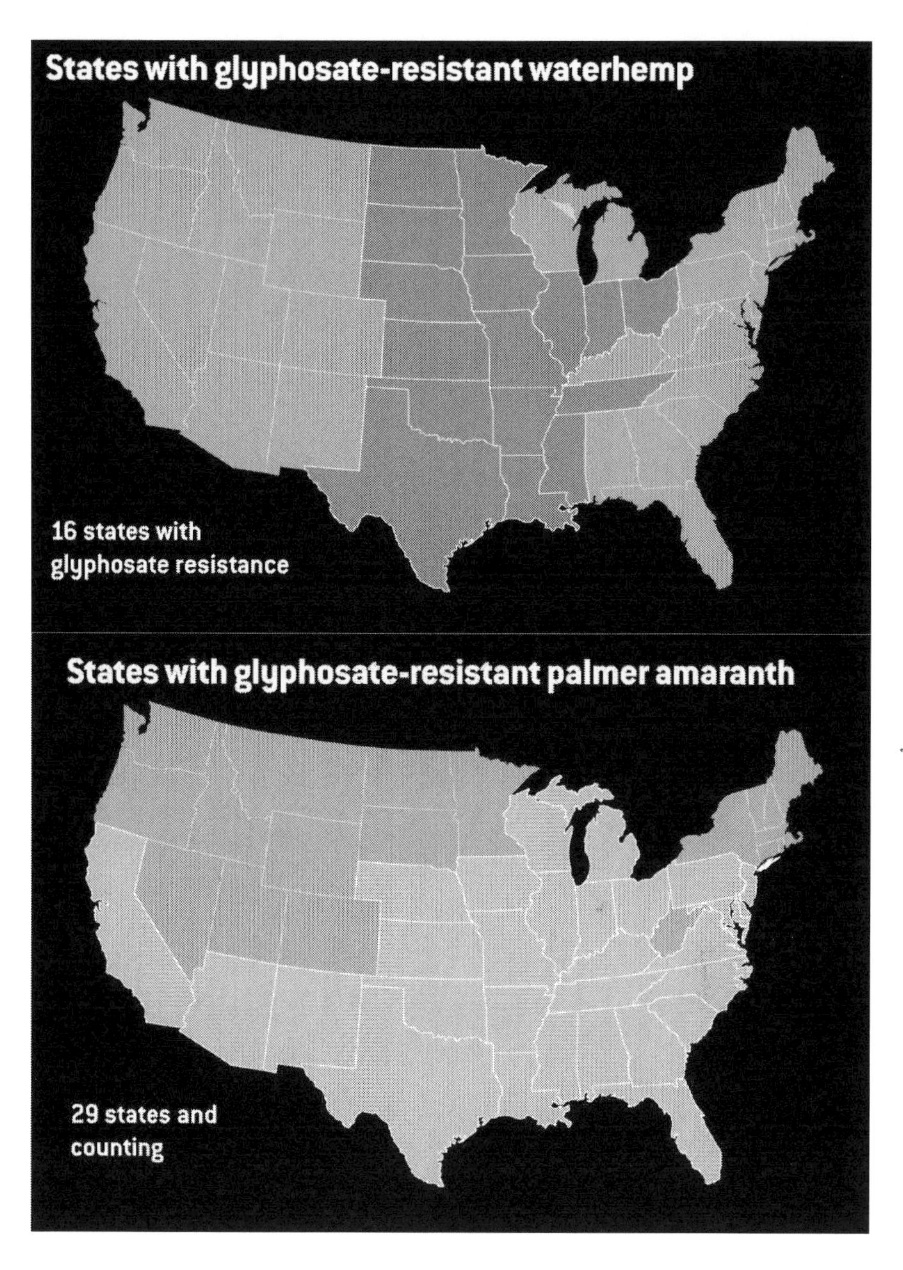

Roundup resistant weeds will infest all U.S. crop land by 2020.

9

The Killing Fields:
A Eulogy for the Bees

"You will probably more than once have seen her fluttering about the bushes, in a deserted corner of your garden, without realising that you were carelessly watching the venerable ancestor to whom we probably owe most of our flowers and fruits (for it is actually estimated that more than a hundred thousand varieties of plants would disappear if the bees did not visit them), and possibly even our civilisation, for in these mysteries all things intertwine."

—Maurice Maeterlinck
Nobel Prize winner for Literature, 1911
From *The Life of the Bee*

On June 15, 2013, ironically at the beginning of National Pollinator Week, 55 trees in a Target Shopping Center parking lot in Wilsonville, Oregon were sprayed to kill aphids with a pesticide named "Safari" -- a nice, outdoorsy, "fun" kind of name. Who doesn't want to go on a Safari? Shortly after, it was noted that it wasn't just the aphids that were no longer having that much "fun," neither were over 55,000 bumblebees, representing more than 300 wild colonies. It was the largest known bumble bee die off in US history.[1]

The commercial sprayer was fined some pocket change, $550, a penny per bumble bee. The chemicals he had used, called neonicotinoids, are typically applied to seeds before planting, allowing the pesticide to be taken up through the plant's vascular system as it grows. The chemical is then expressed in the pollen and nectar of the plant, and hence the danger to bees and other pollinating insects extends throughout the life of the plant and beyond.[2]

In April, 2014, over a four-day span, millions of bees were found dead outside their hives in Delaware, Fairfield, Hardin, Miami, Pickaway, and Ross Counties, Ohio. Beekeepers suspected neonicotinoids because the deaths occurred at roughly the same time corn was being planted using seeds coated with the chemicals.

Bees have been around for about 60 million years. Albert Einstein is often credited with a statement like, "If the bee disappeared off the face of the earth, man would only have four years left to live." Although Einstein may not have actually said that, and he was not an entomologist, the importance of pollinators to modern agriculture is difficult to overstate. Seventy-five percent of flowering plants depend on pollinators. Humans will not likely survive a total collapse of the bee population, but we are headed there. Eighty-seven of the top human food crops, which is about 75 percent of all our crops suppling about 90 percent of the world's nutrition, are pollinated by bees.[3] In fact the two most important human foods could not exist without pollinators—we're talking chocolate and coffee, obviously.

Exposure to multiple pesticides can have a synergistic toxic effect, far more than just additive. Samples of beeswax, pollen, and hives have shown toxic brews including as many as 121 different pesticides in detectable amounts. Some of the individual insecticides were found at concentrations higher than the median lethal dose for bees.[4] Bees are also plagued with gut parasites, but the pesticides made the bees more vulnerable to the parasites.[5]

Human beings have subconsciously "fabricated the illusion that in the 21st century they have the technological prowess to be independent of nature. Bees underline the reality that we are more, not less, dependent on nature's services in a world of close to 7 billion people," said UN Environment Programme (UNEP) Executive Director, Achim Steiner.[6]

If you google a phrase about whether bee colonies are disappearing, one of the first entrees that will appear is the site run by the American Council on Science and Health (ACSH). That sounds like a credible, authoritative body, or at least people who play one on TV. According to them, the problem isn't that bees are suffering from Colony Collapse Disorder (CCD), the problem is science journalists are reporting on Colony Collapse Disorder. And everyone knows if you don't report on something, then it doesn't really exist. In other words, all the angst about bees disappearing is just "Fake News" drummed up by environmental groups as an excuse to solicit donations. However, it turns out that ACSH is basically a corporate front group funded by a "Who's Who" of industrial agribusiness, fast food giants, pharmaceuticals, the fossil fuel industry, and Big Tobacco.[61]

Non-ACSH sources indicate that CCD began to appear in 2007. Since then commercial bee keepers have been reporting an annual winter loss of about 30 percent of their hives on average. Beekeepers, scientists, federal regulators, and the media initially pointed the finger at climate change, poor nutrition, fungus, cell-tower radiation, mites, and viruses. The bees were abandoning their hives, acting disoriented, then just disappearing. Everyone was looking for the culprit under every rock but perhaps the most obvious, like the emperor with no clothes. A new class of pesticides, the neonicotinoids, had just been introduced that systemically infiltrated the entire plant from seed to flower, and attacked the nervous system of insects with devastating efficiency. Neonicotinoids are synthetic derivatives of nicotine, the well-known toxin produced by the tobacco plant. They surged in popularity about the time CCD appeared. It didn't take an Albert Einstein to predict that they would be as toxic to beneficial insects as they were to pest insects.

Although not the only factor, neonicotinoids, now the most widely used insecticides in the world, have risen to the top of the list of contributors to CCD, and the European Union has banned their outdoor use in an effort to protect bees. So far, nothing has been done by the EPA in the US. To find out why, one needs to look no further than a corporate balance sheet. Neonicotinoids are very profitable. Bayer, one of the main manufacturers, makes over $1.25 billion a year on them.

Neonicotinoids dissolve readily in water and migrate into rivers and streams from agricultural runoff. Much to the surprise of researchers, they have been found to emerge again in flowers of plants miles away from the farm where the pesticides were externally applied or were embedded in the crop seeds. The pesticides remained throughout the growing season,[7] exposing the bees for four continuous months.[8]

In fact, these pesticides can persist in the soil for as long as six years after a single application, opening the door for non-treated plants to absorb the toxin. Neonicotinoids, like many other chemicals are metabolized in the plant over time, and some of the metabolites are even more toxic to bees than the original compound.[9]

In a pivotal study, bees were exposed to levels of the pesticides that are found near farm fields where they are used, and then to decreasing concentrations mimicking real life conditions when rain washes away the pesticide. This exposure decreases the life expectancy and reproductivity of the bees, impairs their ability to clean dead or sick bees out of their hive, and causes hives to lose their queens. A hive losing its queen is as much as catastrophe as losing your queen in a game of chess.

The study also showed that the toxicity to bees created by the pesticide was essentially doubled in the co-existing presence of a commonly used fungicide.[10] According to the Xerces Society for Invertebrate Conservation, bumble bees and solitary bees appear to be even more vulnerable to neonicotinoids than honey bees.

David Vogel, professor at the Haas School of Business and in the Department of Political Science at the University of California, Berkeley, points out that between 1960 and 1990, American health, safety, and environmental regulations were more stringent, risk averse, comprehensive, and innovative than those adopted in Europe. Vogel's book, *The Politics of Precaution*, explains that since around 1990, global regulatory leadership has shifted to Europe.[11] With many types of environmental risks, extreme conservative ideologues in the US have brought legislative protection of public health to a screeching halt. America's failure to deal with the climate crisis is probably the most conspicuous casualty (see Chapter 6). But what is happening to the brains of our children from environmental toxins may be just as important and may be related to the disappearance of bees. More about that shortly.

Since 1990, Europe and the United States have gone in two different directions on environmental regulation of toxic substances. Europe now errs on the side of safety, the U.S. errs on the side of corporate profits. Where Europe requires chemicals to demonstrate safety before release and is willing to take products off the shelf at a relatively low threshold of evidence, the United States does virtually the opposite. While that has been true under both Democratic and Republican Administrations, it is especially true of the EPA under Donald Trump. The EPA basically assumes all products are safe and will only withdraw products after they have been conclusively proven guilty of serious harm—obviously long after the fact.

That is not to say that European governments don't protect powerful corporations. The European Commission published a study of overall bee health and mortality rates in the UK and Europe[12] in April, 2014. Professor David Goulson, a biologist at the University of Sussex ,observed, "It does seem odd that the EC spent over €3m on a project on bee health and the words pesticide and insecticide are not used once in the document." "Odd" is certainly one way to describe it. "Sham" would be an even better way.

The chemical companies Bayer, Syngenta, BASF, Dow, DuPont, and Monsanto (Chapter 8) have taken full advantage of this regulatory laxity and waved a smoke screen in front of the bee calamity, claiming the mystery cannot yet be solved, and not even precautionary action can be taken. They say the mounting science that implicates their product is "faulty," just like the supposed "junk" science that exposed tobacco, asbestos, lead, and human caused global warming. But responding to the backlash against their bee poisons, to prove their objectivity and concern, Monsanto, Bayer, and Syngenta have reinvented themselves as "bee stewards." They've even gone so far as to purchase the leading bee research firms, ostensibly to study colony collapse disorder and engage in other bee research. For example, Monsanto recently bought Beeologics, a company whose primary goal was to find a solution to the Colony Collapse Disorder. Again, one does not need to be a descendant of Albert Einstein or even his maid to predict how that will turn out.

Monsanto hosted a "Bee Health Summit," at their St. Louis, Missouri, headquarters in June, 2103, and invited commercial and hobby bee keepers, researchers and Monsanto scientists. To be sure, some smart and dedicated scientists and committed bee keepers participated. However, Monsanto is clearly using the credibility of these outsiders to wrap themselves in a cloak of sincerity, objectivity, scientific credibility, altruism, motherhood and apple pie.

Not to be outdone in the "bee washing" department, on April 15 (just after they got their taxes done), 2014, Bayer held a grand opening for its "North American Bee Care Center," because they wanted to celebrate "Bayer Crop Science's more than 25-year commitment to pollinator health." As part of the grand opening celebration, Bayer launched a "Color Me Bee-autifully" coloring contest, "a learning opportunity for educators, parents and students 12 years of age and under."[13] Well, if that doesn't bring a screeching halt to the demise of bees, and a tear to your eye, I don't know what will.

But Bayer had even more bee salvation hidden up their sleeve. They launched a "Bee Care Tour" — at land grant universities like the University of Nebraska, Washington State, UC Davis, Purdue, Oregon State, and Ohio State—pronouncing themselves as an environmentally sensitive company with a "commitment to bee health" that includes developing a treatment for parasitic mites and advocating for "responsible" neonicotinoid use (as much a non sequitur as responsible cigarette smoking). The tour was promoted by the universities as well. In an attempt to control their skeptics and antagonists, the Bee Care Tour at Oregon State University was by invitation only with an RSVP a week before the presentation in an attempt to control their audience. The attempt failed. Members of the Sustainable Beekeepers Association showed up to challenge the program and the presenters. A dozen protesters showed up in bee costumes with picketing signs. Security for Bayer called the local police, who rejected Bayer's insistence that the protesters be forced to move to the opposite side of the street.

The delivery truck painted by Bayer to prove their commitment to "honest" pursuit of the cause of Colony Collapse Disorder.

During Bayer's presentation, their ecotoxicologist, David Fischer, denied that "neonics" were responsible for colony collapse "when used as directed"—the tried and true escape clause always used to deflect blame from manufacturers to the applicators when things don't go well.

"All studies on neonicotinoids do not show any link to widespread colony losses," he told his audience. "They all say the same thing: Colony losses do not correlate to neonicotinoid use or pesticide residue in hives." Bayer's website, heralding their bee tour, states, "Honey bee colony health is impacted by a great variety of factors including disease, mites and habitat. The Bee Care Tour helps to foster collaboration to improve honey bee colony health and help ensure the availability of pollination services." No mention is made of their highly profitable bee poisons.

If you're still a little skeptical of Bayer's motives, if not their public relations bona fides, Bayer added to their tour a "beehicle" (I'm not kidding)—"a specially wrapped vehicle that's on a three-month journey across the Midwest providing bee stewardship workshops and expert presentations on issues impacting honeybee wellness." Bayer had the audacity to promote to attendees participating in the "Bayer's Pollinator Pledge," an initiative "encouraging consumers and growers to make pollinator health and stewardship a priority when planting."[14] Not sure if taking the pledge required a Nazi-like military salute, but it wouldn't surprise me given that Bayer is headquartered in Germany.

Syngenta is another chemical company profiting heavily from manufacturing neonicotinamides. They also manufacture highly dangerous pesticides like paraquat and atrazine that, hypocritically, are banned in their home country of Switzerland but used largely in poorer countries (as well as Hawaii). Syngenta has threatened to sue individual EU officials for banning neonics in Europe. They are responsible for 18 superfund sites in the US and are accused of hiring armed militias in Brazil, including ordering the murder of Brazilian farming peasant activists who were fighting to protect native seeds and natural agricultural methods from Syngenta. Environmental and human rights organizations have demanded that Syngenta be charged with responsibility for the murder.[15] Syngenta is now funding grants for research into the accelerating demise of honeybees in the United States. I'm going out on a limb here, but I'm willing to guess what the pesticide industry's "independent" research will show.

Currently, the pesticide industry has succeeded in focusing attention on the varroa mite as the principle cause of CCD. That strategy places the burden of unprecedented bee losses on beekeepers while shielding pesticides from responsibility. Most beekeepers and academics, however, note that the varroa mite has been around a long time, predating dramatic bee declines in the US that started in 2006. The varroa mite is indeed a problem, but pesticides weaken bees' immune systems, allowing them to fall prey to secondary, "natural" bee infections, such as parasites, mites, viruses, fungi, and bacteria. Pathogens such as Varroa mites, Nosema, fungal and bacterial infections, and Israeli Acute Paralysis Virus (IAPV) are prominent in honeybee hives on the verge of collapse, and this opens the door for industry researchers to exonerate their employers and blame the colony collapse on "natural" causes rather than pesticides. It's kind of same argument that "guns don't kill people, people do." But people with guns kill a lot more victims than people with knives (see Chapter 12). Natural pathogens kill a lot more bees whose immune systems have been wiped out.

There are ways to manage pests that don't involve indiscriminate use of pesticides. The system of Integrated Pest Management (IPM) maximizes control through crop rotation, enhancement of natural pest enemies, and uses chemicals only as a last resort. There is no commercial incentive for researching non-chemical methods simply because there is no corporate profit involved. Most research into farming methods is focused on high-tech solutions that can be sold by the companies that manufacture them.

Despite the unique and critical function that bees perform, despite what is common sense to everyone else, the EPA appears only too happy to let the chemical industry's obfuscations distract from their responsibility to protect our food supply. They have shunned the researchers who have drawn conclusions critical of neonicotinoids[16] and are poised to make the situation even worse by approving another bee-killing class of pesticides from our friends at Dow--the sulfoximines. I'm sure that the chemical companies are delighted to hear that a team of engineers at Harvard University are now working on a solution to Colony Collapse Disorder—building artificial, mechanical, robotic bees.[17] I wish I were kidding.

Look for the pesticide manufacturers to start pouring money into perfecting robotic bees, which they could sell to farmers after they have killed off the actual bees, and create yet another revenue stream for themselves. Perhaps you will soon see "beehicles" cruising slowly through America's suburbs—like the "good humor" ice cream trucks of old—but instead of selling ice cream, they'll be from Monsanto, selling robotic bees to unleash in your garden and pollinate your robotic, albeit "crunchy" home grown tomatoes.

Pesticides and "Bee Autism "

In Chapter 8, evidence showing a possible connection between autism, inflammatory intestinal disorders, and GMOs was discussed. But there are other connections between autism, pesticides, the decline of bees, and industrial agriculture.

On a recent front page of *The Salt Lake Tribune*, a frightening, oversized headline read, "Highest rate in the nation, 1 in 32 Utah boys has autism." Less well publicized, another national story ran the same day: "New pesticides linked to bee population collapse." If you eat food and hope to do so a few years from now, this should be equally frightening. A common denominator may underlie both stories.

A recent Stanford University study, examining 192 pairs of twins, where one twin was autistic and one was not, found that genetics account for 38 percent of the risk of autism and environmental factors account for 62 percent.[18]

Suggesting an environmental and genetic tag team are other studies showing mothers of autistic children and autistic children themselves have a high rate of a genetic deficiency in the production of glutathione, an antioxidant and the body's primary means of detoxifying heavy metals.[19]

Numerous studies have shown a connection between autism and urban air pollution (see Chapter 5). Furthermore, pregnant mothers who take extra folic acid, which helps increase glutathione production, offset the increased risk of their newborn eventually being diagnosed with autism that comes with air pollution exposure.[20] High levels of toxic metals in children are strongly correlated with the severity of autism.[21]

Low levels of glutathione, coupled with high production of another chemical, homocysteine, increase the chance of a mother having an autistic child to one in three, according to Dr. Jim Adams, director of Arizona State University's Autism/Asperger's Research Program. That autism is four to five times more common among boys than girls is likely related to a defect in the single male X chromosome contributing to antioxidant deficiency. There is no such thing as a genetic disease epidemic because genes don't change that quickly. So, the alarming rise in autism must be the result of increased environmental exposures that exploit these genetic defects.

During the critical first three months of gestation, a human embryo adds 250,000 brain cells per minute, reaching 100 billion by the fifth month. There is no chemical elixir that improves this biological miracle, but thousands of toxic substances can cross the placenta and impair that process, leaving brain cells stressed, inflamed, less well developed, fewer in number, and with fewer anatomic connections with each other, all of which diminish brain function. The opportunity to make up for the resulting deficits later on is limited, almost non-existent.

The list of autism's environmental suspects is long and comes from many different studies that show higher rates of autism with greater exposure to flame retardants, plasticizers like BPA, endocrine disruptors in personal care products, heavy metals in air pollution, mercury, and pharmaceuticals like anti-depressants, and pesticides.[22, 23, 24, 25, 26, 27, 28, 29, 30, 31] (Utah's highest in the nation autism rates are matched by the highest rates of anti-depressant use and the highest mercury levels in the country in the Great Salt Lake.)

Doctors have long advised women during pregnancy to avoid any unnecessary consumption of drugs or chemicals. As mentioned in Chapter 4, the average newborn has over 200 different chemicals and heavy metals contaminating its blood when it takes its first breath,[32,33] with 158 of them known to be toxic to the brain. Little wonder that rates of autism and attention deficit and behavioral disorders are all on the rise.

How does this relate to disappearing bees and your ability to put "food on your family" in the immortal words of Pres. George W. Bush? Numerous studies show that the rapid rise in the use of insecticides are likely responsible for the mass disappearance of bee populations.[34, 35, 36] But in May, 2014, a study that should be regarded by regulators as the definitive answer, was published by the Harvard School of Public Health in the "Bulletin of Insectology."[37] Bees exposed to neonicotinoids at commonly found doses showed a fatality rate of 50 percent, with the deaths showing all the characteristics of

Colony Collapse Disorder. Furthermore, in the control hives, only one of six colonies were lost and that was due to a parasitic fungus. But in that hive, the dead bees remained in the hive. Despite the fact that virtually all the colonies were infected with pathogens, those that collapsed and those that didn't, only those exposed to the pesticides had abandoned their hives. The authors stated, "It is striking and perplexing to observe the empty neonicotinoid-treated colonies because honey bees normally do not abandon their hives during the winter. This observation may suggest the impairment of honey bee neurological functions, specifically memory, cognition, or behavior, as the results from the chronic sub-lethal neonicotinoid exposure."[38]

Researchers from Royal Holloway University of London fitted bumble bees with tiny radio frequency tags, similar to those used by courier firms to track parcels, (and I guess the kind James Bond uses when he wants to know if international bumblebee villains are trying to kill him), and found that neonicotinoid pesticides affected their choice of flowers to visit and impaired their ability to forage and gather pollen.[39]

Scientists affiliated with the Task Force on Systemic Pesticides (a global group of 29 independent scientists) completed a full analysis of all the available literature as of 2014, involving 800 peer-reviewed reports, and stated, "The evidence is clear that neonics pose a serious risk of harm to honey bees and other pollinators. In the case of acute effects alone, some neonics are at least 5,000 to 10,000 times more toxic to bees than DDT." But focusing on only acute effects conceal their true impact. As mentioned earlier, the breakdown products (metabolites) of neonics are often as toxic, or more toxic, than the original active ingredients.

Neonics are systemic pesticides, meaning they permeate the entire plant. The review emphasized that the classic measurements used to assess the toxicity of a pesticide (short-term lab toxicity results), only scratch the surface for systemic pesticides. The combination of persistence over months or years and solubility in water has led to large scale contamination of and accumulation in soils and sediments, ground and surface water, treated and non-treated vegetation. They concluded that neonics pose a serious risk to honeybees and other pollinators, such as butterflies, and to a wide range of other invertebrates, such as earthworms, and vertebrates, including birds.

I recently provided anesthesia for a professional bee keeper who had hundreds of hives. I asked him what he thought was the cause of bee colony collapse disorder. Without hesitation he said pesticides. The nervous system of insects is the intended target of these insecticides. They disrupt the bees homing behavior and their ability to return to the hive. "Bee Autism" is a perfect description of what these pesticides do to bees. Because about 90 percent of native plants require pollinators to survive, most of the world's food chain hangs in the balance.

Let's get back to humans for a moment. But insects are different than humans, right? Human and insect nerve cells share the same basic biologic infrastructure. Chemicals that interrupt electrical impulses in insect nerves will do the same to humans. But humans are much bigger than insects and the doses to humans are miniscule, right? During critical first trimester development, a human is no bigger than an insect, so there is every reason to believe that pesticides could wreak havoc with the developing brain of a human embryo. But human embryos aren't out in corn fields being sprayed with insecticides and herbicides, are they? Recall that we mentioned in the previous chapter a recent study showed that every human tested, including urban dwellers who had no direct contact with any stage of crop or livestock agriculture, had the world's most popular pesticide, Roundup, detectable in their urine at concentrations between five and twenty times the level considered safe for drinking water.[40]

Autism Nation:
America's Chemical Brain Drain

In March, 2014, leading scientists from the Mount Sinai School of Medicine in New York and the University of Southern Denmark warned of a "silent pandemic," citing strong evidence that "children worldwide are being exposed to unrecognized toxic chemicals that are silently eroding intelligence, disrupting behaviors, truncating future achievements and damaging societies."[41] These "brain" toxins—heavy metals, fluoride, chemicals like PCBs, toluene, solvents, flame retardants, BPA, phthalates and pesticides—are found in the furniture you sit on, the clothing you wear, the air you breathe, the food you eat, and soil your kids play in. And this short list of chemicals and compounds is just the tip of a very large, toxic iceberg. "It's time to start looking for the environmental culprits responsible for the remarkable increase in the rate of autism in California," said Irva Hertz-Picciotto, an epidemiology professor at University of California, Davis.[42]

Glyphosate is a potent endocrine disruptor (see Chapter 4), meaning it can interfere with the production, release, transport, metabolism, or elimination of the body's natural hormones, which are the most potent biologic substances known to science. Fetuses and infants are particularly at risk, as any disruption of endocrine systems can affect brain development.

Research published in the *New England Journal of Medicine*[45] compared brain autopsies of autistic children who had died from unrelated causes, to those of normal children. Autistic brains demonstrated abnormal patches of disorganized neurons disrupting the usual distinct layers in the brain's cortex. The primary implication of the research is that the abnormalities almost certainly had to have occurred in utero, during key, short, developmental windows between 19 and 30 weeks' gestation. Even if Monsanto scientist Dan Goldstein (see Chapter 8) was correct, brief exposure in utero could have lasting consequences to brain development. Perhaps even more important than the dose of a toxin is the timing of the exposure, and presence or absence of other facilitators or synergistic toxins.

Other research suggests an even longer list of toxic substances can irreversibly interfere with the delicate process of organizing fetal brain architecture. It is a popular misconception, fed in part by weak government regulations, that toxins produce an all-or-nothing affect. Levels above "safe" doses are acknowledged to be harmful, but below "safe" levels are misinterpreted as harmless. But that's not how the body works, especially the developing brain.

Brain-damaging chemicals can provoke the entire spectrum of outcomes, from imperceptible changes to severe neurologic handicaps. Furthermore, the absence of cognitive or behavioral problems in childhood is not necessarily evidence that an early exposure to a neurotoxin had no adverse effect on brain development. In fact, studies in both animals and humans have demonstrated that some substances cause damage to the brain that is manifested only in the delayed onset of learning problems, attention deficits, and changes in emotional regulation, which can have long-term consequences in teenage and adult years.

The immature brain of an embryo, fetus, or infant is at risk for significant and permanent damage from exposure to chemicals, like pesticides, at levels that may have no detectable impact on adults. Public policies, which too often focus on adults, fail to protect developing brains during pregnancy and early infancy.

Most pesticides work by causing chemical disruption of the brain and nervous system of insects. In fact many pesticides are merely derivations of chemical warfare agents of the WWI and WWII era, i.e., nerve gases. It should be no surprise then that human nerve cells could also be affected, especially when considering that at the critical embryonic stage, the human fetal brain is no larger than that of many insects. Research confirms that mothers more exposed to commonly used, "safe" pesticides bear children with lower intelligence, [46, 47, 48, 49, 50] structural brain abnormalities,[51] behavioral disorders, compromised motor skills,[52, 53] higher rates of brain cancer,[54] and smaller head size[54,55]

In December, 2013, the European Food Safety Authority ruled that neonicotinamides may adversely affect the development of neurons and brain structures in unborn babies. Adult neurologic diseases like Parkinson's and an acceleration of cognitive decline are more common in adults with even modest exposure to "legal" pesticides.[56,57] Adults with high levels of DDT metabolites are four times more likely to have Alzheimer's.[58]

Of the hundreds of chemicals that pregnant mothers are exposed to and that cross the placenta during embryonic and fetal development, many of those undoubtedly reach the brain during critical windows of brain formation. None of those chemicals enhance the natural process of brain growth and maturation, many of them are known to be toxic to neurons and brain tissue.

As alluded to earlier, there is obvious genetic variability in susceptibility to autism among individuals. That autistic genetic profile may have deep evolutionary roots,[59, 60] because a subset of genes found in "antisocial" bees share molecular similarity with a subset of genes in humans that suffer from autism. This group of genes can affect how well bees become integrated into a hive.

With alarming and still rising rates of autism and behavioral disorders in the US, public health officials and politicians should be running around with their hair on fire determined to find out exactly what is happening and why, and most importantly how to stop it. But the current American aversion to holding powerful industries accountable for anything, makes it virtually certain that regulatory agencies will continue to turn a blind eye to most, if not all of the likely environmental triggers of autism. Pesticides almost certainly are playing a significant role in the tragic decline in America's collective intellectual prowess, and the chemical assault on our children's brains is spiraling toward catastrophe.

Through ineptitude, ignorance, apathy, and corruption, we are allowing the same handful of chemical companies that are wiping out pollinators critical to our food supply to also threaten the intellectual potential of our children.

Artificial, robotic bees are being developed by researchers at Harvard University to replace the bees that are being decimated worldwide.

Mass bee die offs are happening throughout the world.

"If We Die, We're Taking You With Us"

10

The Asbestos Industry:
A Century of Choking Us to Death

"It is difficult to get a man to understand something, when his salary depends on his not understanding it."

—Upton Sinclair

If you have occasion to witness a proper asbestos clean-up of a construction or remodeling site you will see workers in hazmat suits, gloves, hoods, and respirators every bit as elaborate as a scene from a sci-fi horror movie where a deadly virus threatens to kill all of mankind. And when all of mankind is dead, that certainly ruins the weekend. Actually a documentary about asbestos would have made a much better horror movie, because the suffering and death was real, as were the monsters that caused it.

Asbestos is a natural mineral that consists of microscopic, flexible fibers that are light weight and have high tensile strength. Individual fibers are too small to be seen. Asbestos is highly resistant to heat, water, degradation, and corrosion. These natural properties and inexpensive cost made asbestos an ideal substance for a wide range of applications, like building insulation, brake pads, gaskets, paints, drywall, flooring tiles, cements, and even fabrics. It is also an ideal substance for wreaking havoc on cells and human organs when inhaled or ingested, causing inflammation, scarring, genetic damage and death.

The lung disease asbestosis and the asbestos caused-cancers, are variations of the same fatal process. Because of scarring and thickening, the lungs progressively lose their capacity to expand and deliver a decent breath, and victims eventually suffocate. It's like being slowly water boarded with molasses. It's a miserable way to die, worse than listening to a Ted Nugent concert.

Archaeological evidence suggests that asbestos has been used since at least 2500 B.C., beginning in what is now Finland, as an additive to clay to make stronger pots and utensils. Since then it was used by most of the world's major civilizations, including the ancient Greeks, Romans, and Persians, principally for its fire-retardant properties.

Reportedly, the medieval king Charlemagne who ruled Europe in the late 700s, used asbestos in a parlor trick. He would throw an asbestos woven tablecloth into a fire and amaze his dinner guests that it would not burn. Asbestos's fire-retardant properties were also useful in making napkins and table runners in castles and homes that were lit by candle light. In the Middle Ages, asbestos was used to insulate suits of armor. I guess suits of armor were a bit too soft and comfortable, so adding some scratchy asbestos took care of that problem.

In 1858 the asbestos industry was launched when the Johns Company in New York began mining asbestos for use as industrial insulation. Asbestos mining soared during the Industrial Era in the late 1800s. It was once called the "miracle mineral" because of its versatility, flexibility, sound absorption, resistance to high temperatures, electricity and chemical corrosion and could be used to insulate buildings, boilers and steam engines.

Asbestos was sprinkled on Christmas trees as a fake snow. It was even used on the film set of *The Wizard of Oz* when Dorothy and her companions, having been placed under a sleep spell by the Wicked Witch, began getting snowed on, or "asbestosed on" whilst lying unconscious in a poppy field.

When Bing Crosby sang for the first time, America's signature Christmas classic, "I'm dreaming of White Christmas," at the end of the 1942 film *Holiday Inn,* the fake snow he was singing in was asbestos. For some reason, the song "I'm dreaming of an asbestos Christmas" never became very popular.

After WWII, surgeons began using asbestos for surgical suture. In the 1950s asbestos made it into cigarette filters, just because cigarettes weren't dangerous enough all by themselves. Women's beehive hairdos of the 1950s might not have been possible—a huge loss of course—without asbestos which was used to line the huge beehive hair dryers in the salons of the era. It even made its way into talcum powder and toothpaste because of it abrasive properties. It seemed like the uses for asbestos were almost limitless.

But the association between asbestos and poor health was noted almost as soon as it was discovered. Pliny the Elder, a Roman author, naturalist and doctor documented a "sickness of the lungs" among slaves that worked in the asbestos mines. He went so far as to not recommend wealthy Romans buy slaves who worked in those mines because they died young.[29] Pliny would have made a great corporate CEO, concerned only about his investment, not about his employees and customers' dying.

By 1898, asbestos had been declared in Great Britain to be an extremely hazardous dust. During an autopsy in 1900, Dr. H. Montague Murray, a physician in London's Charring Cross Hospital, discovered asbestos fibers in the lungs of a 33-year-old man who had worked fourteen years in an asbestos textile factory and died of severe pulmonary fibrosis, which Murray identified as caused by his occupation.

In the United States, in 1917, Dr. Henry K. Pancoast of the University of Pennsylvania School of Medicine observed lung scarring in the X-rays of fifteen asbestos-factory workers. In 1918, the US Bureau of Labor Statistics released a report describing the premature deaths of asbestos workers and the routine practice for insurers to deny workers coverage because of the "assumed health-injurious conditions" in the asbestos industry.[12]

In 1927, a British physician, Ian Grieve, wrote a detailed study of the health of workers at the J W Roberts asbestos textile plant in Leeds, England. Using X-rays to confirm his evaluation, Grieve determined that working with asbestos could cause extensive destruction of the lungs within five years. A government inquiry set up a year later found that a quarter of workers with five or more years of experience in asbestos textile factories had fibrosis, rising to half of those who had worked in the industry for ten years.

By the 1930s, asbestos manufacturers and their insurance companies were aware of the danger and lethality of asbestos in the workplace. Regulations on dust control, medical surveillance, and compensation were introduced in Britain in 1931. Like with tobacco, it was German doctors who were the first to recognize the specific cancer threat from asbestos, and in the 1930s, Germany began compensating asbestos victims for cancer and asbestosis. In 1934, the insurance company Aetna published *The Attorney's Textbook of Medicine*, with a full chapter detailing the consequences of asbestos exposure, including the lung-disease-related disability, that it was irreversible, and that the end result was often ultimately fatal.

The first lawsuits began to be filed against the asbestos industry in the 1920s. Johns Manville then successfully lobbied for national legislation, shunting asbestos workers' claims to workers compensation panels and away from juries. Effectively shielded from costly plaintiff lawsuits, the industry then proceeded to fund medical studies whose published results were falsified, exonerating asbestos as a cause of cancer.

Publicly, asbestos companies claimed there was no evidence people could become sick and die from asbestos exposure. Internal company communications, however showed asbestos executives admitted that the disease process begins as soon as asbestos is inhaled, is progressive, irreversible, and is very advanced by the time it is diagnosed. But their response was to launch a concealment, distortion, misinformation, and outright lying offensive that has continued until the present day.

The reputation of asbestos started to catch up with it by the eve of WWII, and manufacturing of the product was starting to decline. But the war created a demand for asbestos in ships, leading to a revitalization of the industry, and boosted asbestos manufacturing for the next three decades. Apparently images of Nazis in Times Square made most people forget about the health consequences. But the corporations that used asbestos didn't forget. They just didn't care.

Eventually Johns Manville filed for chapter 11 bankruptcy protection. David Oster, the attorney in charge of the Manville trust, said the documents show that corporations knew the dangers of asbestos back in 1934, and that there was a corporate conspiracy to prevent workers from discovering that their exposure to asbestos could kill them. "Manville officers, directors, and employees held secret information, that had it been revealed, would have prevented the deaths of thousands of people."

One employee of Manville, who co-authored a company document, was told by Manville's Chief of Litigation to hire his own lawyer after the document came to light, because it was his opinion that the employee could be indicted for manslaughter.[1] Far more disturbing is that the list of conspirators in the crimes of the asbestos industry goes far beyond Johns Manville. A stroll through the documents of the entire asbestos history, from 1948 through 1988, as was done by the Environmental Working Group,[2] reveals virtually an entire line up of the major industrial corporations of America—Exxon, Dow (Union Carbide), DuPont, Bendix (now Honeywell), The Travelers, Metropolitan Life, Dresser Industries (now Halliburton), National Gypsum, Owens-Corning, General Electric, Ford, and General Motors, just to name a few—that were involved in craven, inhumane abuse and deception, almost unrivaled in American industrial history. All of these companies were shown to know full well what asbestos was doing to their workforce.

A startling example of these documents comes from a physician hired in 1964 by Philip Carey Manufacturing, to produce a report on the health consequences of asbestos for the company. The report stated,

> "There is an irrefutable association between asbestos and cancer. This association has been established for cancer of the lung and for mesothelioma. There is suggestive evidence . . . for cancer of the stomach, colon and rectum also. There is substantial evidence that cancer and mesothelioma have developed in environmentally exposed groups, i.e., due to air pollution for groups living near asbestos plants and mines. Evidence has been established for cancer developing among members of the household. Mesotheliomas have developed among wives, laundering the work clothes of asbestos workers. Substantial evidence has been presented that slight and intermittent exposures may be sufficient to produce lung cancer and mesothelioma. There should be no delusion that the problem will disappear or that the consumer or working

population will not become aware of the problem and the compensation and legal liability involved."[3]

The company's response was to fire the doctor.

The 1966 comments of the Director of Purchasing for Bendix Corporation (now a part of Honeywell) may be the crown jewel of that inhumane attitude, expressed over and over again in company documents spanning the past 60 years.

"… if you have enjoyed a good life while working with asbestos products, why not die from it."—1966 Bendix Corporation letter[4]

Or how about your whole family die from it? Workers exposed to asbestos in the insulation and ship building industries carried enough asbestos home on their clothes and skin to contaminate their own homes and risk the health of their entire families.[9] This "secondary exposure" is just as dangerous as primary exposure. One of the first physicians to describe hazards to asbestos workers linked those hazards to family members as well as early as 1897.[10] One study showed that even 20 years after an asbestos factory shut down, asbestos was still found in the homes of former factory workers.

Louise Williams was a child when her father would come home from work with tiny asbestos fibers clinging to his clothes. Many years later, she was diagnosed with peritoneal (lining of the abdomen) mesothelioma. Six years after that, she was diagnosed with pleural (lining of the lung) mesothelioma. John Panza, an English professor at Cuyahoga Community College, was diagnosed with mesothelioma of the lung when he was 38.[11] He had one lung removed, and has had chemotherapy and 27 radiation treatments. As of 2016, John was still alive, much to the amazement of his doctors. He became stricken with the disease as a result of second-hand exposure to his father's clothing, whose job it was to clean up asbestos dust at the Eaton Airflex brake company. His exposure began when he was a newborn baby and continued for ten years. He was awarded $27.5 million by an Ohio jury. Panza's father died at the age of 52 of lung cancer.

Going back to 1934, letters and company documents from two of the largest asbestos manufacturers, Johns-Manville and Raybestos-Manhattan, revealed their executives engaged in covering up information that asbestos was harming their workers.[5] In the 1930s and 1940s, the industry committed thousands of dollars to research on the human health consequences of asbestos under the direction of Dr.

Leroy Gardner. When results began showing alarming results, they prevented their researchers from publishing their data.[6] As quietly as possible, the companies settled injury and death claims from workers well before they publicly acknowledged the danger of asbestos.

By the late 1940s, asbestos manufacturers, industries that used significant amounts of asbestos in their operations, and their insurance companies all acknowledged internally that asbestos caused an extraordinary health disaster trifecta—lung cancer, asbestosis and mesothelioma (cancer of the lining of the lung, heart or the abdomen). Asbestos is best known for causing lung cancer and mesotheliomas, and it is the only known cause of mesotheliomas, but asbestos also causes other types of cancer.

All these companies had reasonable, affordable, more benign product options. But rather than adopt safety standards, switch to safer products, or provide protections for workers, these companies went to great lengths to conceal the truth about asbestos from workers, the public, and the press. Typically, worker health was not actively monitored. But some corporate officials monitored the health of workers and deliberately withheld the results from them. To every extent possible, information on the toxicity of asbestos was held secret. In other cases, companies interfered with and even rewrote scientific study results, deliberately failed to label their products, or altered the labels.

Internal corporation memos, exposed only because of litigation, reveal a shocking disdain by industry executives for the suffering of the victims strewn along the asbestos highway. Those memos also reveal that the industry knew the magnitude of the number of likely and potential victims—literally in the millions. At Johns-Manville, it was company policy as late as the 1970s to not tell its employees that their health check-ups showed signs of asbestosis, despite the company knowing full well of the diseases from asbestos and their ultimate fatal outcomes if not discovered and treated early.

As put in a memo from Johns-Manville's medical director to corporate headquarters:

"The fibrosis of this disease is irreversible and permanent so that eventually compensation will be paid to each of these men. But, as long as the man is not disabled it is felt that he should not be told of his condition so that he can live and work in peace and the company can benefit by his many years of experience."[7] The lead attorney for Johns-

Manville, Vandiver Brown, was asked this question, "Mr. Brown, do you mean to tell me you would let them work until they dropped dead?" He responded, "Yes, we save a lot of money that way."[12] Mr. Brown was nothing if not charmingly empathetic.

In 1973 an industry insider predicted that approximately 25,000 workers would eventually die from asbestos, and then, incredibly, wrote, "and the good news is that despite all the negative articles on asbestos-health that have appeared in the press over the past half-dozen years, very few people have been paying attention."[13] A 1976 internal industry memo said that protecting workers was too expensive, specifically, that "the drawback of course is several pennies cost."[14]

A 1975 insurance industry memo summarized non-workplace exposure as a major risk facing the industry. One study found that 40 percent of housewives and 50 percent of blue-collar workers had identifiable asbestos fibers in their lungs at death. The author concluded that, "It is now found (that) the public in general is or has been exposed to asbestos products to a far greater degree than previously recognized."[8] In 1977, the insurance industry's "discussion group on asbestosis" unanimously agreed to not admit any liability in asbestosis cases. Another industry memo from 1988 referred to damning internal memos dating back to 1934, showing "corporate knowledge of the dangers" of asbestos, evidence that plaintiffs would be able to cite, revealing a "corporate conspiracy to prevent asbestos workers from learning that their exposure to asbestos could kill them."[15]

Mesothelioma is one of the deadliest cancers, and it is caused almost exclusively by asbestos. It is showing up more and more across all occupations, undoubtedly from much more causal exposure, like sitting at the knee of a father who had worked in the many trades that were directly exposed. Both the lung disease asbestosis and the cancers take decades to develop. As with so many of the health consequences of environmental exposures—such as radiation, chemicals, air pollution and tobacco—the latency period, or lag time between exposure and symptoms, can be so long that corporations who are purveyors of those contaminants can often stonewall regulations by convincing governments and the public that the evidence implicating their product or their emissions is too weak to draw conclusions.

Meanwhile, the companies often viewed their victimized workers as chattel.

Pauline Levesque, who lived in the UK, died of mesothelioma in 2001 at age 72. Her only exposure was washing the clothes of her husband, who was an airplane mechanic. He never got the disease. Gayla Benefield's mother died of asbestosis when her only exposure was cleaning up after her husband when he came home from work in the W.R. Grace vermiculite mine in Libby, Montana. She and her husband both have the disease, as do thirty members of her extended family. Helen Bundrock's husband Arthur died of asbestosis after working in the Libby mine for 19 years. Then Helen got the disease, and then four of her five children.

Libby, Montana, is a beautiful town near the Canadian border, nestled in some of the most picturesque mountains anywhere in the country. But as is so often the case, the beauty is only skin deep. The history of Libby is a slow-moving horror story. It is the site of one of the biggest corporate crimes in American history, and given the competition, that's saying a lot.

In 1919 vermiculite (a soil amendment for retaining moisture) was discovered nearby, and a mine to extract it was soon opened seven miles out of town. But the vermiculite was embedded in deposits of asbestos, and the miners would be covered with asbestos dust, and perhaps worse, bring it back to their homes and expose their families.

As a child, Dean Herreid played baseball on a Libby field near an asbestos mine owned by W.R. Grace. He never worked in the mine, yet in his 40s his lungs started succumbing to the scarring from asbestos. Everyone who lived in the town faced odds of their lungs being destroyed by asbestos, or a cancer caused by it, 40 to 60 times higher than the national average.[19]

The vermiculite deposit in the W.R. Grace mine is particularly dangerous because it is laced with tremolite, the most toxic of the six different types of asbestos. Tremolite's long fibers are like tiny barbed wire. They work their way into soft lung tissue, never come out, and then create an inflammatory scaring of the lung that eventually destroys it.

Vermiculite mined in Libby was processed in a dry mill, a place constantly covered in dust. The mill workers suffered the worst exposure, but the rest of the miners and the townspeople got their share of vermiculite and asbestos dust as well. What wasn't swept out of the dry mill and dumped down the mountainside was released out a ventilation stack and into the air, creating a fine mist that often blanketed the town of about 12,000 people. By W.R. Grace's own estimates, some 5,000 pounds or more of asbestos was released each day.[18] The abandoned tailings pile from the vermiculite mine in Libby is estimated by the EPA to consist of 40 percent tremolite. Left over vermiculite was "donated" to the town for use in gardens, roads and playgrounds, and baseball fields.

In 1955, an internal company memo addressed the dangers of exposing the employees to asbestos. Four years later, Zonolite (who sold the operation to W.R. Grace in 1963) ordered chest X-rays for 130 workers. More than a third of the films were judged "abnormal," with many showing early signs of asbestosis. No one at the company ever told the affected employees the results. It was not from carelessness or incompetence; it was deliberate.

As callous as Zonolite executives were, the new W.R. Grace owners were even worse. Shortly after buying the company Grace executive, P.L. Veltman wrote a letter indicating not only that asbestos was present in the vermiculite compound mined at Libby, "but that the company was already looking for ways to market the stuff."[20]

Court-mandated release of volumes of internal memos, letters, and reports show that W.R. Grace knew full well the health hazards of asbestos when they bought the mine in 1963, and failed to share virtually any of that knowledge with the miners or anyone else in the town.

By 1969, the company concluded that 65 percent of employees who had worked at the mine for 20 years or more had some form of lung disease. Once again, the workers weren't told the results.

Grace ignored the warnings of local doctors that Libby miners were acquiring severe lung disease. Men judged to be ill were moved to less hazardous jobs -- not for their protection, but to keep them working. In 1968, the company's safety officer had speculated that if Grace could keep sick workers from being exposed to more dust, "chances are that we may be able to keep them on the job until they retire, thus precluding the high cost of total disability."[21] Grace actually did its own studies proving tremolite not only produced asbestosis in laboratory animals but led to mesothelioma and malignant lung tumors. And then they hid the results of those studies.

In 1983, the company decided not to spend $373,000 on showers, uniforms, and paid overtime -- the cost of giving miners the chance to clean the dust off their bodies and clothes before heading home to their families.[21]

It's not hard to understand why Don Judge, executive secretary of the Montana AFL-CIO has called Libby, Montana, "America's Chernobyl," and why it is considered one of the nation's worst environmental disasters, and the deadliest superfund site in the nation's history. Hundreds of miners and residents of Libby have died, and at least 2,000 more have developed cancer or lung disease from exposure to the asbestos-containing ore.

W.R. Grace & Co. stock closed at $1.52 per share on April 2, 2001, the day the company filed for bankruptcy protection. On Feb. 1, 2014, the stock closed at $92.28 per share, the day Grace exited bankruptcy, with a market capitalization of over $7 billion. Grace's victims haven't done nearly as well. The bankruptcy action allowed Grace to go to trial over the collective value of the asbestos claims. Strong cases were bundled with the weak, current cases were bundled with expected future cases, rather than deal with each victim one at a time. That strategy allowed them to dramatically reduce their payout to victims.[22]

Many of the plaintiffs died during the more than 12-year bankruptcy proceedings. Many of the most seriously injured were pushed to the back of the line of claimants on the Grace trust, to be dealt with after other, less seriously injured people collect. "Our clients are going to wait many more years to receive something that is, at best, no more than one-quarter or one-third of what they could have gotten if they had been allowed to go to court," said one of the plaintiffs' attorneys.[23]

In 2009, after a ten-week trial, a federal jury acquitted W.R. Grace and three of its former executives, Henry A. Eschenbach, Jack W. Wolter, and Robert J. Bettacchi, of all charges related to knowingly exposing their workers and the residents of Libby to asbestos and covering up their actions. It seemed like the final insult to a town and its people that had been cruelly and contemptuously exploited and hollowed out by a large multi-national corporation for profit.

Libby, Montana wasn't the only town victimized by W.R. Grace. John Travolta played the plaintiff's attorney who sought justice from W.R. Grace in the movie, "*A Civil Action*," which depicted another real

Grace crime--contaminating the water supply of Woburn, Massachusetts.

The asbestos chronicles is a prime example of how corporations routinely succeed in dodging regulations, protecting their profits, and escaping responsibility for the suffering they cause.

Despite the National Toxicology Program having classified asbestos as a known cancer-causing agent, and the World Health Organization had called for a global ban, asbestos is still used in consumer products in the US, including building materials like shingles and pipe wrap, and auto parts like brake pads. The EPA had briefly succeeded in banning asbestos in the US in 1989, but a court of appeals overturned the ban in 1991.

In the UK, as of 2014, 13 people are dying every day because of asbestos exposure, making it the single greatest cause of workplace mortality.[24] In the U.S., asbestos is responsible for around 10,000 deaths a year, meaning it kills close to as many people as gun crime or skin cancer.

In a testament to the toxicity of the mineral, one fourth of asbestos deaths occur in people who were never occupationally exposed. In a report called *Projection of Mesothelioma Mortality in Great Britain* produced for the British government, around 91,000 deaths are predicted to occur in the UK by 2050 as a direct result of exposure to asbestos. The risk of acquiring an asbestos-related disease is lifelong and does not decrease over time.

Nonetheless, as of 2012, worldwide asbestos production increased and international exports surged by 20 per cent. Two million tons of the deadly mineral are still being mined annually by corporations in Russia, China, Brazil, Kazakhstan, and until 2011, even in Canada. Approximately 600 asbestos companies, producing 60,000 asbestos-laden products, operated worldwide in 2011. According to the United States Geological Survey, even the U.S. exported about $27 million worth of asbestos products in 2011. The WHO estimates that worldwide, 125 million people are still exposed to asbestos in the workplace, and over 107,000 people die every year from asbestos-related diseases. According to a new study by the President of the International Commission of Occupational Health (ICOH), Dr. Jukka Takala, legacy use of asbestos is causing nearly 40,000 deaths in the United States every year, double previous estimates by U.S. government agencies.[27]

None of these corporations can claim ignorance about their deadly product. None of the people who run them can claim they don't realize that they make their living serving up a slow, miserable death for others. It's hard to argue with Libby resident Gayla Benefield's assessment, "They have gotten away with murder."[25, 26]

Over 60 countries have completely banned the use of asbestos. The United States is not one of them. The EPA tried to ban asbestos in 1989 after a ten-year study, declaring that asbestos was a carcinogen and "one of the most hazardous substances to which humans are exposed."[16] The US Court of Appeals for the Fifth Circuit overruled most of the EPA's ban in 1991.[17]

Part of Donald Trump's plan to Make America Great Again is to make asbestos great again. Trump has a long history of a having a bewildering love affair with asbestos, seemingly to further his interests as a real estate developer in New York City. Trump extolled the virtues of asbestos in his 1997 book, *The Art of the Comeback*, proclaiming it was "100% safe." He went on:

> "I believe that the movement against asbestos was led by the mob, because it was often mob-related companies that would do the asbestos removal. Great pressure was put on politicians, and as usual, the politicians relented. Millions of truckloads of this incredible fire-proofing material were taken to special 'dump sites' and asbestos was replaced by materials that were supposedly safe but couldn't hold a candle to asbestos in limiting the ravages of fire."[28]

Apparently the researchers who proved diseases were caused by asbestos, the patients who faked all those horrible-looking chest x-rays, and those debilitating symptoms, and the doctors who treated those patient victims, then pulled a sheet over their faces at the morgue, must have also worked for the mob. Trump has gone so far as to blame the collapse of the Twin Towers on 9/11/2001 on a lack of asbestos. His affection for asbestos is almost undoubtedly due to the high cost of removing it. As if there wasn't enough of a tangled web between Trump and Russia, asbestos is another tie that binds. Brazil used to be the main source of asbestos used in the United States, but interestingly enough, a Russian company, Uralasbest, has stepped up to replace them. In fact, a picture of Donald Trump's magnificent face (and hair) are found on plastic-wrapped pallets of Russian asbestos, as posted on Uralasbest's Facebook page, touting that their product is endorsed by President Trump.

Trump's face stamped on plastic wrap covering pallets of Russian asbestos.

Electron microscopic view of "barbed wire-like" asbestos fibers.

11

The "Happy Meal" Apocalypse:
Eating Our Way to Extinction

"For every quarter-pound fast-food hamburger made from Central American beef, 55 square feet of tropical forest -- including 165 pounds of unique species of plants and animals -- is destroyed."

Jeremy Rifkin
Economist, author of *Beyond Beef*

Human life on earth exists at the mercy of intact and healthy ecosystems. The climate crisis that threatens our very survival (see Chapter 6) is often thought of as only the result of our dependency on carbon-intensive, fossil fuel energy consumption. But environmental and climate threats are intimately related to the issues of food production. Industrial agriculture (the livestock industry in particular), deforestation, and land-use practices contribute more to the human-caused greenhouse gas phenomenon than all of our transportation emissions, about 24 percent compared to 14 percent.[51]

Industrial agriculture is itself very energy intensive, in particular, fossil fuel dependent. Synthetic chemical fertilizers and pesticides are produced from fossil-fuel feed stocks. Livestock, crops committed to livestock feed, and animal byproducts, account for over half of all greenhouse gas emissions. Methane emitted from livestock is 86 times more potent as a greenhouse gas over a 20-year time frame than CO_2.[1] That 20-year time frame is critical because that is likely much longer than the window in which we have to avoid the climate tipping points, where self-reinforcing feedback mechanisms ignite an irreversible climate disaster.

Worshipping the Golden Calf

Converting global civilization to clean energy will take decades and an estimated $43 trillion. But everyone can have an even larger, and more immediate impact by not eating meat, and by doing so, improve their health. And they can do that starting tomorrow. But not if the companies that make up industrial agriculture have their way. This industrial model offers multiple profit streams for companies like Monsanto, Cargill, and Burger King, and they will be damned if they will sacrifice a very profitable status quo for something so trivial as an inhabitable planet.

Writing for EcoWatch, John Roulac says the CAFO (confined animal feedlot operation) industrial-meat system is a "cancer-linked, bee-killing, carbon-busting, soil-destroying, nitrous oxide-emitting, air-and-water polluting, ocean-acidifying, human and planetary health disaster."[52] He's actually being too kind. And Monsanto's GMOs and Roundup (see Chapter 8) is front and center in the line-up of guilty parties responsible.[2] CAFOs depend on enormous tracts of monoculture GMO corn and soybeans, with tons of pesticides and synthetic fertilizers. Presently, about 65 billion animals, about nine for every person on earth, are crammed into CAFOs.

The Western meat-intensive diet, and the industrial agricultural system that maintains it, is also responsible for widespread deforestation and draining of wetlands. Rational, ethical, reasonably well educated and informed people (I realize that may be a small audience) know that the looming climate crisis threatens modern civilization if not humankind as a species. Ironically, while humankind's accelerating abuse of our planet is now widely recognized as responsible for the sixth great massive species extinction[3] since the earth was formed, we are rapidly setting the table for our own mass extinction, and due in large part to what we are eating at the dinner table.

TransCanada, ExxonMobil, Peabody Coal, and the Koch brothers are widely demonized as the enemies of meaningful action on the climate crisis. No argument there, and given that their deception and obstructionism will lead to misery and death for billions of people, putting them on trial for crimes against humanity seems reasonable, not radical. But for far too long, what we eat and the corporations that produce what we eat and entice us and our kids to want what they produce, have continued to operate in our climate blind spot. MacDonald's, Burger King, Wendy's, Coca-Cola, Kraft, and the entire "processed, fast food" empire are just as guilty as Peabody Coal of leading us to our own extinction. "Happy Meals" may ultimately become our last meals.

As the countries of the tropics become increasingly integrated into the global economy, and there is steadily increasing demand for ever-limited natural resources, efforts to protect the region continue to be undermined by unsustainable economic demands. Wholesale exploitation and destruction of soil, water, and vegetation in order to provide cheap beef to the developed world is wreaking havoc on the ecosystems that sustain all life, including human life.

The once mighty Amazon Jungle in South America is the largest, most ecologically diverse tropical rainforest on Earth, spanning 2.1 million square miles. It is regarded as the lungs of the planet because of its pivotal role in absorbing atmospheric carbon dioxide and releasing oxygen. Just like humans don't do well when some of their lung capacity is destroyed by smoking, pollution, scarring, or tuberculosis, Earth's atmosphere and climate are suffering because it's lung capacity is being destroyed. The Amazon is also increasingly thought to be the "pumping heart" of the planet because of its role in moving moisture, wind, and weather patterns throughout the globe.[4]

The Amazon is also home to more than half of the world's species of plants and animals. But the Amazon and other tropical forests are under assault because of the pressure to exploit its natural resources. Every day, worldwide, 80,000 acres of tropical, virgin rainforest is destroyed, and roughly the same amount is seriously degraded.[5] That's about ten football fields every minute. At the top of the list of culprits is Western culinary culture, which worships the "golden calf," i.e. , beef, i.e., cheap hamburgers. Just as smoking is harmful to your personal lungs, hamburgers are killing the planet's lungs. And we are losing about 135 plant, animal and insect species every day with the loss of these forests.[5]

Throughout the world, native grasslands, forests, and wild species have been burned, butchered, and brutalized for cattle ranches. Eighty percent of the deforestation of the Amazon has come at the hands of the beef industry. Semi-deciduous forests in Brazil, Bolivia, and Paraguay are cut down to make way for soybeans, which are fed to cows as high-protein soy cake. Since 1960, more than 25 percent of

Central America's forests have been destroyed to make way for cattle ranches, with most of the beef exported to the US and Europe. For every quarter-pound fast-food hamburger made from Central American beef, 55 square feet of tropical forest, with all its plant and animal life, is destroyed.[6]

In the past 20 years, the majority of the rain forest in Costa Rica has been lost to cattle ranches. By the mid 1990s, cattle pasture land in Honduras had swallowed over 40 percent of the country's fertile acreage. Last year, deforestation in the Amazon reached a ten-year high. In a single year, 3,050 square miles of Brazilian rain forest were destroyed.[54]

It is worth drawing a distinction between the behaviors of some of the corporations perpetrating American's meat addiction, especially to beef. Burger King is the second largest international burger chain, selling 11 million meat sandwiches a day. The international watch dog, Mighty Earth, released a report, "The Ultimate Mystery Meat,"[55] describing Burger King's defiant posture as the largest retailer in the fast food industry that still refuses to disclose any information on where or how its meat or its livestock feed is produced. Burger King has refused all requests to address the environmental footprint of their supply chain. The Union of Concerned Scientists has given Burger King a zero grade for their environmental behavior. They released this statement. "RBI's [Burger King's parent company] sustainability plan says it is working towards the elimination of deforestation throughout its supply chain. However, the plan involves only the smallest efforts and amounts to little more than words on paper."[56]

Despite McDonald's and some other corporations having publicly committed to a process that eliminates deforestation from their upstream operations, a report from Greenpeace challenges the idea that they are in fact any better than Burger King.

To supply cattle feed, huge monocultures of soybeans have been carved out of pristine Central and South American forests. Using satellite imagery, drones, on site visits, and data from Brazil's federal agencies, the Mighty Earth report showed that middle-men soy traders, Cargill and Bunge, were the worst companies for environmental destruction in Brazil and Bolivia, financing forest clearing, road building, seeds, and fertilizer for South American soy farmers. Greenpeace says that without multi-nationals' financing the infrastructure, soy farming in the Amazon would be financially non-viable. Bunge says they have committed to ending their participation in deforestation, but they have little to show for it. Outrageously, Cargill has said it will stop participating in deforestation, but not before 2030.[57] MacDonald's is Cargill's largest customer.

Hundreds of defenders of indigenous people and intact rainforest have been literally murdered by those seeking land grabs and forest destruction for cattle ranches and soy plantations. "NGO Global Witness found that more indigenous forest defenders had been killed in Brazil than in any other country."[55] The agribusiness companies that profit directly and indirectly from these crimes certainly cannot be considered naïve, innocent bystanders. Farmers told the Mighty Earth field workers that fires set by soy farmers have spread far beyond areas intended, dried out the landscape, made it vulnerable to subsequent fires, and are reducing rain fall in the Amazon.

Raising beef in the tropics is as short sighted from a global perspective as lighting your furniture on fire to keep your house warm. Land that was once rain forest has only a short lifespan in supporting cattle grazing. Rainforest soil is poor in quality, nutrient deficient, making cattle grazing in the tropics grossly inefficient. Initially, each hectare of cleared land may support one animal, but after 6 to 8 years, each animal may require five hectares. Without the forest, the soil quickly becomes very dry. The grass often dies after only a few years, and the land becomes a barren desert. The cattle farmers then have to move on and destroy more rainforest in a vicious cycle to maintain grazing capacity.

Despite this, destroying rain forest for cattle remains highly profitable. By simply clearing forest and placing a few head on the land, colonists and developers can gain title to the land in countries like Brazil. The land tenure system in many countries promotes conversion of land from a natural productive asset to a "manmade" one by wealthy landowners and speculators. Pasture land prices exceed forest land prices, making land clearing a good hedge against inflation, especially in many Central American countries that have had a painful history of explosive inflation. Recently, when the Brazilian currency, the Real, was devalued, the price of beef in Real approximately doubled, creating a huge incentive for ranchers to expand their pasture area. At the same time, the price of Brazilian beef in dollars fell, which made Brazil's exports more competitive on international markets.[7]

Cattle have low maintenance costs and are highly liquid assets easily brought to market because the Western appetite for beef continues to grow. Cattle don't require much up-front capital and are a low-risk investment relative to cash crops, which are more subject to wild price swings and pest infestations.

Tropical deforestation for cattle and beef production is contributing significantly to the dramatic loss of biodiversity, including species extinction. Preserving species protects the overall biological gene pool, which enhances the chances of survival for all life forms, including humans. All species are interdependent, each one depending on the services provided by other species. Diverse, healthy ecosystems allow all organisms a better chance to recover from a variety of disasters.

Multi-national, Big Ag corporations have resurrected a new era of colonialism, deliberately exploiting the weakness and corruption of under-developed countries to get what they want, and it isn't just in Central and South America. After 2008, they have been moving into the poorest of African countries -- Ethiopia, Liberia, the Democratic Republic of Congo, Mozambique, and Sudan -- to provide meat, livestock feed, and palm oil for the U.S., Europe, Saudi Arabia, and China. According to Human Rights Watch, these foreign agricultural land grabs often involve "arrests, rapes, beatings, and killings of people who resisted leaving villages to make way for foreign projects, as well as starvation among the newly landless."[58]

Metastasizing cattle ranches is not just an issue in foreign countries. The cattle industry has undoubtedly caused more environmental destruction in the Western U.S. than all the highways, dams, strip mines, oil drilling, and power plants put together. Nevada's outlaw cattle rancher Cliven Bundy and his unpaid cattle grazing bills from the BLM don't even scratch the surface of the damage that his and other cattle do to the environment. The U.S. government, through the BLM, has long capitulated to ranchers, making cattle grazing the predominant use of Western public lands. The BLM goes so far as to spray herbicides over large tracts of range, eliminating vegetation eaten by wild animals and replacing it with monocultures of grasses favored by cattle.

Overall, your personal carbon foot print is determined as much by what you eat as it is by how much you drive your car. Avoiding the devastation of the climate crisis may be impossible unless most of us in the developed world change our eating habits. Likewise, feeding the world's billions will become impossible with either current or projected meat consumption.

Grain-fed cattle are responsible for huge quantities of the three main greenhouse gasses: CO_2, methane and nitrous oxide (N_2O). Our dysfunctional food system is responsible for one-third of all manmade greenhouse gases, according to the Consultative Group on International Agricultural Research (CGIAR), a partnership of 15 research centers around the world.[8]

A 2006 report by the United Nations Food and Agriculture Organization (FAO) calculated that our diets and, specifically, the meat we consume, cause 18 percent of manmade greenhouse gases, more than either transportation (14 percent as mentioned previously) or smoke stack industries.[9] It has been said that if you were to choose between riding your bike to work (rather than driving your car), and eating a hamburger for lunch (instead of a salad), it would be better for the climate crisis if you forgo the hamburger and drive your car.[15]

Merely the burning of forests for cattle ranches is responsible for the release of 340 million tons of carbon into the atmosphere every year, equivalent to 3.4 percent of current global emissions.

Nitrous oxide, (N_2O)--laughing gas--is a potent greenhouse gas, and it is hardly a laughing matter that it is 300 times more potent than CO_2. It is a component of traditional anesthetic technics. I abandoned it in my own practice about 12 years ago, and every other anesthesiologist should as well. N_2O is emitted by numerous natural sources as part of the nitrogen cycle, but the amount emitted by human activity is steadily increasing and now amounts to about 40 percent of the total amount in the atmosphere. Agriculture, livestock production in particular, is one of the primary, human-caused sources of N_2O. It is released from the bacterial breakdown of synthetic, nitrogen-based fertilizers in the soil and from the breakdown of livestock manure and urine. To great fanfare, the Obama Administration announced an important plan to reduce CO_2 emissions from power plants, but not even lip service was given to reducing N_2O, which is projected to increase by 5 percent between 2005 and 2020, driven largely by increases in emissions from agricultural activities.

Half of the greenhouse gases associated with the livestock industry come from methane. Over a 100-year time frame methane is at least 25 times more potent than CO_2 as a greenhouse gas and over a shorter life time even more potent. Methane from livestock alone adds up to around 7.1 gigatons per year, or 14.5 percent of all human-caused greenhouse gas emissions, according to the UN Food and Agriculture Organization.[10]

Our farming system of grain-fed livestock consumes resources far out of proportion to its useful dietary yield. Producing half a pound of hamburger for a patty of meat the size of two decks of cards, releases as much greenhouse gas into the atmosphere as driving ten miles in a 3,000 lb. car.[11] Think about that next time you drive to work. Producing that same amount of beef also requires nearly 1,000 gallons of water.[12]

Much of that water comes from non-renewal sources like the Ogallala Aquifer, perhaps the most important body of water in the United States and believed to be the largest body of in-ground fresh water in the world. The farm acreage dependent on the Ogallala yields about 20 percent of US grain production. However, in a brief half-century we have drawn the Ogallala down from an average depth of 240 feet to about 80 ft. This is essentially water "mining," because recharge of this aquifer is not even close to keeping pace with the extraction rate. David Brauer of the US Agriculture Department and the Ogallala Research Service, says, "The Ogallala supply is going to run out and the Plains will become uneconomical to farm. That is beyond reasonable argument. Our goal now is to engineer a soft landing. That's all we can do."[13]

And the disturbing truth is that much of that precious, one-time gift of water is being squandered on one of the worst American dietary staples--Twinkies. I'm kidding, something even worse--cheap hamburgers. Apart from the Ogallala, the most important water resource in the western half of the country is the Colorado River. It has long been recognized as over allocated and vulnerable to much reduced flows with the climate crisis, but a recent study[14] about just how fast that water is disappearing is truly frightening to anyone who lives in the West and should be to anyone that enjoys eating all the produce that depends on that water, i.e., the entire country. The lead author of the study, Stephanie Castle, a water specialist at the University of California at Irvine, said, "We were shocked to see how much water was actually depleted underground." Lake Mead is at its lowest level since the Hoover Dam was built, and Lake Powell is below

50 percent of capacity. And just like the Ogallala, much of the Colorado River water is being squandered at Burger King.

Beef's environmental impact dwarfs that of other meat including chicken and pork, new research reveals. A meat lover's diet requires double the carbon footprint of a vegetarian diet.[16] According to a recent study published in the Proceedings of the National Academy of Sciences,[17] beef requires 28 times more land to produce than pork or chicken, 11 times more water, and results in five times more climate-warming emissions. When compared to staples like potatoes, wheat, and rice, the impact of beef per calorie is even more extreme, requiring 160 times more land and producing 11 times more greenhouse gases. Seventy percent of all US grain production goes to feed livestock, not humans. An estimated 30 percent of the earth's ice-free land is directly or indirectly connected to livestock production.

Ruminant animals (those with multiple stomach compartments and use regurgitation and fermentation to digest plants) like cattle, make far less efficient use of their feed than other animals. Because the stomachs of cattle are meant to digest grass, not grain, cattle raised industrially thrive only in the sense that they gain weight quickly. Only a minute fraction of the food consumed by cattle goes into the bloodstream, so the bulk of the energy is lost. Feeding cattle on grain rather than grass exacerbates this inefficiency. The end result is a cow may emit 1,000 liters of methane and CO_2 per day. That's even more than most wives believe their husbands are capable of. In the average Western diet animal products make up 60 percent of greenhouse gas emissions despite accounting for just a quarter of food energy.[18]

Lindsay Wilson at Shrink That Footprint compares the carbon footprints of various diets in a very enlightening exercise. Assume consumption of around 2,600 kcal of food energy each day, roughly equal to an average American's diet. In each diet, food energy is split up among nine different food groups. The five diets are: The Meat Lover—eats a lot of red meat, white meat and dairy in place of some cereals, fruit and vegetables; The No Beef—just the average diet with all beef consumption switched to chicken. The Vegetarian—switches away from beef and chicken to fruit and vegetables, while also reducing oils and snacks. The Vegan—does much the same as the vegetarian while also eliminating dairy through further switching to cereals, fruits and vegetables.[19]

Red meat is the most carbon-intensive way to get food energy, followed by dairy, fruit, and chicken. Somewhat unexpectedly, cereals, oils and snacks are the least carbon intensive. Oils, snacks, and cereals are each highly calorific and have relatively low losses and waste, which results in their performing very well. The opposite is true of fruits and vegetables, which are less calorific per unit weight but have a very high share of consumer waste and supply chain losses, decreasing somewhat their "carbon advantage." A hothouse tomato can have emissions 5 times higher than one grown in season. Potatoes have tiny footprints compared to many other vegetables, and cheese has much higher emission than milk. High meat eaters contribute an average 7.19 kg of CO_2 equivalent each day, while vegetarians contribute 3.81 kg CO_2 and vegans contribute 2.89 kg CO_2.

In addition to being a land use and climate disaster, the livestock industry is responsible for one of the largest sources of contamination of water and topsoil. Agriculture in the United States—much of which now serves the current demand for meat—contributes to nearly three-quarters of all water-quality problems in the nation's rivers and streams. The EPA allows agriculture exemption from the Clean Water Act even though 60 percent of our rivers and streams are considered "impaired" due to agricultural water wastes. The recent contamination of drinking water for 500,000 people in Toledo, Ohio, is related, in large part, to the livestock industry. The blue-green algae that excrete the deadly cyanobacteria thrive in warm waters (and obviously will be exacerbated by the climate crisis) when there are high levels of nitrates and phosphorus to feed upon. Years ago, states surrounding Lake Erie had banned or severely limited phosphates in detergent, which reduced algae proliferation. But like many other critical bodies of water throughout the country, Lake Erie is the ultimate destination of large amounts of nitrogen and phosphorus released from farm runoff. Fouling our drinking water is another cost of our worshiping the "golden calf."

Plants and animals become extinct when they fail to adapt or evolve quickly enough to threatening perturbations in their environment. Despite our "higher intelligence," humans as a species are mired in the same pit and drifting towards spectacular failure. The climate crisis is on par only with nuclear holocaust as a threat to our collective survival. Warding off either one is thoroughly dependent on the decisions of others and our willingness to cooperate in averting disaster. What you eat greatly affects the viability of my granddaughter's future. If humans are a species of higher intelligence but without a higher moral platform, we will muster little defense against extinction.

Cathedral of Cholesterol

I sincerely hope you are not reading this chapter while chomping on a double cheese burger or a bucket of KFC. If you get to the end, you'll probably want to go somewhere and throw up because you just got a lot more, or perhaps less, than you bargained for. If you don't need to throw up, at least you'll probably become very depressed. But if you feel depressed, maybe eating a bucket of KFC will cheer you up before it gives you cancer. Sound like nonsense from the Mad Hatter in Alice and Wonderland? Let me explain.

No one needs to eat livestock to survive, yet meat is almost universally the focus of meals in a Western diet. When you go to a restaurant and the waiter asks you what you'll have, you respond with the meat or fish entree. You don't say, "The asparagus" or "the rice" or the "mixed veggies." Everything else on the menu is known as a "side dish," or is even regarded as an after-thought. For years Arby's has advertised "Mega Meat Stacks," and "Meats Upon Meats Upon Meats," and "We have the MEATS! And just like the Mad Hatter, this is pure insanity, on a both a personal and global health scale.

The average American eats between two and five times more protein than they actually need. Basically, we eat animals because we want to, or because we're duped into it by the Big Ag Empire.

In the last 50 years, worldwide meat consumption per capita has doubled, primarily because of corporate advertising. Karl's Jr. puts a scantily clad supermodel eating a monster hamburger and dripping it all over herself, and subconsciously men, being the primates that they are, think that eating a hamburger will lead to sex with that supermodel. Women think, just as absurdly, that eating that hamburger will make them look like that supermodel, which might lead to getting a new husband. MacDonald's spends about $1.4 billion a year in advertising.[20] The rest of the meat and dairy industries spend vast sums of money in television and magazine advertising every year to convince Americans that the key to happiness is eating huge amounts of cow meat, cheese, milk, eggs, chicken, and other assorted animal products.

Amino acids, the essential components of protein, can either be synthesized by the body or must be ingested with dietary intake. There are 20 different amino acids in the food we eat, but our body can only make 11 of them. The remaining nine "must have" amino acids, must be obtained on a consistent basis from food sources. Non-meat sources--grains, legumes, and vegetables--can provide all of the essential amino acids. Simply put, no one must eat meat to provide their protein needs. As an added bonus, plant-based proteins don't have the burden of saturated fat and are usually lower in calories.[21]

Furthermore, a high-protein diet strains the kidneys from excess nitrogen absorption, which discharge the waste through the urine. High-protein diets are associated with reduced kidney function. Chronic consumption of large amounts of animal protein risks a loss of kidney function if a person's kidney function is already impaired.[22] The problem is, mild loss of kidney function is usually silent but may affect 20 million Americans.

When cooked at high temperatures, especially grilling or frying, proteins in meat, fish, and poultry, will produce compounds called heterocyclic amines (HCAs) and polycyclic aromatic hydrocarbons (PAHs). These chemicals are toxic. In Chapter 5, many studies about diseases related to PAHs in air pollution were mentioned. But PAHs, when ingested, are just as dangerous and have been linked to multiple cancers, including those of the colon and breast. The manly ritual of barbecuing mountains of meat in the backyard or while tailgating before a football game is a personal health disaster, [23, 24, 25] (and the football game is as well). The average American meat eater pours 100 pounds of animal fats into his/her arteries every year.[26]

Meat consumption, especially red meat, accelerates the development of atherosclerotic vascular disease leading to heart attacks and strokes and also plays a role in the development of cancer. In order to absorb fat, the liver makes bile, which it stores in the gallbladder. When food is digested, the gallbladder releases bile acids into the intestine, where they break down the fat so it can be absorbed. But bacteria in the intestine can convert these bile acids into more toxic, carcinogenic substances known as secondary bile acids. Meats not only contain a substantial amount of fat, they also foster the growth of bacteria that cause carcinogenic secondary bile acids to form.

A long-range study from the Harvard School of Public Health[29] of 110,000 adults over 20 years found that adding just one three-ounce serving of unprocessed red meat to their daily diet increased participants' risk of dying during the study by 13 percent. Adding a hot dog or two slices of bacon increased their risk by 20 percent. On the other hand, replacing beef or pork with nuts lowers your risk by 19 percent, and replacing them with poultry or grain lowers your risk by 14 percent.

Red meat, regardless of whether it's covered in HCAs or PAHs, is also linked to breast, kidney, pancreatic, prostate, and colorectal cancer,[30] and to diabetes.[31] Seventh Day Adventists, who are vegetarians, have about half the normal cancer risk. Simply put, red meat is unnecessary and bad for your health in any amount.[27,28]

The only organisms higher on the food chain than cattle are humans. That means all meat, especially beef, has much higher levels of pesticides, dioxins, and industrial chemicals than any plant food. Not only are the chemicals given to commercially raised livestock a toxic stew, but the overwhelming majority of the grains fed to livestock are GMO, meaning they are soaked in pesticides. The National Research Council of the National Academy of Sciences considers beef the most dangerous food for herbicide contamination and it ranks third in insecticide contamination. The NRC estimates that beef pesticide contamination represents about 11 percent of the total cancer risk from pesticides of all foods on the market today.[32]

Even grass-fed cattle are now eating GMO alfalfa and corn stubble, meaning more pesticides in your hamburgers and steaks. Animal feed that contains animal parts or animal waste (as is often the case) only compounds the problem. For reasons similar to those for meat, the fat in dairy products poses a high risk for contamination by pesticides. Animals concentrate pesticides and chemicals in their milk, fat, and muscle. Growth hormones and antibiotics are also invariably found in commercial milk, cheese, and butter.

Dioxins are perhaps the most deadly group of compounds in our environment after radioactive isotopes. The World Health Organization states that dioxins "can cause reproductive and developmental problems, damage the immune system, interfere with hormones, and also cause cancer."[35]

Dioxins are found throughout the world, and they accumulate in the food chain, mainly in the fatty tissue of animals where their half-life is between 7 and 11 years. Over 90 percent of the average American's exposure to dioxins comes through food, primarily animal fat, meat, and dairy consumption.[36]

Virtually all feedlot-raised cattle are administered growth hormones and antibiotics like penicillin and tetracycline. In fact about 80 percent of the antibiotics sold in the United States go to livestock. The antibiotics are used not only for bacterial protection given the putrid conditions livestock are kept in at CAFOs, but also because they act to fatten them up. That should prompt the question in your mind about what those antibiotics do to you when you eat those same livestock. In fact numerous experiments on humans, dating back to the 1950s have shown that humans also gain weight when fed a steady diet of antibiotics.[37]

This has potential implications for the world-wide obesity epidemic and should provide "food for thought" next time you order a "thick and juicy Karl's Jr." and expect that to be your ticket to becoming a supermodel. No one seems to have studied whether the residual low doses of antibiotics in livestock meat are enough to make you gain weight, but there is evidence that those doses are sufficient to disrupt the normal composition of your gut bacteria, increasing your susceptibility to infections.[38] It's almost an annual occurrence that massive amounts of ground beef are recalled because of E-coli contamination. In May 2014, when the largest recall in the last six years occurred, almost two million pounds of ground beef were recalled from ten different states.

Here's more food for thought. There is now an unfathomable potpourri of toxins fed to livestock. It's almost like livestock, especially cattle, are now being used as toxic waste dumps. An Associated Press article from 1997 cited these hideous examples.

> "In Gore, Okla., a uranium-processing plant gets rid of low-level radioactive waste by licensing it as a liquid fertilizer and spraying it over 9,000 acres of grazing land.
>
> "At Camas, Wash., lead-laced waste from a pulp mill is hauled to farms and spread over crops destined for livestock feed.
>
> "In Moxee City, Wash., dark powder from two Oregon steel mills is poured from rail cars into silos at Bay Zinc Co. under a federal hazardous waste storage permit. Then it is emptied from

the silos for use as fertilizer. The newspaper called the powder a toxic byproduct of steel-making but did not identify it.

`When it goes into our silo, it's a hazardous waste,' said Bay Zinc's president, Dick Camp. 'When it comes out of the silo, it's no longer regulated. The exact same material.'

"Federal and state governments encourage the recycling, which saves money for industry and conserves space in hazardous-waste landfills.

"The substances found in recycled fertilizers include cadmium, lead, arsenic, radioactive materials and dioxins, the Times reported. The wastes come from incineration of medical and municipal wastes, and from heavy industries including mining, smelting, cement kilns and wood products."[39]

Nutrition and Health reported back in 1981 that some ranchers in the Midwest were feeding their steers hundreds of pounds of cement dust to "get their weight up" for sale. The FDA was asked to halt the practice, but after investigation, responded that since there has been no indication of harm to humans, the practice can continue until such time as harm is proven.[40]

Jeremy Rifkin, in his book *Beyond Beef* from 1992, reported this disgusting practice:

"Some feedlots have begun research trials adding cardboard, newspaper, and sawdust to the feeding programs to reduce costs. Other factory farms scrape up the manure from chicken houses and pigpens, adding it directly to cattle feed. Cement dust may become a particularly attractive feed supplement in the future, according to the USDA, because it produces a 30 percent faster weight gain than cattle on only regular feed. Food and Drug Administration (FDA) officials say that it's not uncommon for some feedlot operators to mix industrial sewage and oils into the feed to reduce costs and fatten animals more quickly.

"At Kansas State University, scientists have experimented with plastic feed, small pellets containing 80 to 90 percent ethylene and 10 to 20 percent propylene, as an artificial form of cheap roughage to feed cattle. Researchers point to the extra savings of using the new plastic feed at slaughter time when upward of 20 pounds of the stuff from each cow's rumen can

> be recovered, melt[ed] down and recycle[d] into new pellets.' The new pellets are much cheaper than hay and can provide roughage requirements at a significant savings."[41]

Here's my little love note to the FDA:

To Whom It May Concern (which apparently is none of you):

On the matter of cows being fed a soufflé of cement dust, sewage, oil, plastics and radioactive waste on their way to the slaughter house. There's a biological process that I studied in medical school known as "eating." That describes a process where what a human or a cow ingests is absorbed into their blood stream, and can then be carried throughout their bodies and into every cell in their bodies. So the chemical, heavy metal,[42] and radioactive toxins that you think are Ok for cows to eat will end up in the hamburger, hot dogs, and steaks that Americans think they can't live without. A basset hound could understand that probably wasn't a good idea. But hey, thanks for protecting Big Ag, which is specifically not your job, instead of public health, which is your job.

Yours truly,

Brian Moench, MD

Meat on store shelves can be subject to temperatures too high to prevent bacterial growth from spoiling, so the industry invented "modified atmosphere packaging" or "MAP." That is a euphemism, of course, meaning that the meat was packaged in an artificial atmosphere of carbon monoxide (CO). As much as 70 percent of meat sold in stores is displayed in CO packaging. The oxygen in the package is sucked out and replaced by CO, much like vacuum packaging with an impermeable membrane. This is the same CO emitted from tail pipes, chimneys, space heaters, and charcoal grills that in high enough concentrations can kill people. The CO reacts with the myoglobin in the blood, giving the meat the same artificial bright red color that a victim of carbon monoxide poisoning has when they arrive in the emergency room. CO can keep a piece of meat or fish looking artificially red and fresh for up to a full year, and of course, how a piece of meat looks is the primary consideration of a consumer, like Donald Trump when he's picking a new wife.

Eating CO does not have the same health consequence as inhaling it. But make no mistake, the purpose of CO packaging is to fool shoppers into believing that the meat they buy is fresh. The normal physical evidence of spoilage is masked, almost no matter how old it really is. And therein lies the danger of using CO packaging.[43] This practice is not allowed in many countries, like the European Union, but in another capitulation to the Big Ag Empire, the FDA has approved this practice.

Processed meats like beacon, lunch meat, hot dogs, red meat in frozen prepared dinners, and nearly all red meats sold or served at public schools, restaurants, hospitals, hotels, and theme parks are commonly mixed with sodium nitrite, which acts as a preservative and antibacterial agent. Once digested, sodium nitrites can be converted to nitrosamines, which are carcinogens. The American Medical Association says that sodium nitrites can lead to oral, gastrointestinal, pancreatic, and brain cancer,[44,45] The USDA tried to ban sodium nitrite in the late 1970's but was steamrolled by the meat industry.[46]

It turns out that the slab of meat everyone enjoys is more likely less a slab of meat, and more like a fusion of meat scraps held together with something commonly referred to as "meat glue," officially known as "transglutaminase." The chemical reaction triggered by meat glue also produces ammonia. The amount of bacteria on a steak that has been put together with meat glue is hundreds of times higher than an actual steak. If you cook rare what you think is a steak, you may very well be dangerously undercooking the outside parts of the slab that have the highest bacterial content.[47] That's one of the reasons meat glue has been banned in the European Union. Beef, pork, chicken, and fish are "fused" together to make scraps appear as "prime cuts." Unless you're a vegetarian, the overwhelming likelihood is that you're eating meat glue, no matter how fancy a restaurant you eat at.

The McRib sandwich is the quintessential triumph of meat glue. The McRib only appears on McDonald's menu periodically, whenever pork prices are really low. But the McRib has 69 other ingredients, including bonus features like azodicarbonamide, a flour-bleaching agent that is most commonly used in the manufacture of foamed plastics like in gym mats and the soles of shoes.[48] Wholesalers have plenty of other ways to deceive consumers. For instance, they are known to pump low-grade meat with water and flavoring to make it edible and to weigh more.

If meat glue doesn't get your mouth watering, then how about pink slime? Remember the high profile exposé of Beef Products, Inc., producing pink slime for national hamburger chains using throw-away meat scraps mixed with ammonium hydroxide—an ingredient in fertilizers, household cleaners and some roll-your-own explosives (who doesn't want the option to grill up some exploding burgers if your family barbecue could use some excitement?). Remember Beef Products closed three of its four shameful pink slime plants? Well, now, shamelessly, they are reopening.[49] And pink slime's new name is, "lean finely textured beef." Sounds delicious! And ground beef containing pink slime doesn't need to be labeled, thanks to a ruling by a USDA official who later stepped down and immediately joined the board of Beef Products, Inc.

But this charade involves more than just beef. To make inferior, factory-farmed chicken look like high-quality, pasture-raised chicken, many large-scale chicken producers add various dyes and additives to chicken feed to make their meat appear more yellow and golden. Likewise, "Atlantic" is another name for "farmed salmon" (there are no wild salmon in the Atlantic Ocean). Farmed salmon are fed pink dyes to make their flesh look more appealing than the disgusting gray they would otherwise be.

Chicken are routinely contaminated with roxarsone, a form of arsenic found in their chicken feed. A 2004 study from the Institute for Agriculture and Trade Policy showed that more than half of the store-bought and fast-food chicken contained elevated levels of arsenic. Roughly 2.2 million pounds of it are being used every year to produce 43 billion pounds of poultry. The poultry industry has been using roxarsone to fight parasites and increase growth in chickens since it was approved by the FDA in 1944.[53]

Chickens are also fed a frightening elixir of drugs that includes caffeine, banned antibiotics, Benadryl, Tylenol, and even Prozac. Prozac was added to feed because stressed out chickens produce tough meat, and brutal conditions often mean a constantly nervous bird. Chickens are fed coffee pulp and green tea powder to keep them up longer so they eat more food, according to a story in *The New York Times.*[50]

By now you're probably badly in need of some Prozac yourself. There is an answer to combating the deception and the health consequences of the meat industry. Stop eating it. You don't need it, none of us do. Your waistline, your arteries, and your kidneys will thank you for it. And we just might preserve a livable climate if enough people follow your example.

Huge swaths of the Amazon are being burned down for cattle ranches and soy farms to feed cattle.

Aerial view of deforestation of the Amazon in Bra

Pink slime is back under a new name,
"lean finely textured beef."

The Union of Concerned Scientists' campaign against Burger King's disregard for their environmental foot print

12

Gun Manufacturers:
In Cold Blood

"The right to profit from the massacre of children shall not be infringed."

—Unknown

In the middle of the holiday season, 11 days before Christmas, a first grade class room in a suburban school in Newtown, Connecticut, filled with bright, smiling, innocent children, exploded in seconds into the most heinous of crime scenes. Twenty first graders and six adults who had tried in vain to protect the children, all dead in the most gruesome way possible, their bodies literally torn apart by weapons of war. A new school has been built in a town that will henceforth be known as a "cradle of sorrow," but the pain and anguish for many will never end.

On a pleasant Sunday evening in October, 2017, Stephen Paddock, holed up in a Mandalay Bay hotel room above a country music concert filled with 22,000 fans, calmly lined up an arsenal of 23 automatic weapons, some with tripods and telescopic sights, and started indiscriminately murdering whoever he could, as fast as he could. It became the deadliest mass shooting in US history committed by a single individual. The massacre left 58 people dead and 546 people wounded. All of the murder weapons had been legally purchased in several nearby states.

Early in the morning of June 12, 2016, Omar Mateen, a 29-yr old, homophobic ISIS sympathizer, entered the *Pulse* nightclub in Orlando, Florida, with an AR-15-style assault rifle and a handgun, and started killing as many of the 350 patrons of the club as he possibly could. He was eventually stopped and killed by an 11-member SWAT team. In the end 49 people had died, and 58 others were wounded.

I know you're tired of reading about all these tragedies, and it's sickening to even list them. What's more sickening is, just like the movie Groundhog Day, all the cowards in Congress (all the

Republicans and a few Democrats), will crawl out of their holes, offer their "thoughts and prayers" for the victims' families, and within days, sometimes within hours, the nation's conscience will again go into hiding, public outrage will vanish, resolutions will be forgotten, and the next day, the cycle repeats itself and the killing continues.

After one of their fellow members of Congress, Gabby Giffords, a Democrat, was shot in the head at a public event, Congress did nothing. Gifford survived but still suffers obvious brain damage. After another mass shooting in 2017, where congressman Steve Scalise was shot, nearly bled to death on a baseball field, and was unconscious in critical condition for four days, Congress did nothing. Scalise returned to Congress months later, and after nearly losing his life, and with the Las Vegas shooting occurring just days before, Rep. Scalise said his opposition to gun control had been "fortified." He's still proud of his A+ rating from the National Rifle Association (NRA), and in addition to his thoughts and prayers for the Las Vegas victims, his answer for addressing America's epidemic of massacres is for all of us to give blood.[1]

"Houston, we have a problem." So does Chicago, LA, Dallas, Atlanta, Boston, New York, and every other city and town in the US. "American exceptionalism" is an elite term for the concept that America is "number one," i.e., the greatest, best, most pure, righteous, and powerful nation on earth. America is indeed number one in many things, and one of them is civilian gun massacres. The Philippines is a distant second with one fifth as many. The advocacy group, Everytown says that an average, of 96 Americans are killed every day by guns--accidents, suicides, and homicides. Seven of those are children and teenagers. Nearly 13,000 per year are victims of a gun homicide. Adding suicides and accidents, in 2016, over 38,000 people in the US were killed by a gun. For every one person killed, two more are injured. An average of 760 people are killed every year with guns wielded by their domestic partners, the overwhelming majority of those shooters are men, and the victims, almost always women. Gun homicides in the US occur at a rate 25 times higher than the average of all other developed countries.[2]

While the U.S. has a dramatically higher rate of gun deaths, in contrast, it does not have a higher rate of crime than other developed countries. However, property crimes are 54 times as deadly in New York as in London, where gun use is still rare.[3] Even in countries with overall murder rates far above those in the US, mass shooting events are extremely rare.[4] The US is also number one in gun ownership per capita. Three million Americans carry a loaded handgun with them on a daily basis.[5]

Any discussion about gun deaths in the United States begins and ends with the NRA. The NRA was originated in 1871 by two former Union officers who had criticized the shooting skills of their Civil War era recruits. They wanted to upgrade marksmanship. In the early 20th century, the NRA cooperated with the federal armed forces, who often donated surplus equipment. New York state helped the NRA buy its first shooting range.[6]

Once upon a time, the NRA was a low profile, reasonable organization furthering the interests of hunters and sportsmen, even priding itself on its independence from corporate influence. But after four American Presidents had been shot and three of them killed, public conversation about the advisability of private gun ownership bubbled to the surface. The controversy deepened in the era of machine-gun-toting bank robbers and gangster wars, the country was seemingly overwhelmed by lawlessness. The first-ever gun control legislation in 1934 and again in 1938, was passed in Congress and signed by President Franklin Roosevelt, with the help and support of the NRA.

When further gun restriction legislation was passed after the 1963 assassination of President Kennedy and again after the Robert Kennedy and Martin Luther King Jr assassinations in 1968, the NRA was supportive, not antagonistic.

Public concerns over crime also rose in the late 1960s in concert with the spread of Vietnam War protest violence, and many citizens turned to gun ownership to soothe their feelings of vulnerability. Younger members of the NRA began to call for the organization to embrace political activism.

In 1971, federal ATF agents killed a member of the NRA who was holed up with a large cache of illegal firearms. This was the "shot heard round the NRA" and culminated with a coupe within the organization in 1977. After a political blood bath at their annual convention, NRA leadership was handed over to hardliners who rallied around an opposition to all forms of gun control in Congress and state legislatures. NRA president Harlon Carter pronounced that the NRA would be "so strong and so dedicated that no politician in America, mindful of his political career, would want to challenge our legitimate goals."[6]

Within three years, Ronald Reagan became the first winning candidate for the Presidency to be endorsed by the NRA. Ironically, the attempted assassination of Reagan was the major impetus for the most significant gun control legislation ever since, the centerpiece of which was known as the Brady Bill.

Wayne LaPierre became the Executive Vice President of the NRA in 1991. When they picked LaPierre as an administrator, he was described as an "absent-minded professor." LaPierre's personal transformation into a rabid gun rights attack dog was even more extreme given his background—working as an aide to former, ultra-liberal presidential candidate George McGovern. With the guidance of PR firm Ackerman McQueen, LaPierre ramped up his speaking style, and the NRA's agenda, and injected them deeply into the country's political debate. It was Ackerman McQueen who concocted the NRA's strategy of leaping on to the public podium after every mass killing and disingenuously railing against any and all gun legislation as the government's first step on the slippery slope of "banning your guns."

"When evil knocks on our doors, Americans have a power no other people on the planet share: the full-throated right to defend our families and ourselves with our second amendment. The NRA is "freedom's safest place.""[7]

As the public face of the NRA, LaPierre is certainly qualified to lecture us about evil, because it would seem to many, and I confess I'm one of them, that he is its personification. With the help of other NRA and gun industry officials, like Chris Cox, David Keene, George Kollitides, and James Baker, LaPierre operates a perpetual death machine that shifts into high gear after every massacre, almost before the victims have reached the morgue, blaming everything but the gun manufacturers and the absurd lack of gun control laws. LaPierre famously and tirelessly rails, "The only thing that stops a bad buy with a gun is a good guy with a gun."

Immediately after every mass shooting, just like clock-work, the NRA-owned members of Congress crank up the chorus of offering their "thoughts and prayers" and phony indignation that "now is not the time to talk of gun control, it is insensitive to the victims, and politicizes the tragedy."

The problem with LaPierre's statement, his world view, and just about every foaming at the mouth word he speaks, is it's all mythology or outright lies. And it's all in the service of corporate profits.

If LaPierre is the NRA's face, Cox is its brain, specifically its chief lobbyist and political strategist. He is credited with masterminding the NRA's victory of a gun rights candidate over Rep. Ron Barber, a democrat who was elected to succeed Gabby Giffords after she was wounded and brain damaged during a gun massacre in Tucson, Arizona. Cox led the NRA in 2004 to their successful block of the assault weapons ban renewal and played a major role in the re-election of George W. Bush. Cox's most enduring "achievement" in fighting gun control was to link unfettered gun rights to the much broader cultural wars and right wing political movement of fighting off a 'sinister and suffocating federal government's intrusion" into all aspects of American life. God, guns, and (anti)gays, played to the cultural backdrop of country western music, became the definition of the Republican Party, newly supplanted by the Tea Party, and now the Trump Party. And among the many questions that raises is: just what type of assault weapon would Jesus recommend?

Invoking constitutional rights and a worshipful reverence for the "second amendment" is relatively recent in the rhetoric to justify opposition to gun control, and deliberate marketing by the NRA. Gun ownership is portrayed as not just an essential constitutional right, but the pre-eminent American value.

At the NRA's 2016 annual convention, Cox pulled the trigger on the NRA's rhetoric against Hillary Clinton with, "There's no limit to the destruction of individual freedom she'd cause with four or eight years in office." David Keene, ex-NRA president, compares gun control advocates to Hitler's taking away the guns of German citizens to stifle Nazi resistance. Keene even defends the NRA's push for legislation to stop military commanders from trying to prevent military suicides, because it involves questioning service personnel about weapons that they have access to.

James Baker has led the NRA to shore up unity and fealty to fanatical gun rights among some gun manufacturers by having to persuade them not to agree to any gun control concessions, ever, no matter how high the body count rises.

On a handful of occasions during the last 20 years, noise has been made by individual gun manufacturers that they would accept more sensible gun laws. The American Shooting Sports Council (ASSC), a gun industry group, signaled to then President Bill Clinton that they would accept restrictions of automatic weapons and magazine sizes. They met with Clinton in 1999 to discuss restrictions in the wake of the Columbine school massacre. Big mistake.

Baker led an NRA counter offensive which resulted in the ASSC dissolving. It's former director, Robert Ricker said, "There are gun companies out there willing to be responsible and ready to stop illegal use of guns—more than willing. But the NRA, and Baker in particular, will not let them. Any gun company who ventures outside the fold is kicked back into line. They go along with the NRA to save their factories."[7]

In 2000, when Smith and Wesson (S&W) was appearing to capitulate to pressure from the Clinton administration to adopt safety technology on its hand guns, like trigger locks and smart gun technology that would prevent anyone but the gun's owner from firing it, Baker, the NRA ring master, led a public boycott of S&W that reduced their sales by 50 percent and nearly drove them out of business. The handgun manufacturer Glock saw what was happening to S&W and withdrew from similar discussions with the Clinton White House.

S&W's "reasonable" management was accused by Baker of "running up the white flag of surrender" and was forced out. The new company president let all the nation's gun dealers know that they had indeed "run up the white flag of surrender"--but not to the Clinton Administration or to common sense, but to the NRA. They restored their sales, and became a major donor to the NRA.

The ascendency of the NRA was launched by its eventual marriage to the gun industry, and they have evolved (devolved actually) into the lobbying arm of gun manufacturers. Less than half of the NRA's operating funds come from membership program fees and dues; the rest is tied to the gun industry in one way or another.[8] Some companies, like Crimson Trace, maker of laser sights, donate 10 percent of sales to the NRA.[9] "Taurus buys an NRA membership for everyone who buys one of their guns. Sturm Rugar gives $1 to the NRA for each gun sold, which amounts to millions."[8] The Violence Policy Center estimated that since 2005, gun manufacturers have given nearly $40 million directly to the NRA.[10] But because the NRA does not have to disclose donor information, that number could be, and likely is, much larger.

NRA operations are inexorably tied to gun sales, and not just to gun manufacturers, but also to the down-stream industry, makers of after-market gun accessories, retailers and shooting ranges.

The NRA exploits their past by air brushing their self-portrait in the historical frame of just being the collective voice of (now 4 million) innocent, patriotic gun owners, sportsmen, and lovers of the constitution. But at the NRA's annual meeting, where's there always a "grassroots workshop," it is funded by gun industry members, who can also sponsor the meeting's annual "Prayer Breakfast." Worshipping the glory of God and guns in the very same breath may seem like something you would only hear from an ISIS fanatic. But it is core ideology for the NRA.

Robert Spitzer, author of *The Politics of Gun Control,* writes that the current debate about individual gun rights began after Congress' passing of the Gun Control Act of 1968, in response to the nation's shock and grief over the Kennedy and King assassinations.[11] The NRA gradually adopted alarmist and victimhood rhetoric as a tool to mobilize and fanaticize their base, which translates into spectacular sales results for the gun industry. Their railings reached a fever pitch, and were given the boost of a racist undercurrent when Obama ran for office. In fact, the Obama Presidency was a godsend to the gun manufacturers, with the NRA demonizing him as wanting to take away the guns of every law-abiding American. LaPierre declared Obama's run for re-election in 2012, "the most dangerous election in our lifetime."[9]

He wasn't talking about impending nuclear war, or the disappearance of mankind to climate apocalypse, he was talking about something much more serious--guns. The very survival of the second amendment was at stake. This despite the fact that Obama had accomplished nothing in the way of gun control in his first term. In fact, Obama signed a bill in 2009 that allowed guns to be carried in national parks. He even managed to earn an F from the Brady Center. No matter. In advance of the 2010 election, the NRA warned its members that failure to elect a pro-gun Senate would allow Obama to handpick a Supreme Court that "puts democracy in peril."[9] The actual legislative gun fights in Congress during the 1990s had long since ended. The NRA had achieved decisive victories over just about every legislative threat, but the reality on the ground was not going to interfere with the message, because that message meant real profit.

While there is not, and never has been, a serious threat to the second amendment, serious money is at stake, "money that the NRA banks with each new member, that the gun makers earn with each new gun."[9] The politics of gun control is, at least on one side of the aisle, the politics of money. The NRA and the gun industry make millions ginning up fear--fear of the political left taking away their guns, and fear of an imaginary American society where violent criminals are lurking behind every bush. Crime rates have in fact steadily fallen for decades (see Chapter 1). The US murder rate dropped 44 percent from 1995 to 2012. And if the political left ever aspired to truly abolish the second amendment, they have proven convincingly incapable of doing so.

In the midst of an overall recession in the beginning of Obama's first term, the nearly 100,000 people employed in the gun industry enjoyed an economic oasis, with a 31 percent increase in employment numbers by 2012. By 2011, the number of guns manufactured in the US had increased 63 percent, and handguns increased 104 percent.[9] Some gun manufacturers were enjoying such a strong burst in sales they were unable to fulfill new orders. From 2009 to 2016, the stock prices of S&W, and Sturm, Ruger each increased more than 900 percent. Louis Navellier who edits several investment newsletters called President Obama "the best gun salesman on the planet."[12] Confiscation paranoia, and the prospects of a Hillary Clinton presidency pushed gun sales to another record again in 2016.[13]

Gun manufacturers are victims of their own products. A gun reasonably well taken care of should last a lifetime. So the industry that makes such a product will either stagnate or decline unless new business is created. The solution is not just new customers but more guns for each one, meaning an increasingly radicalized small number of people.

Concentration of the nation's private cache of weapons is increasingly found in the hands of a dwindling few. Just 3 percent of Americans own half of the nation's privately held guns, and those 3 percent owned an average of 17 guns each.[14] Those 3 percent are known by the gun manufacturers as "super owners." They should call them "super heroes" a la Marvel Comics, because they are saving the financial balance sheet of the gun industry. While the percentage of

Americans that own guns has dropped in the last several years, the number of guns sold has increased.[15]

Promoting fear of crime and confiscation, and the juxtaposition of the two, is crucial to sustaining gun profits. Industry publications and TV shows like *Guns & Ammo, Shooting Times, Rifle Firepower, and Personal Defense,* offers readers and viewers a bullet-laced buffet of stories of home invasion, panic room essentials, how to stop attackers, and assembling your multi-gun arsenal.[16]

At the 2014 annual NRA convention, LaPierre thundered, "We know, in the world that surrounds us, there are terrorists and home invaders and drug cartels and car-jackers and knock-out gamers and rapers, haters, campus killers, airport killers, shopping-mall killers, road-rage killers, and killers who scheme to destroy our country with massive storms of violence against our power grids, or vicious waves of chemicals or disease that could collapse the society that sustains us all."[17]

Wow! Threat of nuclear war is nothing compared to our acute, malignant gun shortage. It was not just that America was a haven for "bad guys;" LaPierre's country had become a "Mad Max" purgatory, and every moral person needs to become a "Rambo vigilante," carrying a gun to protect himself or herself from marauding bands of hideous barbarians. Taking it a step further, he roared, "Do you trust this government to protect you? We are on our own, that is a certainty, no less certain than the absolute truth—-a fact the powerful political and media elites continue to deny, just as sure as they would deny our right to save our very lives. The life or death truth that when you're on your own, the surest way to stop a bad guy with a gun is a good guy with a gun!"[17]

He was whipping up the audience into a frenzy of government sedition and paranoid victimhood with his (non)science-fiction dystopia. The "bad guys" included the government, the IRS, non-compliant politicians, and anyone--anyone who spoke in calm tones about gun control. In so many words, he was telling them that it was "you, standing heroically with the NRA" against the world, and that by arming themselves, they could right all the economic wrongs, all the cultural degenerations, all the government overreaches, and all their personal grievances. LaPierre is smart enough to know it's all a deadly lie and savvy enough to put on such a convincing and murderous act.

As disgusting, terrifying, and disconnected from reality as LaPierre's public speaking is, it is nothing compared to the hideous, fire-breathing TV advertising campaign launched with NRA spokesperson and arch conservative radio talk show host Dana Loesch,[87] that depicts the NRA as the last refuge from an America that has already descended into a full on "George Orwell 1984." We must all grab our weapons right now to do battle against the amorphous "them."

The NRA's inflammatory rhetoric pays dividends in shaping public attitudes. A poll in 2013 found that 65 percent of Americans believe that the purpose of the second amendment remains "to make sure that people are able to protect themselves from tyranny."[18] Never mind the question of just how gun ownership protects one from 21st century tyranny, consider this disconnect. The demographic that clings to their guns to defend us from creeping tyranny is the same demographic that voted for Donald Trump, widely recognized as the greatest authoritarian threat to democracy in the modern history of the country.

NRA shouting about crime is similarly rewarded. A Gallop poll in 2005 found that in 40 percent of American homes, someone owned a gun, and when asked why, the most common response was for self-protection from crime.[19]

Public opinion polls regularly indicate that people believe crime is increasing even when it's not.[20] There are large geographic variations in crime rates. Alaska's crime rate is three times as high as Virginia's. Not surprisingly, gun homicides tend to be concentrated in cities, certain neighborhoods, and even within certain social networks. Chicago is the NRA's and President Trump's whipping boy when it comes to ginning up fear of crime for political purposes. Chicago police department data showed that 70 percent of non-fatal shootings and 46 percent of gun homicides happened within a social network that included just 6 percent of Chicago's total population.[21]

A tried and true NRA public relations tool is the inevitable attachment of the moniker "law abiding" to any reference to gun owners, as if the two are inseparable, almost as if owning a gun itself bestows the mantle of "law abiding." Of course, that is a deliberate marketing ploy, but also misleading to the point of dishonesty. Research shows that in most mass shooting, many gun homicides, and many more gun suicides, the killer is, "until that moment, a law-abiding firearm owner, pulling the trigger on a lawfully held gun."[100] In Europe,

Australia, New Zealand, and the United States, most perpetrators of mass shooting were legal gun owners, using legally purchased weapons.

But most gun deaths are not mass shootings, but single events involving domestic disputes and gang violence. Are licensed gun owners synonymous with good citizens? They can certainly be troubled citizens, and troubled citizens with guns, in the heat of the moment, are responsible for far more gun deaths than hardened criminals. The percentage of gun "events" perpetrated by legal gun owners ranges from 15 percent in England and Wales, to 43 percent in Canada, and 50 percent in New Zealand. In Australia, half of the police killed in the line of duty were murdered by a legal gun owner.[101] We should know more about the data in the United States, but thanks to the NRA, we don't even collect such data. More about that later.

Completely ignored by the NRA is that most gun deaths are self-inflicted—suicides and accidents.[102]

The demographics of who owns most of the guns is not surprising. It's older, Republican, white men living in the South and Midwest. Digging deeper, the psychological profile of the typical gun owner is someone with an authoritarian personality who mistrusts other people, organizations and institutions.

Political polarization certainly defines America in the era of Donald Trump. Polarization is another, likely subconscious reason gun owners feel driven to purchase ever larger caches of weapons, beyond any rational constraints, thus the "super owners."

A gun gives many owners a subconscious feeling of wielding considerable, perhaps ultimate power over other people, something that is especially appealing to someone that feels weak, inferior, vulnerable, or underachieving. That psychological asset diminishes in a culture where guns are common. So in order to compete, those individuals are lured into buying more and bigger guns. That men are gun owners much more often than women suggests that it is also a confirmation of their masculinity (toxic in many cases).

The "gun ownership as a tool for self-preservation" mentality creates a feed-back loop, or literal arms race. If you think you are surrounded by people packing heat, it would be logical to conclude you need to carry a weapon yourself. None of this is lost on LaPierre, the NRA, or the gun manufacturers.

It's much more difficult to maintain a constant state of paranoia about violent crime during a Republican administration, because maintaining that kind of drum beat would hardly be flattering to those who are the NRA's political handmaidens. Furthermore, anxiety over gun confiscation during a Republican administration is similarly a non-starter, and not surprisingly, gun sales have dropped since the election of Donald Trump, for some companies as much as 40 percent.[22] Given this, what's a gun manufacturer to do?

Well, President Trump and the Republicans to the rescue. Within five weeks of taking office, the primacy of gun ownership above all other considerations was once again confirmed when President Trump signed a bill passed by Congress, that allowed 75,000 people, adjudicated as mentally incapable of managing their own finances, to avoid being listed in the National Instant Criminal Background Check System (NICS) used for gun purchases.[23]

While mental incompetence does not uniquely equate to violence or criminal behavior, is it sensical or necessary that someone who can't grasp their own finances still have the right to purchase, use, and be expected to store safely a loaded lethal weapon? It makes no more sense than to offer driver's licenses to the blind.

In Sept. 2017, the Trump Administration indicated that it would loosen restrictions of gun sales overseas.[24] Specifically, they intended to move the regulation of overseas sales from the State Department, which is supposed to weigh national security and human rights consequences, to the Commerce Department, which doesn't.

As part of the NRA's "freedom and American values" identity, "supporting the troops" is a bedrock of their messaging. Well this is how the gun lobby supports the troops. Opening the door for more gun sales overseas will almost undoubtedly increase the fire power of those fighting against our troops--criminals, gang members, paramilitaries, despots, and terrorists. The body count of our own troops will increase, but so will gun sales.

In addition to Democrats in the White House, the other thing that gun makers can always count on to boost sales is more gun massacres. A definite pattern has emerged that within hours after the latest mass shooting, the stock prices of gun makers spike up. It has happened so consistently that the financial world has taken notice and is betting that it will continue.[25] Because of the surge in gun sales following these events, the cruel irony emerges that public talk of any gun control after the latest massacre actually leads to more deaths. Research published in *Science* magazine suggests that, for example, the spike in gun sales following the Sandy Hook massacre actually led to the deaths of an additional 20 children and 40 adults in the five months that followed.[26]

Gun control advocates cannot match the financial juggernaut of the NRA. In 2010, the NRA spent $240 million more than the Brady Centre to Prevent Gun Violence, the largest gun control advocacy group.[27] But the NRA's clout in Washington, D.C., and the 50 state houses throughout the country goes far deeper than just defeating gun control. They claim responsibility for the defeat of enough House Democrats to cost them control of the House in 1994. They credit themselves with defeating Al Gore in 2000 and John Kerry in 2004. But the NRA's involvement in elections reached a new high (low?) in 2016, spending at least $30 million on behalf of Donald Trump or against Hillary Clinton, and $52 million on behalf of Republicans in House and Senate races.[28]

The NRA spent three times more money helping elect Trump than they spent trying to elect Mitt Romney in 2012. But given that campaign disclosure laws currently do not require disclosure of internet advertising or get-out-the-vote efforts, it's likely they spent much more than that, upwards of $70 million.[29]

The NRA's omnipotent, mythical image may be as much responsible for their outsized influence as the reality itself. I remember few concepts from my undergraduate courses, but one I do remember from Political Science 101. When it comes to foreign policy, a country is as powerful as their enemies think they are. The NRA is as powerful as their enemies think they are.

In much the same way that the tobacco industry funded smokers' rights groups like the National Smokers Alliance, the pharmaceutical industry funds patients' advocacy groups, and the fossil fuel industry funds climate denial, the gun manufacturers fund the NRA as their front group. For the manufacturers of guns to be nakedly at the forefront of the public debate on gun use, legislation, and high profile mass murders would seem odd, if not perverse and self-serving.

It is likely that without the NRA, the gun industry would be suffering the same ignominy that beset the tobacco industry and their executives when it became incontrovertible that they were in the business of killing people by the tens of millions and that their model including getting children addicted. Granted, it took 40 years longer than it should have, but eventually tobacco executives were publicly hauled before Congress, shamed, and humiliated. Something similar for gun manufacturer executives will never happen, thanks to the NRA.

But polls show that gun owners are at odds with the NRA. Only about 5.5 percent of gun owners are NRA members. As a group, they are just as supportive of many stricter gun laws as the general public.[30] For example, 85 percent of gun owners support required background checks for purchases at gun shows, and 90 percent are in favor of closing the gaps in the government data bases intended to prevent the mentally ill, drug abusers, and terrorists from buying guns.

The NRA's phony sportsman image allows gun manufacturers to hide behind the flag of freedom and a "divine" constitution, and thereby cleanse themselves of culpability and callousness when the inevitable consequences of their products bear murderous fruit. The NRA can dress up its lobbying and campaign contributions as merely ideologically driven, and the gun industry can stay behind the curtain and shield themselves from the charge of facilitating mass murder for profit.

In the wake of the Las Vegas massacre, the worst mass killing in US history, the response of congressional Republicans was a few days of some muffled, sheepish talk about possibly thinking of, perhaps, maybe, someday, giving consideration to saying in fine print that it's naughty to own a bump stock, the after-market device that allowed the Las Vegas gunman to turn his semi-automatic weapons into fully automatic "machine gun" type rifles. Even that obscenely tepid response quickly disappeared, and then they advanced two bills that will make it even easier for a horrific Las Vegas sequel.

One of those bills would have deregulated the use of gun silencers, making it more difficult for law enforcement, first responders, and the public to quickly identify the source of the shooting. The other would allow people to carry concealed weapons across state lines, from states that allow such practice to states that do not.

Of course, other countries have had mass shooting not related to terrorism. Australia is a country with a high rate of gun ownership, conservative politics, and a culture that prizes rugged individualism, much like the United States. In 1996, in the popular tourist town of Port Arthur, Tasmania, a 28 year old man used a semiautomatic rifle to cut down patrons eating lunch in a cafe, leaving 35 dead and 23 wounded in the worst mass shooting in Australia's history. But in contrast to congressional paralysis in the US, the political parties from across the spectrum united to pass legislation that sharply reduced the availability of guns nationwide within a few months. They "banned automatic and semiautomatic firearms, adopted new licensing requirements, established a national firearms registry, and instituted a 28-day waiting period for gun purchases." [31]

The government also bought and destroyed 600,000 privately owned guns, about 20 percent of the nation's total, and raised taxes to pay for the $500 million cost. This comprehensive strategy even attracted the scorn of the NRA in the US, who added their incendiary rhetoric and political muscle to gun rights groups in Australia in fighting the program, but this time to no avail. In the 18 years prior to the Port Arthur massacre, 13 episodes of mass killing had occurred. After the 1996 legislative reforms--zero mass shootings. Between 1995 and 2006, gun deaths (homicides and suicides) in Australia dropped about 60 percent.[32]

Moreover, a more detailed analysis of Australian states and territories revealed that where the most guns were surrendered or confiscated saw the most reduction in intentional gun deaths. And the NRA contention that people denied guns will find other weapons to kill themselves and others and that you can't stop mass shootings is contradicted by Australia's experience.[33] A study of the reforms estimated that the gun buyback initiative alone saved about 200 lives a year.[34]

While Australian gun rights groups have managed to whittle away somewhat at these reforms, the legislative action remains largely intact, and the benefits and lives saved are difficult to refute, despite the NRA's persistent attempt to do so.

The absurdity of American gun laws is illustrated by a 2010 Government Accountability Office report that found that "during the past six years, individuals on the terror watchlist were able to buy firearms or explosives from licensed US dealers 1,119 times."[35] *Mother Jones* magazine has compiled a data base of the weapons used in all United States mass shooting from 1982 to 2017. They found that 82 percent of the weapons used were purchased legally under current, weak U.S. law.[36]

The gun industry has done far more than strangle every effort at modest legislation. Mark Rosenberg, a gun violence expert who, in 1996, directed the federal Center for Disease Control and Prevention's National Center for Injury Prevention and Control, explains that his center came under fire from the NRA for research that showed having a gun in the home markedly increased the risk of that gun being used for a homicide.[43] The NRA was not going tolerate anything that smeared the holy writ of gun ownership, so they went to work on the Republicans they owned in Congress, and in 1997 they passed the "Dickey Amendment" (Republican Rep. Jay Dickey of Arkansas, self-proclaimed "point man for the NRA"), which effectively ended research by the federal Center for Disease Control and Prevention (CDC) into gun deaths. It did so by pulling $2.6 million, virtually the entire budget, from research earmarked for gun violence.

The second front of the attack on the CDC came with the passing of an NRA-sponsored bill that prohibited the CDC from spending any funds ever, to "advocate or promote gun control." CDC directors accepted ever since then that any research into gun violence had become politically impossible. Despite the recent massacres in Las Vegas, Sandy Hook, Orlando, San Bernadino, Sutherland Springs, Charleston, and on and on, no research is being done to find answers.

After the movie theatre shooting in Aurora, Colorado, in 2012, Rep. Dickey had a change of heart and wrote an Op Ed for the *Washington Post* admitting his mistake. [44]

Perhaps the NRA's crowning legislative achievement was the passage of the Protection of Lawful Commerce in Arms Act of 2005. It is perhaps the most egregious and unprecedented government act of special interest protectionism of modern times. The law prohibits federal lawsuits being brought by municipalities against gun manufacturers for their role as accomplices in the death wrought by the use of their products. This legislative coddling essentially saved the entire industry.

In lock step with the NRA is the out of the spotlight work of ALEC, the American Legislative Exchange Council, the Koch Brothers–sponsored organization that orchestrates a marriage, or at least "live in" arrangement between large corporations and conservative legislators (see Çhapter 5). Anti-gun-control legislation is near the top of the ALEC agenda. For many years, the NRA actually co-chaired the ALEC "Task Force on Public Safety and Elections,"[37] producing bills opposing bans on semi-automatic rifles, opposing waiting periods for background checks, promoting concealed-carry laws, "oppos[ing] efforts by law enforcement to use their purchasing power [to] get gun manufacturers not to market guns or ammo likely to be used against police, like 'cop killer bullets' that pierce armor,"[37] and in various ways encouraged the carrying of guns on campuses and shooting activity for teenagers.

The same task force, chaired by Walmart (the largest retailer of long guns in the nation), came up with the infamous "stand your ground" laws that emerged in the national spot light after the Trayvon Martin killing in Florida.[38] These "shoot first" and "make my day" laws upended centuries of traditional self-defense doctrines, granting a presumption of immunity to killers if they could claim they feared for their safety, even if they had the safe option of merely walking away in a public setting. The gun-control group, Everytown, points out that "Stand Your Ground laws give every-day, untrained citizens more leeway to shoot than the United States military gives soldiers in war zones."[39]

The best that could be said for the laws is that they invited vigilantism, and the worst is that they are a license to kill. ALEC and the NRA succeeded in spreading these laws from Florida to 32 other states. The aftermath is that supposedly justifiable homicide has dramatically increased.[40] For example, in Florida, a 32 percent increase in gun deaths has been linked to the law, and a 300 percent increase in homicide ruled justifiable by law enforcement.[41, 42] Pure and simple, these laws replaced the sanctity of life with a sanctity for guns.

The NRA's on the forefront not just of holding sensible gun control hostage, but of trying to shut down public debate about gun control, and any empirical evidence that challenges the NRA narrative. The NRA's signature slogan, "Guns don't kill people, people kill people," is the headline for the NRA laying the blame for gun deaths on everything but guns--criminals, Hollywood movies, video games, and failures of the public to address mental illness.

There is much we don't know about gun violence, its causes, and how to reduce or stop it, and that is just the way the NRA likes it. They would much prefer we make public policy based on their Wild West anecdotes than on empirical evidence, facts, or science. But the NRA wants to snuff out science every bit as much as they want to snuff out gun control. They have strong-armed Congress into not even allowing research into the problem.

A few studies have escaped the NRA's bull's eye and their ability to blind the public. A 1998 study in the *Journal of Trauma and Acute Care Surgery* found that a gun in the home is "22 times more likely to be used in a criminal assault, an accidental death or injury, a suicide attempt or a homicide than it is for successful self-defense."[45] In other words, a gun in the home makes that family dramatically more likely to experience a tragic death, intentional or otherwise, than it is to prevent one. Another study in the world's premier medical journal, *The New England Journal of Medicine*[46] found that a gun in the home increased the risk of death by homicide between 40 and 170 percent. Another study similarly found a 90 percent increased homicide risk. [47]

Carrying a gun increases a person's risk of being shot during an assault about 4.5 times.[48,49] For every 1 percent increase in gun ownership in a state, there is a nearly identical increase, 0.9 percent, in gun murder rates.[50] Having a gun in your car is associated with aggressive and dangerous driving. Such drivers are 44 percent more likely to get into a road rage episode with another driver and 77 percent more likely to aggressively follow another driver attempting to intimidate the other party. It doesn't take a brain surgeon to grasp that the presence of a gun could escalate a petty, trivial, macho pissing contest into a deadly tragedy.

Most women who are murdered by their spouses or live-in partners are murdered with a gun. While the NRA insists that gun restrictions do not stop "bad guys" from getting a gun, in states that prohibit men who have been served domestic violence restraining orders from owning a gun, homicides of female domestic partners by guns are down 25 percent.[51] The NRA encourages women to carry guns to protect themselves. But a gun in the home of a family with a history of domestic violence increases a woman's chance of being killed by 500 percent. [52] In states with higher gun ownership rates, women are 4.9 times more likely to be the victim of a gun homicide than women in states with lower firearm ownership.

A gun in the home is twice as likely to be stolen as it is to be used successfully in self-defense. [53] The FBI looked at 160 active-shooting civilian events from 2000 to 2013. They found that only one was stopped by a person with a legal gun permit. In sharp contrast, 21 events were stopped by individuals without guns.[54]

The NRA preaches that mass shooters intentionally target gun-free zones because they are easy targets. There is absolutely no evidence to support that claim. Of 156 mass shootings between 2009 and 2016, only 10 percent occurred in "gun-free zones.[55] This tenet is contradicted further by the fact that a likely a high majority of mass shooters typically expect to die in a "blaze of glory" as the center of the nation's attention in the maelstrom they created.[56,57,58]

Of course, all guns can kill, but they are not created equal in their lethal potential. The NRA calls the AR-15 "America's rifle." And that label, on that weapon, speaks volumes about America and about the NRA. It is the weapon of choice for mass murderers and you can buy one for as little as $500. In Florida, you can buy one more easily than a handgun. Anyone over 18 with nothing on their record can buy one within minutes of walking into a store. It's inventor, Eugene Stoner, designed it to have an advantage in war over enemy troops using the Chinese AK-47.

Physics and anatomy are not often part of the gun control debate. They are both far beyond the concerns of the NRA or the comprehension of most super owners. But the physics behind firearms and the anatomical consequences they create tell perhaps the most important story about whether gun battles are the solution to American-only mass murders.

The kinetic energy (and therefore destructive potential) of a bullet is equal to one half of the mass of the bullet times its velocity squared. Bullets come of out an AR-15 at 3,300 feet per second, three times the velocity of the average handgun, producing kinetic energy ten times greater than bullets from a handgun. Furthermore, AR-15 bullets "tumble" as they penetrate flesh, widening the area of tissue damage. Peter Rhee, a trauma surgeon at the University of Arizona compares the damage from a handgun bullet to the AR-15 this way. "One looks like a grenade went off in there. The other looks like a bad knife cut."[88]

While a bullet from a handgun creates tissue damage confined to roughly the diameter of the bullet, unless it hits the heart or a major artery, the injuries are usually non-fatal, and the bullets are often stopped by bones or lodged in tissue. But the high-velocity, high-energy bullets from an AR-15 are specifically designed to cause massive collateral damage in the human body, leaving a much wider path of destruction, utterly destroying organs, not just putting a hole through them. Despite its rather small size, an AR-15 bullet can literally obliterate three inches of the largest bone in your body, the femur. Likewise, it can destroy a major artery even several inches away from the point of impact. It can turn the largest organs in your body, your liver and your brain, into mush. The exit wound can be as large as an orange. Most victims who die from AR-15 bullet wounds bleed to death on the spot, with no chance for survival. [89]

If a victim is "lucky" and survives, removing the dead tissue created by an AR-15 bullet will require a series of multiple, sequential operations, where that from a handgun usually requires only one. Survivors are often left debilitated, with impaired organ function, requiring more surgeries throughout their lifetimes, and they are often relegated to a life of chronic pain and permanent disfigurement.

The AR-15's range is more than 1,300 feet compared to a Glock hand gun at 160 feet. Because of the destructive potential mentioned above, accuracy is not particularly important for a shooter to wreak massive damage to the body. There's no recoil "kick" to an AR-15. It is light and easy to hold, making it extremely easy to pull the trigger, rapidly and endlessly. With a standard magazine that holds 30 bullets and is designed for quick change-out, a shooter can get off hundreds of rounds within a few minutes, especially if it's equipped with a bump stock that makes it fully automatic, requiring only one pull of the trigger to deliver all the bullets in the magazine. Every feature of the AR-15 is part of an overall package designed to kill many human beings as quickly and efficiently as possible. It is a weapon of mass destruction if there ever was one.

Eugene Stoner, inventor of the AR-15, not only didn't have an AR-15 for sport, hunting, or for self -defense, he did not own one at all. Although Stoner has long since passed away, his family says he would be horrified that civilians have access to this tool of mass destruction.[90]

On Valentine's Day, 2018, a former student burst into Marjory Stoneman Douglas High School in Parkland, Florida with a blazing AR-15 and countless magazines of bullets. He ended up killing 17 students and teachers within a few minutes, well before there could have been any opportunity for a defense to be mounted by anyone.

President Trump's answer was to arm more teachers, offering this fact-free, pearl of ignorance and misinformation: "This would be obviously only for people who were very adept at handling a gun, and it would be, it's called concealed carry, where a teacher would have a concealed gun on them. They'd go for special training and they would be there and you would no longer have a gun-free zone. Gun-free zone to a maniac--because they're all cowards--a gun-free zone is 'let's go in and let's attack because bullets aren't coming back at us.'"[59]

An entire book could be written on the irrationality and inconsistencies of this strategy. But with most mass murderers intending to commit suicide or go down in a blaze of glory, they would care nothing about battling armed math teachers. It could actually enhance their thrill of a gun battle where hand-gun wielding teachers would be badly overmatched by automatic, or semi-automatic rifles.

Trump's response used almost the exact same language as a speech given by Wayne LaPierre the same day to the Conservative Political Action Committee. And Trump drew the exact same conclusion. The answer to mass shootings is more guns, and of course, unsaid, is more profit for the gun manufacturers.

More NRA dogma is the refrain that "when guns are outlawed, only criminals will have guns." Those 10 states with the weakest gun laws have an aggregate level of violence three times greater than the ten states with the strongest gun laws.[60]

From the book, *Reducing Gun Violence in America: Informing Policy with Evidence and Analysis,* edited by Professors Daniel Webster and Jon Veronica at the Johns Hopkins Bloomberg School of Public Health, comes this insight. "Strong regulation and oversight of licensed gun dealers—defined as having a state law that required state or local licensing of retail firearm sellers, mandatory record keeping by those sellers, law enforcement access to records for inspection, regular inspections of gun dealers, and mandated reporting theft or loss of firearms—was associated with 64 percent less diversion of guns to criminals by in-state gun dealers."[61] Expanding concealed carry laws has also been shown to backfire. Those states that enacted these laws saw an increase of 13-15 percent in violent crimes rates.[62] That is particularly noteworthy in the context that other states continued to see a steady decline in violent crime.

Contrast the history of gun violence between two states that went in opposite directions on required background checks and permits for gun purchasing. When Connecticut implemented such a law, gun homicides fell 40 percent. When Missouri repealed a similar law, their gun homicide rate spiked 25 percent.[63]

The NRA has succeeded in depicting America's gun homicide epidemic as a problem of mental health rather than gun access. As a testament to their success, the majority of Americans believe that mental illness and the failure to identify the mentally ill who can be predicted to engage in violent behavior is an important cause of gun violence. But only about 4 percent of violence perpetrated by people upon each other can be attributed to serious, diagnosable mental illnesses like schizophrenia and bipolar disorders.[64] Furthermore, mental-healthcare professionals have not demonstrated that they are particularly adept at predicting which patients are going to be violent, and there is no reason to believe their assessments will improve in the future.[65]

If rampant mental illness were the answer to America's gun homicide, then one would expect this country to have much higher rates of severe mental illness than those countries with much fewer gun deaths. But the demographic prevalence of schizophrenia is revealing. Men perpetrate the vast majority of gun violence, but schizophrenia occurs nearly equally among men and women.[66] Mental illness is a strong predictor of gun-wielding suicide, but better identification of those patients offers nothing in terms of protecting gun violence perpetrated on others.

Psychiatrists Liza Gold and Robert Simon edited a book, *Gun Violence and Mental Illness,* that exposes the myth that mental illness is the root cause of gun violence and mass shootings. When the mentally ill shoot someone else, it is far more likely to be someone they know, not total strangers. The truly mentally ill are far more likely to be the targets of gun violence rather than the perpetrators.[67]

But relevance to exploring the contribution of mental illness to gun massacres rests upon whether preemptive diagnosis of potential shooters is actually possible. Given that, continuing to focus on America's gun violence as a mental health issue is, pure and simple, a ruse intended to distract. Improving society's ability to successfully diagnose and predict people likely to commit future gun violence would mean a new level of mass intrusion into the lives of millions of people—ironically, the exact kind of government overreach railed upon by the NRA.

Imagine the Pandora's box of sequestering, perhaps forcing some kind of treatment, counseling, or medication upon people like America's most prolific mass murderer Stephen Paddock, who was a 64-year-old successful real estate businessman with no history of violent behavior, mental illness, or criminal record, who had a long-term girlfriend, and gave no reason for his family or friends to be concerned. He was the NRA's poster boy for a "law abiding citizen," even while he amassed an arsenal to shoot at 20,000 concert goers. He was completely law abiding, showing no sign of mental illness, until the second he started firing into the crowd. It speaks to the glaring inadequacy of our gun laws, the complete inability to anticipate who are potential perpetrators of gun massacres, and the ruse that it is all a mental health issue.

Sorting out dozens of future murders among the hundreds of thousands of angry white male gun owners with aggressive, psychopathic tendencies would be impossible. And just how would a society take meaningful action against citizens who cannot be adjudicated as mentally ill, haven't yet committed a crime, but are suspected by a society to be capable of it? Even if these people were identifiable, hypocritically, the NRA is the last organization that would tolerate having those so identified stripped of their right to own a gun despite the risk they represent.

The psychological profile of gun owners has been well studied. The 310 million privately owned guns in the US are disproportionately owned by people who have baseline anger issues and are prone to impulsive, often violent behavior. Nine percent of American adults demonstrate poor anger management, impulsive behavior, and have gun access, according to self-reporting.[68] It reminds me of one of the best lines in *The Simpsons* TV show where Homer goes to a gun store, attempts to buy a gun, and is told there is a five-day waiting period. His brilliant response is, "But I'm mad now!"

Furthermore, the tendency to that kind of behavior is even directly proportional to the number of guns owned by an individual. And although it is obvious that these are exactly the people who shouldn't be allowed to own guns, of course these are the quintessential ground troops of the NRA, seduced by the fiery rhetoric of its leaders.

People with that type of personality disorder are hardly ever involuntarily hospitalized for a mental illness and are not subject to current mental health-related legal restrictions on owning and using firearms. This fact eviscerates the argument that better diagnoses, and/or restricting gun access to those diagnosed as mentally ill is the panacea or even a likely contributor to fewer gun deaths.

The leader of the study, Jeffrey Swanson, a professor of psychiatry and behavioral sciences at Duke University School of Medicine said, "Gun violence and serious mental illness are two very important but distinct public health issues that intersect only at their edges."[69] The authors of the study point out that the arrest history of an individual is more relevant than the mental illness history. One study showed that merely holding a gun can make you more likely to conclude that others are also wielding guns, "gun paranoia" if you will.[70]

Not surprisingly, racism has also been shown to influence whether a person owns a gun and is associated with their views on gun control legislation.[71] Gun-owning college students are more likely than their peers to drink excessively and engage in other behavior that put themselves and others at high risk for injury.[72] The authors of this study also found that "having a firearm for protection is also strongly associated with being threatened with a gun while at college."

Guns afford an owner a sense of power, intoxicating invincibility almost like nothing else. As a physician, I was attracted to the specialty of critical care medicine and anesthesia. Anesthesia can be fairly described as, "bringing patients as close to death as they will ever come, yet still keeping them alive." It can be exhilarating and occasionally terrifying, but it embodies the cliché about doctors playing God. Owning a gun allows just about anyone into a similar game of playing God, holding the strings of power over other people's lives. One of the Columbine High School killers said, "I would love to be the ultimate judge and say if a person lives or dies—be godlike."[73]

For all that is said about mental illness, motives of personal protection, symbolic defiance of imagined tyranny, and constitutional rights, I believe the most fundamental, and perhaps unconscious reason why people own guns, including those who use them irrationally, is because it makes them feel powerful, and in far too many tragic cases, they wield that power over the life and death of others.

Gun massacres that seem senseless to a rational society are usually carefully planned and make perfect sense to someone riddled with narcissism, inflamed by grievances, seeking revenge or ultimate control over others. NRA bombast by LaPierre and his cohorts deliberately elevates the thermostat of persecution, paranoia, and "righteous" retribution that drives its membership to purchase guns in hopes of teaching their antagonists a lesson.

The gun worship promoted by the gun manufacturers corrupts rational behavior among those who view gun ownership as a status symbol. My father was a forensic psychiatrist. I remember him telling me that he was asked by a court to give a psychiatric evaluation of parents who had had two teenage sons kill themselves in two separate accidental shootings. My father asked them, after losing the first son, had they considered removing the guns from their home? Their stunning reply was, "No, we would never deprive our children of their right to use guns." The parents weren't mentally ill according to any conventional diagnostic criteria, but their fascination with guns was more important to them than the lives of their children.

On the evening of June 17, 2015, parishioners were gathered at the Emanuel African Methodist Episcopal Church in Charleston, South Carolina, for a prayer service. If the NRA's prayer service at their annual convention involves praying for a society without reasonable gun control, then their prayers are certainly being answered. The prayers at the Episcopal Church, however, were not.

Dylan Roof, a 21-year-old "law abiding" white supremacist, broke into the church and murdered nine people with a Glock 45 caliber handgun. One of them was the pastor, Rev. Clementa C. Pinckney. In the NRA's world, the reason that occurred is because those praying should have been armed, locked and loaded, finger on the trigger while in prayer. As disgusting as that thought process may be, that is not the worst of the NRA's responses to the tragic murders. In a post that was later taken down, NRA director Charles Cotton blamed Rev. Pinckney, because he had once served as a South Carolina state legislator supporting gun control.[74] No mention of blame for the shooter, or for a system where a maniacal racist had no problem buying a gun.

On Feb. 14, 2018, at Florida's Marjory Stoneman Douglas High School, a modern Valentine's Day massacre took the lives of 17 adults and children. The most laudable response in the aftermath came from students at a nearby high school, who staged a walk out in protest, furious about political paralysis on gun control, and holding signs like, "the NRA is a terrorist organization."[75] But history reveals that in Republican-controlled states like Florida, in the year after a mass shooting, the tragic event leads to the exact opposite of what you would expect--on average, a 75 percent increase in legislation passed to loosen gun restrictions.[76]

The NRA loves to throw around a statistic that American gun owners use their weapons for self-defense 2.5 million times a year.[77] The source of that "statistic" has long been discredited.[78] Five years of data from the federal Bureau of Justice Statistics' National Crime Victimization Survey produces a number much lower, 67,740 times a year, i.e., 50 times lower.[79] Furthermore, only slightly more than half of those involved strangers (the NRA's bogey man, "bad guy with a gun"); the rest were acquaintances or family members. Another study found that the majority of self-reported, self-protection gun uses were illegal and/or counter to the overall interests of the common good.[80]

When threats to public health emerge (think Zika, Ebola, HIV, traffic accidents, asbestos, lead), Congress has responded with public policy to better understand and manage the threat. Not so with gun violence.[81] Even with robust scientific evidence to contradict NRA dogma, America takes the NRA's word for how to manage gun violence and accepts their solutions. Common sense and empirical evidence indicate more guns only make gun violence worse.[82]

Allowing the NRA to be their spokesperson, the gun industry sees the world in only two colors, black and white, "good guys" and "bad guys." There are no shades of gray or any other color. The only good guys are those devoted to gun ownership. The bad guys are everyone else--Democrats, liberals, even any Republicans who are deemed weak-kneed on gun control. Even the victim of a gun homicide, Rev. Pickney, was a bad guy because he had supported gun control. But if all this makes you think that the gun industry may not be the champion of "good guys" they claim to be, we haven't yet discussed their most revealing behavior.

In the wake of the Supreme Court's *Citizen United* ruling, The NRA has emerged as a most valuable player on the "dark money" team of the Republican Party. As a supposedly "social welfare" organization, the NRA is not required to disclose its donors, and a Karl Rove Super PAC and the Koch Brothers have used the NRA to funnel money to GOP candidates. But it is illegal to use foreign money on US elections. As I write this, the FBI is actively investigating strong evidence that the Russians used the NRA as a conduit for buying into US elections.[83]

Let that sink in for a moment. The most self-aggrandizing icon of American patriotism, the NRA, was apparently a co-conspirator in the most significant act of treason in modern American history, with our greatest foreign enemy, the Putin autocratic regime in Russia. *Rolling Stone* magazine rightly asks the question, "Does Russian influence have anything to do with the [violent], fascistic[84] turn in NRA messaging?"

State legislatures throughout the country enact volley after volley of NRA brain cramps, like open carry, arming teachers, guns on campus, stand your ground, all of which solidify American society as a Wild West gun culture, steadily raising the cumulative number of guns everywhere and the risk of death and tragedy to everyone. In the world of the NRA and the gun manufacturers, there are never enough guns in the streets, at church, at school, in your home, or under your pillow.

Wayne LaPierre's angry shouting that "the only way to stop a bad guy with a gun is a good guy with a gun" is worse than willful ignorance; it is the cynical, disingenuous mission statement of an entire industry that considers violent loss of human life just the price of doing business and making a profit.

The NRA's defiance of gun control in the face of the never-ending stream of gun massacres carries a societal cost far beyond the thousands of murders themselves. Its sends a proud and very public message that empathy, human decency, and reverence for human life are no longer desirable characteristics of American society. The NRA is hastening the arrival of the grotesque, dehumanizing hellscape that they claim to be fighting against.[87]

For American citizens to see that their government, their elected officials, business interests, and their fellow citizens will all stand idly by while gun manufacturers, through the NRA, brazenly facilitate and profit from the murder of thousands of their neighbors, friends and family members, establishes an official national theme that life itself is cheap. Even the lives of our school children are cheap. Pleas for gun control by grieving and tearful family members of victims plastered all over the news are met with cold indifference, or even worse, ridicule and contempt by the perverse occupants of social media's dark corners. It took only hours before the surviving students of the Stoneman Douglas massacre, who were speaking to the media and organizing protests, were maligned and mocked by right wing media and their devotees for their heartfelt outbursts following the unspeakable horror of seeing their friends riddled with bullets and bleeding to death.

The aforementioned Dana Loesch appeared on TV a week after the Stoneman Douglas massacre, accusing the "legacy media" of cheering for gun massacres because grieving, "crying white mothers are ratings gold."[85] The small kernel of truth in her diatribe is that in the era of 24-hour television and internet news, the attention devoted to mass shootings does incentivize the perpetrators, plays into their narcissism, and stimulates copycat murders.[86]

Everyone knows that the media aphorism, "if it bleeds it leads," is a valid criticism and has been since before the days of William

Randolph Hearst. And to be sure, although mass shootings get the most media attention, they represent only a small portion of the gun violence perpetrated in this country. But without an ounce of shame, or self-awareness of the hypocrisy involved, Loesch, the firebrand provocateur, attacked the media for prostituting themselves for ratings by covering the massacre. The NRA even released a video in the same time frame that slammed the mainstream media for being the "casting call for the next mass shooting." Of course, virtually no mention from either Loesch or other NRA brass of the role of guns in the tragedy and no mention of any sympathy for the victims and their families.

Conspiracy theories in the era of social media, Alex Jones, and Donald Trump have infiltrated virtually all sources of information. In no arena has this been more damaging than that of gun massacres, and they have been integral to desensitizing the public toward the deaths and violence of guns. After nearly every mass murder, conspiracy theories ooze like puss from an infected wound, claiming that the tragedy never happened, it was staged by the government, or was a collusion between George Soros and the Democrats. Even those mourning families are really just "paid actors" who travel from one tragedy to another. Ted Nugent, the fanatical right wing music star and long-time board member of the NRA, and Donald Trump Jr. gave life to at least one of the most malevolent conspiracy theories--that the Florida high school students were "crisis actors," paid to show up at mass shootings to rally public sympathy for an anti-gun agenda.

While these conspiracy theories are embraced by the gullible and poorly educated, they are circulated by millions through social media. They not only deceive and misinform but erode the normal empathy that people would otherwise feel for the tragedy and suffering of others. These hideous conspiracy theories provide many people an excuse for apathy toward ongoing the slaughter. When physical and emotional pain of others can be dismissed as "fake news," little is left to maintain the social and psychological bonds between human beings that are the bedrock of civilized society.

In cold blood, gun manufacturers have provided the tools for the murder, suicides, and traumatic accidental deaths of hundreds of thousands of men, women, and children. Moreover, they can be rightly condemned for contributing even more broadly to America's descent into a cruel and callous society. But this chapter is being written as high school students in Florida and throughout the country have risen up, organized protests and marches, and seemly are determined to bring a new chapter to this national plague. They have boldly condemned the NRA, the politicians of both parties for selling their souls to the NRA, and American adults more broadly for their extraordinary indifference.

Politicians and parents throughout the country are learning some very important lessons from these young high school students. They have become the nation's conscience and the flag bearers of common sense and compassion that is so badly needed to dismantle the power of gun manufacturers. It's far too early to say, but these high school students have brought a ray of hope that we all may be witnessing the twilight of the political omnipotence of the NRA and the gun industry

The victims of the Sandy Hook Massacre.

Alex Jones repeatedly claimed that the victims of the Sandy Hook massacre were child actors and no one was actually killed there.

Dana Loesch, the new fire breathing face of the NRA, accused the "legacy media" of cheering for gun massacres because grieving, "crying white mothers are ratings gold."

13

The Poisoning of Bhopal:
They Were Just Poor People

"Now I am become Death, the destroyer of worlds."

—Bhagavad-Gita

"Business is war" is a philosophy sometimes credited to Carl von Clausewitz, a Prussian general and military theorist. It should not be surprising, then, that businesses managed by that philosophy could leave war-like casualties in their wake. In 1984, Union Carbide made a business decision that killed victims on a mass scale in much the same horrific and agonizing way as the "war to end all wars."

In the 1970s, as part of the "Green Revolution," Union Carbide saw tremendous profit opportunity in feeding the masses of the undeveloped world. They chose Bhopal, India as the site to build a factory to manufacture the pesticide Sevin. The Bhopal site, in the rapidly expanding capital of Madhya Pradesh, was near the center of 'Old Bhopal,' at the heart of a metropolis of one million people. The factory site is surrounded by areas of tightly-packed slum housing and shacks and is within two kilometers of the main bus and train stations. The promise of work at the factory and the supporting businesses around the plant, brought many more families to the area adjacent to the factory.

In the early morning hours of December 3, 1984, more than 40 tons of methyl isocyanate (MIC) gas leaked from the Union Carbide plant. In the surrounding area, people were awakened by screaming that filled the night air, much like the way the destroying angel is depicted in the classic movie the Ten Commandments. The gas had attacked their eyes, noses, and throats, burning like fire. Then fluid started filling their lungs as they coughed and suffocated. Vomiting and diarrhea became uncontrollable. The gas hung close to the ground. Owing to their height, children and other people of shorter stature inhaled higher concentrations. Many people were trampled to death trying to run in the opposite direction of the plant.

The morning sun rose a few hours later on streets filled with corpses. At least 3,800 people had been killed immediately. Within the next 72 hours, 8,000-10,000 had suffered more prolonged, torturous deaths. Many were blinded before they succumbed from cardiac and respiratory arrest.[1] Overwhelmed by all the dead, the army began just dumping bodies in the forest and the Betwa River, which became so clogged with corpses that the river began to back up when the corpses became stuck between the arches of the bridges. Entire communities were wiped out. Two thousand buffalo, goats, and other animals succumbed and were buried. The leaves on all the nearby trees turned yellow and fell off within days.

In a later survey of 865 women who lived within a kilometer of the plant and who were pregnant at the time of the gas leak, 43 percent of the pregnancies did not result in live births. Of the 486 live births, 14 percent of babies died in the first 30 days compared to a death rate of 2.6 to 3 percent for previous deliveries of pregnant women in the area in the two years preceding the accident.[2]

All told, the eventual death rate for babies in utero at the time of the leak was about 50 percent. Within weeks of the disaster, doctors were aware that dozens of still births and spontaneous abortions were happening, and many fetuses were not even recognizably human. In March 1985, four months after the leak, a crowd of frightened pregnant mothers besieged a government hospital with bottles containing their urine samples. The women asked for the samples to be tested to check whether their babies could be born damaged, and to ask for sodium thiosulphate injections to rid their bodies of toxins inhaled on that night of horror.

Instead of treatment, tests, medical advice, sympathy or kindness they were driven away by police with sticks. But even as these scared women were being chased away, the Indian Council of Medical Research (ICMR), a government agency, was carrying out a double-blind clinical trial to test the efficacy of sodium thiosulphate injections as a detoxificant for those who were gas-exposed. Results of the ICMR studies took 22 years to be published. They revealed, a whole generation too late, that the injections could indeed have saved tens of thousands of lives.

Credible estimates are that eventually 30,000 people died from the chemical leak. And the dying continues, but the official agency for monitoring deaths has been closed since 1992. Another 50,000 victims are permanently disabled, and another 120,000 face significant, chronic illnesses. In a study of people who lived from one to ten kilometers from the Carbide factory, 12 months after exposure 71 percent of the exposed population showed evidence of chromosomal damage.[3]

Cerebral palsy, horrific birth defects, and cancer are ubiquitous in the people unfortunate enough to still live in the area. "Children are being born with deformities like cleft palate, three eyes, all fingers joined, one extra finger, one testicle, different skull shapes and Down's syndrome," says N. Ganesh, a researcher at the Jawaharlal Nehru Cancer Hospital and Research Centre (JNCHRC).[4] The ICMR concluded that over 520,000 people were sufficiently exposed to the chemical poisons from the Union Carbide leak to have had some damaging effect on their health.

Ironically, the plant wasn't even in operation at the time of the leak, because pesticide sales in India had been far less than forecast. Indian farmers weren't buying Union Carbide's products, but they became tragic victims nonetheless.

In 1981, three years before the catastrophe, Union Carbide formed the "Bhopal Task Force," a team of executives that, after failing to find a buyer for the plant, left no stone unturned in finding ways to make the Bhopal factory profitable once it was clear that the bottom had fallen out of the Sevin market. In 1981 UC's global profits were on the order of $800 million. By 1984 they had dropped to just $79 million.

By 1982, the Task Force economy drive cut the work force in half, and 150 employees were laid off from their regular work schedule and used as "floating" labor. The work crew for the MIC plant was cut in half from twelve to six workers. The period of safety-training for workers at the MIC plant was trimmed from 6 months to a mere 15 days. Faulty pipes were replaced with cheaper metal. Malfunctioning critical equipment like gauges and safety valves weren't replaced. Safety systems were allowed to be disengaged and weren't repaired. Several vent gas scrubbers had been out of service as well as the steam boiler, intended to clean the pipes. The critical refrigeration system was switched off in order to save $37 a day in Freon.[15] The MIC tank was filled to a much higher level than their own loose rules allowed.[5]

The Indian workers at the time of the leak had virtually no training, spoke no English and were expected to maintain the plant using only manuals written in English. Minor accidents became so common place that the factory's alarm siren was turned off so as to stop bothering the neighbors constantly. US engineers had audited the plant in 1982 and found nearly 61 safety violations, half of them critical, 11 of which involved the handling and storage of the MIC that eventually leaked.

Two years before the accident, a journalist for the local newspaper had written a series of articles based on interviews with alarmed employees at the plant. He wrote that "Bhopal was about to be annihilated" by the plant. The employees told him, "It will take just an hour, at most an hour-and-a-half, for every one of us to die."[15] Factory staff had even gone to the extent of putting up posters in the community warning about eminent disaster. The journalist, Raj Keswani, tried to get the minister of the province to investigate the factory, before Bhopal "turns into Hitler's gas chamber"[15] The last in his series of articles was published two weeks before the disaster.

The warnings were precipitated after the death of a fellow employee was killed by a spill of phosgene, the deadly WWI chemical warfare agent. Phosgene was being used at the plant in the manufacture of MIC. But the MIC was far deadlier than phosgene, extremely volatile, and wickedly unstable to the point where engineers recommended it not be stored at all, but only in small quantities if necessary, under optimal conditions. The MIC was being stored in a tank the size of a semi-truck. The conditions at Bhopal were pretty much the opposite of optimal.

UC had also instituted a complete double standard in design, safety and operations, adopting inferior and inherently more dangerous conditions than it chose to install at its sister factory in West Virginia. These differences in design, along with the savage cutbacks overseen by the Bhopal Task Force, are the reasons that on Dec 3, 1984 there was "no immediate detection of a problem, when the problem became apparent it couldn't be found, when it was found it couldn't be contained and when the gas leaked it couldn't be neutralized."[6]

UC was warned by American experts who visited the plant after 1981 of the potential of a "runaway reaction" in the MIC storage tank. Local Indian authorities had warned the company of the problem as early as 1979, but essentially nothing was done. Numerous more minor accidents had occurred in the few years leading up to the big leak. Some sent as many as 24 workers to the hospital.

UC has long claimed the blast was triggered by employee sabotage. But it was clearly gross negligence and butchering of safety standards at the plant that was directly responsible.[7] It was corporate, free market capitalism at its worst, and a callous disregard for human life that caused the worst industrial accident in human history.

UC immediately tried to dissociate itself from legal responsibility. Eventually it reached a settlement with the Indian Government through mediation with that country's Supreme Court and accepted "moral" responsibility. It paid $470 million in compensation, a total of only $370 to $533 per victim - a sum too small to even pay for most of the medical bills, let alone provide any just compensation. Union Carbide ultimately merged with Dow Chemical, who refuses to assume these liabilities in India - or clean up the toxic poisons left behind.

But the disaster left behind by the plant is far more than just the leak in 1984. Union Carbide's Sevin factory was a "disaster" for fifteen years before the gas disaster. Company officials had allowed a routine of dumping highly toxic chemical wastes inside and outside its factory site. Some were buried, some simply lay dumped on the soil, open to the buffeting of nature, scattered by the wind, and seeping into ground water.

According to the leader of one of the survivors' organizations, Rashida Bee: "As early as 1972, they had discussed various proposals to stop it happening – but they ignored all of them. Instead, knowing the dangers, they okayed the dumping of thousands of tons of solid and liquid chemical waste in and outside the factory. They knew it would poison our water and our daily lives and they did it anyway."[13] "Thousands of tons of pesticides, solvents, chemical catalysts and by-products lay strewn across 16 acres inside the site. Huge chemical evaporation ponds, covering an area of 35 acres outside the factory, received thousands of gallons of virulent liquid wastes."[14]

By 1980, evidence that chemicals were leaking from the factory emerged when cattle in nearby fields began to die.[8] UC denied responsibility, but nonetheless paid compensation to the animals' owners. After the catastrophic gas leak, the factory gates were locked and the facility was abandoned. Not even a token attempt was made to remove the toxic chemicals and hazardous wastes still on site.

Union Carbide's parting gift to Bhopal was to leave the dilapidated factory and its surrounding property a perpetual environmental and public health disaster.

Since 1984, monsoon rains beat upon the decaying plant causing toxic waste ponds to overflow their banks. Lethal chemicals seep down through the loose soil, and migrate into the ground water contaminating the only water supply for the people living in eighteen surrounding townships. Among them is JP Nagar, the community most devastated by the gas leak of 1984.

More than 40,000 people still live near the factory which sits at a slightly higher altitude than the surrounding residential areas. That means the contaminated water flows downhill. Disturbingly, some people actually moved into the area after the murderous disaster because the real estate became extremely cheap. The residents are now chronically and persistently exposed to toxic chemicals through groundwater and soil contamination, and the suffering, disease, and tragedy continue. The residents complain that their water burns when they drink it, but they drink it anyway because there is no other choice. A long list of illnesses plagues the residents, including cancers, birth defects, kidney disease, depression and numerous other mental illnesses, as well as everyday aches and pains, skin rashes, blisters, fever, headaches, stomach pains, nausea, dizziness, vertigo and constant fatigue.

At least three generations of people will have had their lives cut short, ruined and or severely impacted by Union Carbide. Many of these people have been doubly abused, once by the horrible gas leak, and twice by the chronic contamination of the abandoned site before and after the leak.

In 1989 UC conducted secret tests and learned that its abandoned factory site was still lethally contaminated. Water from the site, and water in which soil had been soaked, produced instant 100 percent mortality in fish. The places where the samples had been taken were often just inside the boundary wall and on the other side where villagers lived, drawing water from wells and above-ground pipes. A later 1997 study, commissioned by UC, warned that pollution of the underground aquifer which feeds local drinking water wells could be happening at a rate far faster than anticipated. Still the company kept silent and did not warn people, nor did it lift a finger to clean up the contamination.

Company managers, in India and the US, knew the families devastated by the gas leak drank, cooked, and bathed their kids in the water they had poisoned. When environmentalists expressed concern about soaring rates of cancer, birth-defects and early deaths, UC denounced them as "mischief-makers," and even wrote to Indian government authorities suggesting some vague action be taken against them.[9]

Finally, in 1999, 2002, and 2004, Greenpeace tested soil and water in 14 neighborhoods near the factory. In some places they found mercury levels 6 million times higher than background levels and a potpourri of 30 chemicals in the water, many known to cause cancer and birth defects. Mercury had been used in the factory as a machinery lubricant. A 2002 study by the "Fact-Finding Mission on Bhopal," found lead, mercury and organochlorines in the breastmilk of nearby nursing mothers.

A 2009 study by Bhopal Medical Appeal, a non-profit trying to help victims by operating free medical clinics, revealed chloroform concentrations exceeding guideline values by a factor of between 4 and 7 times and carbon tetrachloride between 900 and 2400 times. The study also found at least 15 other highly toxic chemicals present in the drinking water samples, at levels that greatly exceeding the WHO safety guidelines, and many other chemicals present for which there are no safe guideline values. This report not only supported the findings of previous reports, but also demonstrated that the contamination of the local ground water was in fact getting worse.

As if Bhopal residents hadn't suffered enough, a conspiracy of silence was enforced by local government officials worried about their culpability more than they were about the victims of the disaster. The Indian Council of Medical Research (ICMR) stopped all research into the health effects of the gas in 1994 and has yet to publish the findings of the 24 research studies it had carried out up to that point involving over 80,000 survivors. A ban was placed on their publication by the Indian Ministry of Chemicals and Fertilisers until 1996. Reporter Tim Edwards found only one doctor that would talk to him. "The government doctors won't say a word about gas problems. They are under orders not to talk about these things. If they talk to the press they will say there are no problems."[10]

Drugs for only temporary symptomatic relief have been the mainstay of medical care by the local doctors ever since the morning of the disaster. Basically it has consisted of indiscriminate prescription of steroids, antibiotics and psychotropic drugs which probably does little to help the victims and may even compound their problems.

The Bhopal Medical Appeal makes this observation in their solicitations for donations: "The Bhopal Memorial Hospital & Research Centre (BMHRC), which is meant to prioritize gas victims, has a fully-fledged nephrology department, but the gas-affected are routinely denied dialysis, crucial in saving lives and halting further damage, to give the priority to fee-paying patients. Gas victims – contemptuously referred to as 'gasees' at BMHRC—are told to come back in three months. While waiting, many die." Government doctors often refused to touch the victim patients.

Edwards raised these very disturbing questions in 2002: "Why would a government so greatly weaken its own position in a settlement dispute? Why would it prevent medical information on the gas being publicly available, information that could benefit the efforts of physicians working with the aftermath of gas poisoning? Why indeed, even 18 years later, was there an official climate of secrecy and cover-ups concerning what a foreign company had done to the very people that the same officials are assigned to represent and protect?"[10]

Why the Indian government took such little interest in helping Bhopal victims deserves its own book. But friendliness to corporations certainly loomed large in the decisions to remain largely passive bystanders. In 2006, a victims' rights campaign organized by the International Campaign for Justice in Bhopal, planned a march to travel from Bhopal to Delhi to demand the federal government of India force a clean- up of the contaminated site and fairly compensate the victims. The distance is 500 miles. About 70 people walked for 36 days to get there.[12] One of them was Sanjay Verma. His parents and five of his siblings died in the disaster. Sanjay was orphaned, and like many of the other marchers who suffered from various bizarre and hard-to-explain illnesses, Sanjay suffered a stroke at the age of 20. When they finally arrived in Delhi, the prime minister, Manmohan Singh, refused to meet them. Then Sanjay and some of the other marchers went on a hunger strike, which eventually produced a meeting with Singh. Despite many promises made, nothing of substance actually came from the meeting.

The "Polluter Pays Principle" applies in India as in the US, but Union Carbide and its present day parent Dow Chemical, flatly refuse to clean the factory site themselves, or pay a cent towards someone else cleaning it up, or towards the care of those it has already damaged.

These people remain unofficial victims, denied compensation or medical help. Their only care comes from the Sambhavna Clinic in Bhopal. Sambhavna means "possibility." It was founded in the mid-1990s by survivors of the disaster. All who come, no matter how destitute, are welcome. The majority of the clinic staff were injured themselves by the now vacant plant or live where the water remains poisoned.

The clinic sees patients like 14-year-old Adil. He must crawl to go anywhere because his legs are too weak to walk. Suraj, another patient, cannot stand or walk because his legs are paralyzed. He was born with severe brain damage, and cannot speak a word. All he can do is "roll over and over and smile a wide, brilliant smile." Suraj's mother drank from the well poisoned by Union Carbide during her pregnancy and while she nursed him, even though it smelled and tasted bad. She had no other choice; no other water was available.

In 1987, a Bhopal District Court charged Union Carbide officials, including then CEO Warren Anderson, with culpable homicide, grievous assault and other serious offenses. Anderson skipped bail and fled to the United States. In 1992, a warrant was issued for his arrest, but he was never extradited to face criminal charges for thousands of deaths. It took the Indian government 19 years to make a formal request for his extradition to face trial. After that the Indian government made multiple attempts to extradite Anderson for trial, but President Obama, like other presidents before him and the US Departments of State and Justice all willingly harbored one of the world's most wanted criminals for 30 years.

On Sept. 29, 2014 Anderson died a free man in a nursing home in Vero Beach, Fl. Survivors of the Bhopal gas tragedy assembled outside the now-defunct pesticide factory and placed a large portrait of him. Then, one by one, they spat at the photograph.[11]

Anderson, of course never lived in one of the typical Bhopal shacks made of flattened metal cans, rubber tires, wooden planks and disintegrating plastic sheets that most of his victims live in. Until his death he enjoyed a very comfortable, privileged life with three homes —in the Hamptons, Florida and Connecticut—to help him soothe his troubled conscience. And Dow Chemical is doing just fine for those of you who are concerned.

Dow Chemical bought Union Carbide in 2001, including all their assets and liabilities. Dow, itself tainted by its role in manufacturing Napalm for use in the Vietnam War, has repeatedly denied that they are responsible for any more compensation to victims or for any further clean-up of the ghostly, hellish abandoned plant site. In fact Dow Chemical has shifted the blame for the ongoing humanitarian crisis in Bhopal to the Indian government and to the Indian battery manufacturing company Eveready Industries.

The world has largely forgotten Bhopal and all of its victims. Only when a movement to hold international corporations accountable by protesting their participation in the 2012 London Olympics did the issue re-emerge. Dozens of disabled kids affected by the 1984 Bhopal disaster held their own Games to protest against the London Olympics sponsorship by Dow.

The "Bhopal Olympics" kicked off with an opening ceremony a little less lavish than that in London. Children suffering from cerebral palsy, partial paralysis and mental disabilities paraded in wheelchairs and walked only with the assistance of others around an outdoor stadium in the shadow of the old pesticide plant. One of the events was called "the crab walk"--three children who were unable to stand propelled themselves down the 25-meter racecourse with their hands. Not a single political leader, or any prominent social worker turned up during the event or even to sponsor refreshments to the participants. There was no TV coverage or multi-million-dollar endorsement contracts following the medal ceremony, and none of Bhopal's victims will ever appear on a box of Wheaties, People magazine, or the Tonight Show.

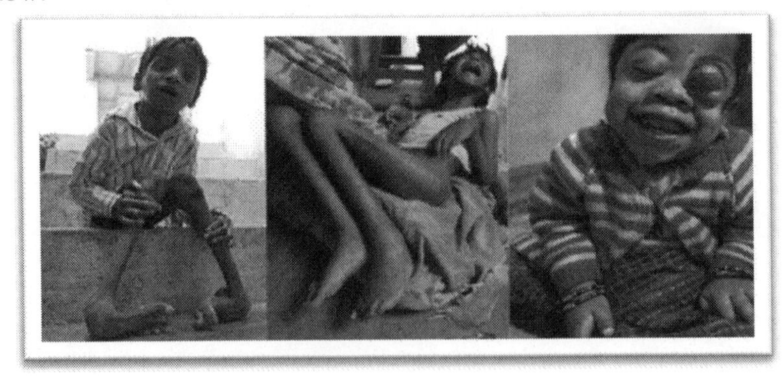

Children with birth defects from the Bhopal tragedy

Protestors against Dow Chemical sponsoring the 2012 London Olympics

One of the child victims at Bhopal

More Bhopal victims

Warren Anderson, Union Carbide CEO in 1984

14

A King Size Lump of Coal For Christmas

"A corporation often means the protection of a few to the detriment of the many."

—F. A. Klein,
attributed, Day's Collacon

On Dec. 22, 2008, just before 1:00 am, the nation's largest utility, the Tennessee Valley Authority, gave Roane County, Tennessee, the worst possible Christmas present--an enormous lump of coal, actually already burned coal--coal ash to be more precise. A dike holding back billions of gallons of a coal ash slurry broke, sending 1.1 billion gallons down the Clinch and Emory Rivers near the city of Kingston. In just a few hours 5.4 million cubic yards of a ghoulish, gray slime smothered 300 acres of land nearby and over 20 houses to a depth as high as six feet. The volume of the coal-ash slime was 40 times greater than the Exxon Valdez oil spill of 1989. It was the largest environmental disaster in US history, until the infamous Deep Water Horizon oil spill in the Gulf of Mexico in 2010.

The ash is a byproduct of coal combustion in electrical power plants, and it is deadly, hazardous waste. As residue after coal combustion, the ash is riddled with concentrated toxins that don't combust, like heavy metals--arsenic, lead, cadmium, vanadium, selenium, zinc, cobalt, chromium and mercury, and radioactive metals like uranium, radium, thallium, and thorium. Although the levels of radioactive elements in coal are rather small, when coal is burned those residual elements become concentrated by a factor of about ten in the fly ash that ascends up the smoke stack.

The ash sent to landfills comes from the fine particles of carbon captured in pollution control devices in the smoke stack, and from the bottom ash which is primarily larger, coarser particles that fall to the bottom of the furnace, similar to what is left over after burning wood in a fireplace.

The toxic potential of coal and other solid fuel combustion was highlighted in a landmark health report in 1775 by Sir Percivall Pott, an English surgeon, who found high rates of scrotal cancer in chimney sweeps. It is regarded as the first observation of occupationally caused cancer. Young boys of the time, whose body frames would allow squeezing into small spaces, were routinely sent naked up narrow chimneys to scrub them out. (Don't laugh, we've all done it). The chimney sweeps were also known for having "soot warts," ugly skin lesions, sometimes hundreds of them on their scrotums and other parts of their bodies, that often preceded the scrotal cancer. Multiple studies from the 1980s and 1990s have shown an increased risk of mortality among chimney sweeps of about 30 percent, from ischemic heart disease, respiratory diseases, and just about every type of cancer.

Exposed to a toxic brew similar to what chimney sweeps would have been exposed to, over 900 blue collar workers were hired to clean up the coal ash disaster in Kingston. Most of the workers labored without any protection for 12 to 14 hours a day, some for just weeks, others for months or years on end. Ten years later, of the 900 clean-up workers, 30 of them are dead, and more than 250 are sick or dying[1,7]

The list of diseases provoked by exposure to the ash is very similar to the diseases suffered by the chimney sweeps--various types of cancer, lung diseases and skin disorders. And much like the chimney sweeps, the Kingston clean-up workers sacrificed their health and life expectancy for a paycheck, and the profitability and comfort for someone else.

Harry Hemingway was one of those 900 workers exposed to the coal ash for two years. Although he was allowed to wear clothes, unlike the chimney sweeps, Harry was not allowed to wear the right clothes, much like Cinderella (the name literally means 'little ashes"). Perhaps because Harry had no "fairy godmother" to intervene, he was not allowed any protection other than a hard hat. In that work area, a hard hat was about as useful an accessory as a pair of glass slippers because the real occupational danger was the soft, fine powdery coal ash once it had dried out.

Hemingway was a "dredger" and general clean up worker constantly exposed to the coal ash. Every day he would leave work covered with the fine gray powder. In March 2017 he broke his arm while swinging a hammer. It turned out to be a pathologic fracture, meaning the strength and integrity of the bone had been eroded by an underlying disease. Despite chemotherapy and radiation, he died of a blood cancer, multiple myeloma, in August 2018.[2]

An investigation into the spill and the clean-up revealed that the TVA, and a contractor hired by the TVA, Jacobs Engineering, repeatedly lied to the workers about the toxicity of the ash, and "refused to provide them protective gear, threatened to fire them if they brought their own..."[1] Jacobs and the TVA apparently didn't want the media to see workers with respirators and protective clothing because they didn't want the public to think that the coal ash was toxic. They brushed off the workers' symptoms, pulling completely out of thin air, a claim that they were related to high pollen counts.[3]

They went so far as to mock the workers' complaints of health problems, and refused doctors' orders to supply the workers with masks and respirators. The TVA claimed that the coal ash was no more toxic than regular household garbage. Independent testing done immediately after the spill showed very high levels of radium and arsenic, hardly a surprise given what is well known about the witches' brew that is coal ash slurry. But a few weeks after the spill, the TVA turned control over the clean-up to Jacobs, including testing of the toxin levels in the area. After that, the tests showed much lower levels, consistent with what is "legally permissible" according to the involved state and federal agencies. Suspiciously, the EPA ruled those initial tests showing higher levels of toxins, were not reliable.

Meanwhile workers' testimonies and secretly taken videos indicate a multi-pronged strategy by Jacobs to manipulate the test results, all to show less worker exposure. Secretly taken videos by workers showed that Jacobs would dump out the residue from testing monitors before they were sent out for quantitative analysis, so that the levels of coal ash inhaled by the workers would appear artificially lower. Water trucks would spend most of their time near the stationary monitors to reduce the amount of ash settling on the filters. Personal monitors worn by the workers were supposed to be issued at random, but the workers noticed that those who worked in the dustiest areas of the clean-up were seldom given those monitors to wear, and that the monitors were used primarily on rainy days when the dust would have been minimal anyway. Not surprisingly Jacobs' attorneys deny all these workers allegations.

Jacobs Engineering is a multi-billion-dollar construction firm, with annual revenues of around $12 billion. The TVA paid Jacobs and other contractors over $1.2 billion for the clean-up. The contract with Jacobs allowed the TVA to bill rate payers $27.7 million supposedly to cover Jacobs' costs to keep workers and the community safe and protected during the clean-up. But little of that $27.7 million was spent for "decontaminating" the workers at the end of their shifts, as they were only given a bucket of water and a brush.[4] But Jacobs did manage to devote over $1 million of that allowance to decontaminate their own vehicles during the clean-up.

Five months after the spill and the clean-up had begun, Mark Kovak, an EPA national safety manager, informed the TVA and Jacobs that the workers who were most exposed to the coal ash should be given at least protective coveralls. Jacobs and another government contractor, The Shaw Group, pushed back and eventually pressured the EPA to overrule Kovak's request. The EPA even acquiesced to the TVA in removing the words "hazardous waste" from warning signs for the public to avoid contact with the coal ash.[5]

Not surprisingly, more than 70 of the workers have filed lawsuits against the TVA and Jacobs. Court documents reveal e-mails from the "safety" officer at Jacobs, Sean Healey, to Jacobs managers directing them to convey to the workers, the public, and even the EPA that they could "eat a pound of coal ash a day," and not suffer any ill effects.[6]

Attorneys for Jacobs claim that despite coal ash being full of toxins, the plaintiffs cannot prove that the ash was the cause of their suffering and death. That defensive strategy must be put in context. If a person who smokes cigarettes and eventually gets lung cancer, despite the known and proven link between smoking and lung cancer—87 percent of lung cancer is related to smoking—no one can say for sure that the smoking caused the cancer in any particular person. Smoking increases a person's risk of lung cancer between 15 and 30 times. But only about one in seven smokers will get lung cancer. And there are other causes of lung cancer—air pollution, asbestos, tuberculosis, and radon in buildings for example. Despite the uncertainly inherent in these statistics, there's no question that smoking causes lung cancer or that arsenic or radium also cause cancer. Whether Harry Hemingway's multiple myeloma was provably caused by his exposure to coal ash depends on your definition of "proof." What is not in question is whether Hemingway should have been exposed to coal ash without any protection, and that his risk of a wide range of serious and fatal illnesses was increased.

Six years after the spill, when the illnesses and deaths of some of the clean-up workers was already evident, the American Coal Association still achieved a particularly sinister lobbying triumph in pressuring the EPA and Congress to water down regulations on how coal ash is to be handled, authorizing it to be treated as household garbage.

On November 7, 2018, after hearing three weeks of testimony, a federal jury deliberated only five hours to render a guilty verdict on Jacobs Engineering for negligence in causing the deaths of their clean-up workers.[7] At the time of this writing, damages have not yet been set, and likely appeals will carry an ultimate resolution and compensation for the victims far into the future (see chapter 10 on asbestos settlements). But meanwhile those workers' families are currently getting another Christmas present from the TVA and Jacobs- - they are having to pay for their funeral expenses, and their own medical care and legal bills. Adding a final insult to the injury for everyone that is not Jacobs Engineering, ratepayers will also end up paying for Jacobs' legal fees.

One of the homes destroyed by the 1.1 billion gallon river of coal ash

Clean-up worker covered with toxic coal ash

15
The Financial Industry:
Death by Accounting

"Corporations are people, my friend ... of course they are. Everything corporations earn ultimately goes to the people. Where do you think it goes? Whose pockets? Whose pockets? People's pockets. Human beings my friend."

Mitt Romney
speech in Des Moines, Iowa, Aug. 11, 2011

"No, Governor Romney, corporations are not people. People have hearts. They have kids. They get jobs. They get sick. They thrive. They dance. They live. They love. And they die. And that matters. That matters. That matters because we don't run this country for corporations, we run it for people."

Elizabeth Warren
speech at Democratic National Convention, Sep. 5, 2012

The global financial crisis of 2008, which, at an economic loss of over $20 trillion, caused about 34 million people to lose their jobs, homes and pensions in the worst recession since the Great Depression. It was an avoidable financial and humanitarian disaster caused by widespread failures in government regulation, corporate mismanagement and heedless risk-taking by Wall Street, according to the conclusions of a federal inquiry.[9]

The financial institutions that caused the crisis were largely spared the consequences of their fraud with the largest global bailout in history, and the individuals responsible remained virtually untouched.

The financial crisis became a human crisis. The World Bank estimated that 53 million people worldwide were thrown into poverty because of the recession. A UNESCO report calculated that the financial crisis caused a loss of 20 percent of the per capita income of Africa's poor and that between 200,000 and 400,000 babies died annually, directly because of it.[1] Millions of children in sub-Saharan Africa alone have suffered severe malnutrition and long term irreversible brain damage as collateral damage from the financial disaster.[2]

Anxiety, fear, depression, crime, domestic abuse, drug abuse and addiction, murder and suicide all increased world-wide because of the financial crisis, all of which would have increased national mortality rates. Rates of depression increased between four and five fold in the wake of the recession. Anxiety and depression are known risk factors for numerous chronic diseases, like cardiovascular disease, diabetes, stroke, and Alzheimer's, all of which can lead to premature death.

Suicide rates rise and fall with the state of the economy. Unemployment and foreclosure are the largest triggers in increased suicide risk. About 35,000 Americans die every year from suicides, up about 28 percent since 1999.[3] Researchers from the University of Oxford and the London School of Hygiene and Tropical Medicine estimated that the suicide rate in the US rose 4.8 percent between 2007 and 2010. Suicide rates in Europe, where the recession was even more severe, rose even higher, by 6.5 percent. A study published in the British Medical Journal calculated that in the first year after the banking and financial crisis of 2008, there were 5,000 extra suicides in Europe and North America.[4]

Overall there were 10,000 suicides in North America and Europe triggered by the 2008 recession. The authors of this study warned that the 10,000 suicides may be a conservative estimate and that, "Suicides are just the tip of the iceberg" regarding the true extent of the misery and mental health toll inflicted on the general public by the corporations responsible for the financial collapse.[5] Indeed, it is estimated that for every successful suicide, 30-40 people have made unsuccessful suicide attempts.[6]

Ervin Lupoe from Wilmington, Calif., shot his five children and wife to death before turning the gun on himself. Lupoe was deep in debt, behind on his mortgage and had been fired from his hospital job. In Italy, Giovanni Schiavon, 59, a contractor, shot himself in the head in the office of his construction company. As he faced the inevitable reality of having to order Christmas layoffs at his family firm that had survived two generations, he wrote a last message: "Sorry, I cannot take it anymore."[7]

On December 5, 2008, Ricky Guseman of West Palm Beach, Florida, became another victim of the Wall Street engineered global financial crisis. Ricky was going to be evicted from his mobile home. Instead, local officials told the South Florida Sun-Sentinel, he "barricaded himself in a mobile home… set the place on fire and then shot himself in the head with a shotgun."[8]

The 2008 recession increased unemployment, and unemployed people have a higher overall mortality rate than employed people. Millions of people lost their health care because of the great recession, and that loss of health care was fatal for hundreds of thousands of people, perhaps millions. A study published in the prestigious British medical journal, *The Lancet*, calculated that because of the recession related loss of health care, between 2008 and 2010, an extra 500,000 people died worldwide of cancer alone.[12]

It was the private market -- not government programs -- that made, packaged, and sold most of the wretched loans responsible for the 2008 financial collapse without regard to their quality. The packaging, combined with credit default swaps and other esoteric derivatives, spread the contagion throughout the world. That's why what initially seemed to be a large but containable US mortgage problem touched off a worldwide financial crisis. The speculative binge was abetted by a giant "shadow banking system" in which the banks relied heavily on short-term debt. The credit rating agencies were essentially bought off by their clients for "services rendered." Regulators lacked the political will to scrutinize and hold accountable the institutions they were supposed to oversee.

The financial industry spent $2.7 billion on lobbying from 1999 to 2008, while the individuals and committees affiliated with it made more than $1 billion in campaign contributions. One would think that after such a worldwide debacle, those who caused it would be chastened, or laying low until the memory of it had largely vanished. And that expectation is incredibly naïve. The banking industry began pressing a full frontal assault on repealing Dodd-Frank, a small, inadequate attempt to prevent them from destroying our economy again, almost as soon as it was passed, and they continue to this day.

But to appreciate just how unchastened, how callous, how pathologic Wall Street titans are, one must give credit to financial journalist Kevin Roose for crashing the annual induction gala for the Wall Street secret fraternity, Kappa Beta Phi. As told in his book, *Young Money*, and in *New York* magazine, [10] Roose describes a truly amazing, other-worldly experience of slipping undetected into this secret celebration of two hundred of Wall Street's barons. Kappa Beta Phi includes many of the richest and most famous investors in the world, including executives from nearly every "too-big-to-fail" bank, private equity megafirms, and major hedge funds. This secret fraternity of financial elites, founded at the beginning of the Great Depression, acts as a sort of Friar Club/Opus Dei hybrid for the one tenth of one percenters. Each year their gala dinner features comedy skits that make fun of liberal politicians, gays, and the poor, musical acts in drag, off-color jokes, and the code of silence was always, "What happens at the St. Regis stays at the St. Regis."

The financial bacchanal secretly observed by Roose in 2012 featured disgusting missives of the victims of the financial crisis, and boasting about their financial triumphs at everyone else's expense. After ruining the lives of millions of people, the disregard they revealed for the collateral damage left in the wake of their excess was literally breathtaking.

As Roose said, "Here, after all, was a group that included many of the executives whose firms had collectively wrecked the global economy in 2008 and 2009. And they were laughing off the entire disaster in private, as if it were a long-forgotten lark. (Or worse, singing about it — one of the last skits of the night was a self-congratulatory parody of ABBA's 'Dancing Queen,' called 'Bailout King.') These were activities that amounted to a gigantic middle finger to Main Street." Inhumane, chilling psychopaths by any measure. This was no less a fraternity of Norman Bateses, ruining and ending lives with computers and financial statements instead of knives.

Employees of the notorious criminal enterprise, disguised for years as the Enron corporation, for a time had set the standard for businesses to callously disregard human life. California's 2000-2001 electricity crisis resulted in numerous days of rolling blackouts over the course of several months, a $5.7 billion hike in energy prices for consumers, and $11 billion being drained from the public treasury of the state of California. This was a direct consequence of skyrocketing of electricity prices that went from $40 a megawatt to as much as $1,000 a megawatt in 2000-2001, paving the way for severe fiscal austerity in the California state budget, leading to massive cuts in public education, the state's university system, health care, and virtually all social programs.

In secret deals with power producers, Enron traders deliberately conspired to drive up California electricity prices by, among other things, ordering power plants shut down. Enron trading maneuvers were brashly named by the traders "Death Star," "Fat Boy," "Ricochet," and "Get Shorty." When a forest fire shut down a major transmission line into California, cutting power supplies and raising prices, tapes recorded Enron energy traders cheering, "Burn, baby, burn. That's a beautiful thing."[11]

After California's disastrous experiment with energy deregulation, Enron energy traders were heard on audiotapes gloating and praising each other as they helped create, and then exploit, an electricity supply shortage in the Western US. Tapes record traders making fun of their victims. "Yeah, grandma Millie, man…now she wants her f------g money back for all the power you've charged right up, jammed right up her a------ for f------g $250 a megawatt hour."[17] Charming dialogue to be sure, but also a raw portrait of the chilling psychopathy of the financial wizards behind this corporate plundering of citizen assets.

Several studies suggest that financial elites are more likely to feel that rules and societal constraints don't apply to them. Donald Trump's entire business and political career, especially his approach to paying taxes, is just one of the most conspicuous examples of this personality trait. CEOs live in bubbles of privilege, entourages and private jets, lifestyles that defy the rest of society, and breed a disdain for the rules that others must live by.

In numerous experiments, rich people and even people only made temporarily "rich" during a game of monopoly, tend to attribute their success to their own personal skills and talents, ignoring other contributing factors to their good fortune, like luck and the work of other people. Self-absorption's ugly twin sister is lack of empathy.

The "upper class," as defined by a research project involving multiple experiments published in the *Proceedings of the National Academy of Sciences*, are more likely to break the law while driving, lie during negotiations, endorse unethical behavior at work, cheat to increase their chance of winning a prize, and literally take candy from children.

Researchers Dacher Keltner of the University of California-Berkeley, Michael W. Kraus of UC-San Francisco, and Paul K. Piff of UC-Berkeley write that the life experiences of the wealthy make them less compassionate and empathetic than the lower class. People who come from a lower-class background have to depend more on other people. Wealthier people don't have to rely on others as much or at all. These disparities show up in psychological studies. People from lower-class backgrounds are better at reading other people's emotions. They're more likely to act altruistically. "They give more and help more. If someone's in need, they'll respond." When poor people see someone else suffering, they have a physiological response that is missing with the wealthy.[13]

Dr. Dale Archer, a psychiatrist and frequent guest on "FoxNews.com Live" of all places writes, "Physically, studies have shown that the brain chemistry is different in powerful politicians, leading to sensation-seeking and risky behavior. They have lower levels of the brain chemical monoamine oxidase-A, which means they have higher highs when they engage in risky behavior and that they get bored much more easily than the norm."[14]

Financial psychopaths (see chapter 18) commonly found in the trading industry and other financial services corporations are also often as addicted to thrill-seeking as compulsive gamblers. They thrive off of increased neurotransmitters like endorphins and serotonin released by deals that give them a sense of power, control, invincibility, and grandiosity. They may even brag about their losses and care little about what other people feel or think.

Consider the 71 CEOs behind the recent corporate-backed Campaign To Fix The Debt, formed in 2012. These were CEOs making the media rounds and spending $30 million dollars[15] pounding the table on achieving federal deficit reduction exclusively by dismantling the social safety nets--Medicare, Medicaid, and social security--while they sit on their own massive retirement funds averaging $9.1 million. These are the same CEOs who have contributed mightily to, and benefited personally from, the deficit they now want closed. Their companies have received trillions in federal war contracts, subsidies and bailouts, as well as specialized tax breaks and loopholes that virtually eliminate the companies' tax bills--like Goldman Sachs, Honeywell, AT &T and Boeing. And no, they are not offering to reduce their feeding at the public trough, instead they want us to turn away the poor, the disabled and the vulnerable, calling government support for them "low priority spending."

The senior directors of many a defunct Wall Street firm walked away with huge amounts of money, but even more disturbing, they walked away with clean consciences, seemingly oblivious to those who lost their jobs, their life savings and investments, their hope and means for a future. Clive Boddy writes in his essay, *The Corporate Psychopaths Theory of the Global Financial Crisis,*[16] that psychopaths largely caused the crisis, having risen to key senior positions within modern financial corporations, influencing "the moral climate of the whole organization." Their "single-minded pursuit of their own self-enrichment and self-aggrandizement to the exclusion of all other considerations, has led to an abandonment of the old fashioned concept of noblesse oblige, equality, fairness, or of any real notion of corporate social responsibility."

To say that there are people in the world blinded by greed, and willing to wreak havoc with the lives of others to satisfy their greed, is certainly not a new insight. To say that there are even a few greedy people willing to jeopardize the lives of millions, perhaps the future of mankind itself, to acquire and horde their wealth is also nothing new. But corrosive greed doesn't tell the whole story here.

Boddy suggests this further explanation. When the era of corporate takeovers and mergers arrived, company stability began to crumble. "Jobs for life" disappeared and mutual loyalty between company and employees became an anachronism. "Employees increasingly found themselves working for unfamiliar organizations and with other people that they did not really know very well. Rapid movements in key personnel between corporations compared to the relatively slower movements in organizational productivity and success made it increasingly difficult to identify corporate success with any particular manager. Failures were not noticed until too late and the offending managers had already moved on to better positions elsewhere. Successes could equally be claimed by those who had nothing to do with them. Success could thus be claimed by those with the loudest voice, the most influence and the best political skills," which played perfectly to the skill set and motivation of corporate psychopaths.

One direct result was that senior level compensation became completely untethered to that offered to the rest of the company's work force and to the company's profitability. "Corporate psychopaths are ideally situated to prey on such an environment and corporate fraud, financial misrepresentation, greed and misbehavior went through the roof, bringing down huge companies and culminating in the Global Financial Crisis." Boddy considered this a recipe for disaster, for the financial world and society in general. Indeed, this phenomenon has seeped into other industry board rooms as well. But instead of the casualties including the companies themselves, the casualties include the lives and health of millions of people, and the livability of the climate itself.

Psychopaths are notoriously refractory to treatment or behavioral modification, another trait they share with business and political elites. As F. Scott Fitzgerald wrote in *The Great Gatsby*, "Let me tell you about the very rich. They are different from you and me...They think, deep down, that they are better than we are."

Our nation's responses to the climate crisis, the federal deficit, our economic stagnation, extreme wealth inequality, disparities in health care, and many of our other serious challenges are still being held hostage by people who manifest a detachment from reality as profound as that of schizophrenics. We are still allowing a powerful elite, who behave like psychopaths, to steer our government towards protecting their interests at the expense of everyone else. The greatest threat to the United States will never be Al Qaeda, Russia, China or Iran. It will be our failure to wrest control of public policy from the inmates of our own corporate insane asylum.

Wilbur Ross, current Commerce Secretary, and "Grand Swipe" of the secret Wall Street fraternity, Kappa Beta Phi. Ross was bewildered about why federal employees couldn't just get loans to bail them out during the Jan. 2019 government shut down.

16

The Third World:
Dumping Ground for Corporate America

"The largest 100 corporations hold 25 percent of the worldwide productive assets, which in turn control 75 percent of international trade and 98 percent of all foreign direct investment. The multinational corporation . . . puts the economic decision beyond the effective reach of the political process and its decision-makers, national governments."

—Peter Drucker

Heriberto Obando was a 15-year-old boy living in a remote village in Costa Rica . He died in 1988 after using his bare hands to spread a bag of Counter, an American made pesticide so toxic that the EPA would not allow its sale to anyone in the US other than trained and protected workers.[1] But illiterate farmers in Costa Rica were able to buy Counter off the shelf despite the fact that they were not capable of reading the warning labels and couldn't afford any protective safety equipment.

Reports of unsafe food and products from China, including toxic toothpaste and toys with lead-based paint, garnered wide spread media attention in 2007. The Federal Consumer Product Safety Commission (CPSC) acted quickly to prevent consumers from being exposed to these kinds of dangers. However, neither the CPSC, nor any other federal agency has much interest in protecting foreign consumers from dangerous products manufactured by US corporations. And US corporations have plenty of interest in selling them.

A common theme throughout this book is the willingness of corporations to peddle their deadly goods, and/or the manufacturing of those goods, to third world countries once the inherent danger of their products becomes apparent, widely publicized or illegal in the developed world. It's called "dumping," a euphemistic, corporate term for unloading toxic inventory. And often the US government does little to stop it, and sometimes has even encouraged or facilitated it. Companies are required by law to notify the CPSC whenever they intend to export any manufactured goods that violate US safety standards. In 2007, corporations notified the CPSC 97 times of "dumping" they were planning, almost twice as many as in 2002,[16] involving chemicals, toys, fireworks, clothing, lighters, carpets and pacifiers. The recipients of these "dumped" goods were consumers in Belgium, Ireland, New Zealand, Colombia, the Czech Republic and the Philippines.

An investigative series published in the late 1970s, by a team led by Mark Dowie at *Mother Jones*, called out the "Corporate Crime of the Century." It showed that the practice of successful dumping required a cooperative conspiracy between manufacturers, retailers, exporters, black marketeers, the US Export-Import Bank, and even the Commerce, State, and Treasury Departments. You see, no one in the US government wants to do anything to increase our trade imbalance, even if it means the death or suffering of millions of non-Americans.

But it's not just non-Americans harmed by dumping. In many instances hazardous products dumped overseas have ended up being smuggled, or openly imported back into the US. For example, American consumers often end up eating produce covered with pesticides banned by the US government decades ago simply because they are grown in a foreign country where those chemicals were conveniently dumped. Mother Jones called it, the "Boomerang Dump." With less than 2 percent of imported food shipments sampled by the FDA, don't assume that your federal government protects you from this absurdity. As I write this (early 2019) the federal government is shut down, and no food inspection is being done.

A child burning in a fire is a frightening image ripe for exploitation. Pajamas covered in a flame retardant started out as a marketers dream and within a few short years over 200 million pairs of pajamas were sold having been soaked in Brominated Tris (2,3-dibormoprophy). But the manufacturers' dream ended abruptly with their own nightmare.

Tris was found to put children at risk for kidney cancer, at the rate of 300 cases per million exposed male children. When the toxicity of Tris was revealed, the CPSC acted swiftly, banning their sale and even ordering them all to be recalled. CPSC went so far as to require that the sleepwear couldn't even be thrown away. It ordered them to be burned (a misguided directive itself because combustion emissions would be toxic), or buried. So the manufacturers of Tris-soaked pajamas went from financial heaven to financial hell in only a few months. Once again, salvaging corporate profit became the imperative.

Soon, like sharks circling their next prey, exporters like Paul Rothman Industries, and Cord Exporters smelled a tidy profit and moved in for the kill, literally. Their ads began appearing in Women's magazines and they offered to buy up the carcinogenic pajamas at 10 to 30 percent of the wholesale price, then dumped them on the non-domestic market by the tens of millions.[2] The Tris pajamas saga led to the first serious attempt to prohibit US corporations from dumping dangerous products overseas.[3]

After intense public debate and congressional investigations, just before he left office, President Jimmy Carter signed Executive Order 12264 on Jan. 15, 1981, to prohibit companies from exporting toxic chemicals, pesticides, drugs and other products the federal government won't allow to be sold domestically. One month later, in one of his first acts as the new President, and heralding the post-Nixon Republican Party's new position to nakedly abandon consumer, public health, and environmental regulations that interfered with corporate profits, Ronald Reagan withdrew Carter's Executive Order.

Tris is only one of an almost endless line of dangerous products "dumped" overseas by American corporations. The list includes 450,000 baby pacifiers known to cause death from choking. In 1972, 400 Iraqis died and 5,000 more were hospitalized from eating wheat and barley tainted with a mercury based fungicide that had been exported after being banned in the US. "An undisclosed number of farmers and over 1,000 water buffaloes died suddenly in Egypt after being exposed to leptophos,"[4] a pesticide not allowed for domestic use by the EPA but that was then exported to 30 different countries.

In 1970, because of concerns about consequences of estrogen consumption (see Chapter 2) the FDA warned physicians to prescribe only birth control pills with the lowest possible estrogen dose, which at the time was 50 ug (micrograms). Three years later, millions of higher dose, 80 ug pills remained unsold because of the multiple dangers that had been identified. Syntex corporation was one company with lots of unsold 80 ug pills. So they offered a discount to the US Agency for International Development (AID) who was eager to offer them to third world countries to help with population control. So with help from the federal government, Syntex unloaded millions of dollars of unsold, but dangerous 80 ug pill stock on millions of unsuspecting third world women.

Upjohn was another pharmaceutical company that improved its bottom line by dumping dangerous birth control drugs on the overseas market. Makers of Depro-Provera (DP), a hormone injection designed to provide contraception for three to six months, Upjohn was prohibited from selling DP in the US after it was found to cause an arm length list of side effects in animals, some of them potentially fatal in humans--breast masses, reproductive cancers, uterine bleeding disorders, immunosuppression, permanent sterility and birth defects when it failed to prevent pregnancy. After being turned down by the FDA for domestic approval in 1971, Upjohn managed to ship plenty of DP to 70 foreign countries, even with the help of the United States' government-sponsored population control programs. And Upjohn continues to do so, including making it available over the counter without a prescription.

The NY Times described the "sad legacy of the Dalkon Shield" as a story of "loss and suffering and bitterness and pain … individual heart break and corporate defeat."[5] A. H. Robins sold the intrauterine birth control device to 2.5 million women in the early 1970s. More than 350,000 women eventually filed injury claims against the company. A major study found that the device increased the incidence of pelvic inflammatory disease about 500 percent over other intrauterine contraceptives, and at least 18 women died as a result. When the device failed to prevent pregnancy, as it did 5.5 percent of the time, the rate of miscarriage was 60 percent. It was estimated that for every $1 million profit the company made, $20 million was spent by the women and their insurance companies trying to deal with the complications from the device. A. H. Robins set aside $1.75 billion to settle all the lawsuits that followed. The corporate "defeat part," however, was significantly reduced by Robins who then restored their balance sheet by selling the device to 42 under-developed countries with an important assist from the U.S. Treasury Depart via a foreign aid program.[6] For at least five years after the device was recalled by Robins, it was still being inserted in women overseas.

Winthrop Products made a pain killer, dipyrone, that was discovered to cause a fatal blood disorder called agranulocytosis (failure to produce white blood cells). When the FDA banned dipyrone in the US in 1977, Winthrop took it overseas and it is still sold in Russia, Spain, Brazil, South America and Africa. A synthetic male hormone, Winstrol, which was found to stunt the growth of American children, is still sold in Brazil as an appetite stimulant for children and as a body-building steroid.[4]

Dow-Elanco, a joint venture between Dow Chemical Company and Eli Lilly, was eventually renamed Dow AgroSciences when Dow bought out Eli Lilly. Dow-Elanco had been producing an herbicide, Galant, that the US EPA withdrew from the American market because of a demonstrable link to cancer. Dow-Elanco then sold Galant in Costa Rica causing hundreds of farm workers to become sterile. In 2013, the EPA's own data showed that pesticides banned, restricted or unregistered in the United States, were manufactured in 23 states to be exported overseas, and most of those pesticides will return to the US as contaminants of food imported from those foreign countries, ultimately eaten by Americans.

Perhaps the worst example of companies plying their dangerous products overseas when the US market is forbidden or curtailed comes from the tobacco industry (see chapter 7). For years, the same federal government that has warned American citizens about the dangers of smoking, has actively pressured Taiwan, Thailand, Japan and South Korea to import American cigarettes or be penalized with trade sanctions. Twenty years after the US banned cigarette commercials on television they were pressuring Japan to allow those kinds of commercials, many of which ran during Japanese children's programs.

Even when the export of dangerous products is forbidden, companies still have options to ply their wares, and boost profits. They can change the name of the product, or export the ingredients to a plant in a foreign country, and then have it manufactured off shore

The Race to the Top Means a Race to the Bottom

Manufacturing dangerous products and then dumping them overseas is only half of the corporate exploitation of foreign populations. The other half is the imperative of cutting costs to the bone and letting the chips fall where they may. The race to the top of the profit mountain frequently means a race to the bottom for worker health, safety and survival.

In November, 2012, a fire in a garment factory took the lives of 112 Bangladeshi workers, most of them women. Tazreen Fashions, Ltd., owned the factory, and sold the clothing to United States companies like Disney, Sears, Wal-Mart, Enyce, and the US Marine Corps. An entire laundry list of safety compromises was obvious and complained about by the workers, including no emergency exit doors. But the worst was that factory bosses shut the main exit when the alarms went off, and ordered everyone back to work.[15] The American companies who all bought from Tazreen denied any responsibility, and in some cases any knowledge, of the fact that the manufacturing of their goods were endangering and killing the workers. As the LA Times editorialized, "Ignorance is a poor excuse. As long as American companies press for quicker production of goods at lower prices while failing to monitor the conditions under which those demands are met, corners will be cut in dangerous, unfair and inhumane ways."[15]

It's not like it couldn't have been anticipated. In several years prior to that fire, at least 300 other workers died in Bangladeshi garment factories. Whether or not clothiers would suffer consumer backlash from slightly more expensive clothes that were manufactured under safer conditions is perhaps an unknown. But it is simply not responsible for any corporation to even give consumers that option.

Recently K-mart forced one of their contractors in El Salvador to limit their production costs to $1 per dress. El Salvador then raised the minimum wage, but K-Mart refused to adjust the price. So the only option for the contractor was to increase production, which increased the risks and danger to the employees.

Women are especially victimized by the overseas garment industry. Supervisors typically view women as virtual slaves. They can be paid much less, and are easier to subdue and oppress. They are essentially viewed as a disposable means to maintaining tight production targets and profit margins and not worth the expense or bother of investing in appropriately safe working conditions. In many Asian countries it is not surprising that many poor women prefer working in the sex trade, because at least the pay is much better. Pregnant women are particularly discriminated against, and many factories have done prescreening pregnancy tests so that companies would not have to deal with allowing time off for pre-natal doctor visits or maternity leave.

"Distressed" jeans are currently in fashion, and the way that those jeans get distressed in the factory is "distressing" to the health of the workers. It is often with the use of sand blasting. Sand often has crystalline silica in it and chronic exposure can lead to the debilitating and fatal diseases of silicosis and lung cancer. And many of those garment workers toil all day inhaling crystalline silica without face masks or any other protection. Clothing that is wrinkle-free or stain resistant became that way because some factory worker was exposed to a chemical embedded in the fabric, often the carcinogenic formaldehyde.

Hu Fengchao worked in a Panasonic DVD plant in Dalian, China. His job was to spray paint the DVD plastic cases. In doing so he was exposed daily to some of the most toxic industrial chemicals known--benzene, xylene, and toluene. At the age of 28, after just three years on the job, he developed aplastic anemia, a failure of the bone marrow to produce blood products, which is often untreatable, and ultimately fatal.

Designer leather originating in Bangladesh provides the raw material for fancy clothes worn everywhere from Times Square in Manhattan to Mormon churches in Moscow, Idaho.

But it is a bitter irony that some of the most luxurious personal goods worn at the most extravagant places on earth are made in an apocalyptic hell scape,[22] by some of its poorest and most exploited people. The world of high fashion clothes, hand bags, and car seats exacts a steep price from consumers, but a much steeper price for many workers in the tanning industry, who live in the filthiest corners of destitute countries like Bangladesh.

The name Hazaribagh means "Thousand Gardens," and most of Bangladesh's 206 tanneries are crammed in Hazaribagh, a slum of Dhaka, the nation's capital city. In fact, "Thousand Gardens" likely doesn't have a single flower in it, maybe not a single plant. It is one of the most polluted, toxic places on earth. Bangladeshi leather is a billion- dollar industry, and about 12,000 children and teenagers who work 12-14 hours a day, sometimes 20 hours a day, 7 days a week,[18] are the real sacrificial lambs. Sixteen-year-old Muhammad Azim is one of those workers, and glad to be earning $1.50 a day in one of these tanning factories to help augment what his father makes as a rickshaw driver.[19]

In Hazaribagh, untreated waste water runs through canals, whose banks are lined with rotting hides. Photo courtesy of Larry C. Price/Pulitzer Center on Crisis Reporting/Undark.

The animals whose skins are used are smuggled in from India usually in wretched condition. Their slaughter is a hideous endeavor, but not by any means the worst part of the ordeal. Sometimes their hides are peeled off while the animals are still alive.

Softening, pickling, bleaching and coloring the leather is a process of unimaginable pollution and filth. Many children stand barefoot in huge vats of chemicals and animal waste stomping on the hides to soften them up. In the vats is a toxic brew of chemicals and metals that can include sulfuric acid, benzene, formaldehyde, chromate and bichromate salts, aniline, butyl acetate, ethanol, toluene, chromium sulfate, arsenic, lead, magnesium oxide, ammonium hydrogen sulfide, and sodium sulfide. The chemicals burn the children's skin, their eyes and the lining of the respiratory tract, causing itchy, peeling skin, generalized body aches, headaches, dizziness and nausea. The long-term exposure causes cancer, respiratory and cardiovascular diseases, disfigured and sometimes eventually amputated limbs. The World Health Organization has estimated that the life expectancy of tanning workers has dropped to the point where 90 percent of them don't see the age of 50.[17] Cancer rates are 20-50 percent higher than in control groups from the surrounding communities, which are themselves swimming in environmental toxins.

Given that there are no waste treatment plants in places like Hazaribagh, these tanneries inevitably contaminate the local water supply with their waste sludge stew of chemicals, animal tissue and refuse that runs into open gutters, packed streets, past slum housing en route to primary city rivers and streams used for drinking water, like the Buriganga River that runs through Dhaka. A report from Human Right Watch found that children were bathing in water that was black with contamination.[20]

Many of the people who use these waterways for drinking, bathing, and washing report acute rashes and respiratory symptoms. Chronically the consequences will be much greater. Local governments usually do not monitor the water, air or soil for toxins from these tanneries, and the pungent odor alone makes living in these communities almost unbearable. In fact the smell of tanneries has been described as the most horrifying, sickening smell on earth. Even in medieval European cities, laws required tanneries be cited outside of city walls because of the smell.

Investigative reporters found that Bangladeshi leather ended up with companies that manufacture shoes, belts, and purses--like Clarks, Coach, Kate Spade, Macy's, Michael Kors, Sears, Steven Madden, Timberland, Harbor Footwear Group, Apex Footwear, and Genesco.[21] It is inconceivable that clothing corporations that use Bangladeshi leather do not know how much their profits are predicated upon the grotesque conditions and human sacrifice involved in bringing their product to market. The nightmare at tanneries are not unique to Hazaribagh. Similarly deplorable conditions exist in the Philippines and India.[21]

Americans do not know, and to a certain degree choose not to know, the working conditions in the overseas factories where many of their consumer goods are made. The country of origin, "Made in (where ever) " labels on the finished products, can be very misleading or outright dishonest about where the products actually originate. "Made in Italy" leather evokes luxury, style, and sophistication, the polar opposite of the filth, degradation and tragedy likely built into the clothing that carries such labels.

If you think you can ignore this fetid, human disaster because it provides you with fancy clothes at the cheapest possible price, and is largely happening to someone else, far, far way, consider this.

Further downstream from Hazaribagh, in the Bay of Bengal, fed by the same Buriganga River, are ponds where prawns are farmed for export, which may appear on your dinner plate next time you eat at a fancy American restaurant. No one in Bangladesh is monitoring the chemical and heavy metal contamination of those prawns, or any other seafood from Bangladesh. And there is only a small chance it will be monitored by any agency in the United States.

A tanning worker stands in the same soaking solution intended to soften the hides. It likely contains toxic chemicals, like chromium. Photo courtesy of Larry C. Price/Pulitzer Center on Crisis Reporting/Undark

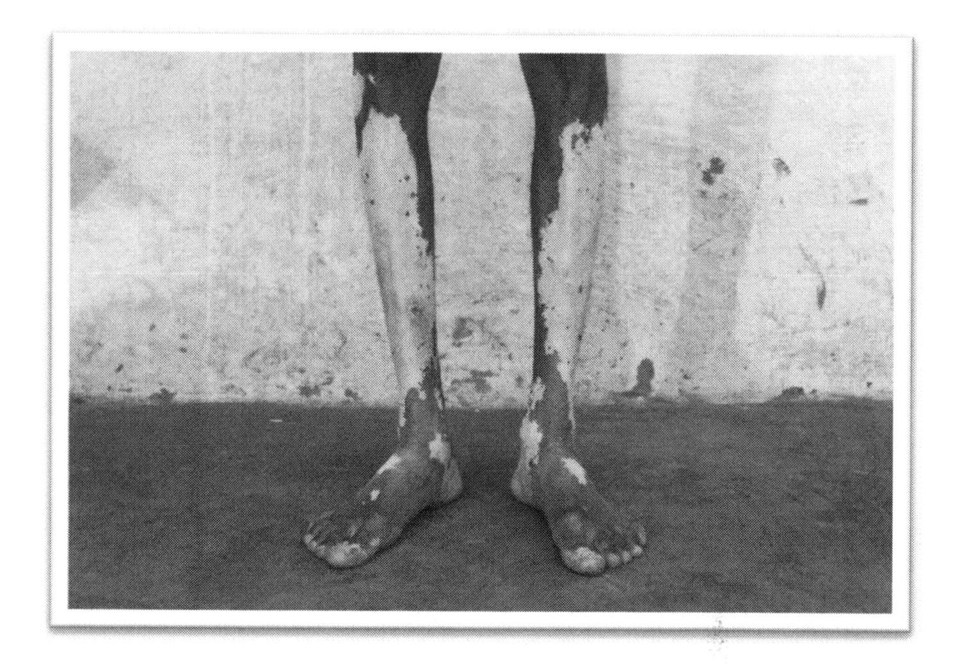

This farmer's field is irrigated by water not far downstream from a tannery in India. The skin on his legs has been excoriated by the toxic chemicals in the water.

17

Dirty Dieselgate

"Multinational corporations do control. They control the politicians. They control the media. They control the pattern of consumption, entertainment, thinking. They're destroying the planet and laying the foundation for violent outbursts and racial division."

—California Gov. Jerry Brown

In 2006 top brass at Volkswagen struggled with developing a strategy to reinvigorate their stagnant US car sales. Europe was in the throes of a love affair with diesel because they prioritized fuel efficiency in the name of reducing greenhouse gases and carbon emissions.

Their predicament was that diesel engines produce far more nitrogen oxides (NOx) than gasoline engines, and that American regulations allowed only one-sixth of the NOx emissions allowed in Europe. Standing in the way of solving the problem were several engineering limitations. Removing NOx meant reduced fuel efficiency, impaired engine performance, bulky hardware, increased cost and frequent servicing, or a combination of all of them.

For seven years Volkswagen ran ads on TV, social media and print directed at environmentally conscious consumers to convince them their new "clean diesel" vehicles were the latest "must have" for people who cared about clean air, the climate and the planet. Their "clean diesel" cars commanded a premium price because of the claims of superb fuel efficiency, sustainability, longevity, and markedly less pollution than historical diesel engines. One ad showed "green police" stopping a long line of cars belching smoke out their tail pipes, then waving through a single, virtuous Audi diesel with an approving, cheerful, "You're good to go, sir." In truth it was Volkswagen and Audi that were the ones blowing smoke out their tail pipes. A 2014

Super Bowl ad from Volkswagen showed a father telling his trusting young daughter that a German engineer "gets his wings" every time a Volkswagen reaches the 100,000 miles driven. In actuality, a German engineer is getting something less glorious than wings--hand cuffs, leg irons, and a jump suit to be exact.

John German (no relationship to the country or Volkswagen) is a modestly paid engineer for a small international non-profit dedicated to helping reduce vehicle emissions, the International Council on Clean Transportation (ICCT). It was not government regulators, but German's team that in 2015 ultimately lead to the discovery that one of the largest and most iconic automakers in the world, Volkswagen, had orchestrated a deliberate and deadly hoax upon consumers and the world at large.

Volkswagen was equipping at least 15 models of their "groundbreaking, clean diesel" engines, sold under the banner of VW, Audi and Porsche, with a secret, sophisticated engine software algorithm that would turn off exhaust pollution control devices as soon as the cars exited regulators' test labs. It was a magic act worthy of Houdini. Eleven million vehicles worldwide, including 580,000 in the US, were equipped with the device allowing them to pass emission certification tests despite producing up to 40 times the amount of nitrogen oxides than the legal limit once they hit the road. Volkswagen's motivation for the fraud was the realization that their ambition of achieving, simultaneously, superb fuel efficiency and pollution control from a diesel engine was too good to be true.

Nitrogen oxides are a "double whammy" atmospheric contaminant in that they can act as precursor gases to both of the high volume pollution components in typical urban smog--particulate matter and ozone. These toxic gases can travel thousands of miles, increasing the air pollution, and provoking all the health consequences detailed earlier in chapter 5, in cities and countries far removed from where they were originally emitted.

German was so surprised by the results of his tests, but rather naively was not suspicious of the cause, that he had notified Volkswagen of the results as a courtesy because he wasn't sure of their significance. He said, "That is actually the single most inexplicable thing about this whole business. VW had a chance to fix the problem, and they continued to try and cheat and do what they had done. That's just amazing. It was not an accident. A lot of work has gone into this."[1] The conspiracy lasted at least a decade, and the defeat devices had been installed for at least seven years before the rigged scheme was uncovered.

Volkswagen's first response after being exposed was that the fraud was perpetrated by "rogue" engineers. They might as well have announced they were helping O.J. Simpson search for the real killers. The company eventually backed away from that assertion, but did everything possible to protect top level management.

Investigators in the US and Germany eventually identified more than 40 people they believe were part of the conspiracy. One has to admire the exquisite engineering behind the cheating devices. An important part of regulators' emission tests in the US lasts exactly 1,370 seconds. Audi's defeat device was programmed to emit the legal amount of NOx for exactly 1,370 seconds. On the very next second the software switched over to the non-lab setting, emitting nine times more NOx. The German carmaker had indeed achieved a triumph in engineering, just not one that they could brag about in their advertising.

On December 6, 2017, "Dieselgate," described by Fortune magazine as "one of the most audacious corporate frauds in history,"[4] yielded its first conviction, 48 yr. old Oliver Schmidt, a former Volkswagen engineer. With a bald head and a baby face he was led into a Detroit courtroom in a bright read jumpsuit, handcuffs and leg irons. His wife struggled to control her tears. After pleading guilty to helping Volkswagen with their emissions scandal, he was given one of the stiffest sentences ever for a white collar, non-violent crime--seven years in prison. But in many ways he was also a corporate fall guy, because those even more responsible are not likely to face justice.

Volkswagen CEO Martin Winterkorn, made about 100 times the salary that Schmidt made, $18.6 million in 2017, the highest of any CEO in Germany. It is as rare as a lunar eclipse that corporate bosses are held accountable for their company's misdeeds. But, in May 2018, criminal charges were finally filed against the former CEO. Winterkorn was charged with "four felony counts, including conspiracy to defraud the United States, wire fraud and violating the Clean Air Act."[2] In Winterkorn's indictment is also the charge that VW employees went so far as to advise the company to seek approval for their 2016 models from EPA and California regulators, without disclosing the existence of the fraudulent software. In fact it had been Schmidt who was given the assignment of flying from Germany to

Michigan to meet with the California Air Resources Board (CARB) to convince them to approve the cars despite their contradictory emissions testing. Yet still, neither he nor any other Volkswagen officials disclosed the deceit. Schmidt said his dishonest sales pitch was approved by at least four senior Volkswagen officials. Finally, another VW engineer confessed to the CARB about the defeat devices and the fraud went public.

Those who are cheering for justice however must know that the charges against Winterkorn are largely symbolic as he is unlikely to be extradited to the United States and it is unclear at this point what his punishment will consist of in Germany. Other members of Volkswagen's board would have undoubtedly known about the fraud, but have not been charged, or even removed or demoted in the company. Nine other lower level VW employees have been charged, with Schmidt and another German engineer pleading guilty and serving prison sentences. Five VW executives and senior managers indicted by the US are believed to be in Germany and will also not likely face extradition, nor apparently will they be charged in Germany. Schmidt wrote a letter from his prison cell to his sentencing judge stating, "I've learned that my superiors that claimed to me to have not been involved earlier than me at VW knew about this for many, many years. I must say that I feel misused by my own company."[4]

Volkswagen agreed to a payment to the US Dept. of Justice of $4.3 billion, and will lose much more in settling with individual vehicle owners. In total Volkswagen has agreed to spend $25 billion as compensation, spread among regulators, state governments, dealers, and individual vehicle owners. My family owned two of these "not so clean diesel" vehicles.

Two months after Volkswagen admitted to their diesel emissions scandal, Audi, one of their subsidiaries, admitted they had done the same thing. Despite admitting company guilt, Audi's CEO, Rupert Stadler, remained in his position until 2018 when he was arrested by German authorities and charged with complicity in the cheating scandal.[3] The German authorities had reason to believe that Stadler was trying to suppress evidence in the case. Audi's company ethics were lambasted by financial analysts who decried the fact that Stadler had not been forced to step down, and was only removed from the company by virtue of his being arrested by German police.

Volkswagen's punishment in Europe will be much different than in the US for a complex of social, legal and political reasons. So far in Europe, where Volkswagen sold 16 times as many of the tainted diesel cars as they did in the US, they have not paid a single Euro in penalties, and have not offered its European customers any compensation. Cars with their emissions certified in any European country must be accepted in all other EU countries, so by default, the lowest common denominator in emissions becomes basically the standard throughout Europe.

Hurdles for plaintiff's cases are much higher in Europe, and the regulatory atmosphere much weaker. Volkswagen maintains that its cheating software is not illegal in Europe. German national regulators were loath to punish carmakers within their own country, and put their domestic manufacturers at a competitive disadvantage. Volkswagen was right to assume that they could expect a mere slap on the wrist, and that's all they did get in Europe.

To understand why Volkswagen thought they could get away with this, one needs to further understand the personality of the company. It was originated in 1937 by none other than Adolph Hitler and the Third Reich, who tried to duplicate for Germany what Henry Ford had done for America. The name Volkswagen literally means "folks' wagon," in other words, "the people's car." It was nationalized after WWII and despite private ownership is still run almost as a state business, perched on a unique pedestal in Europe and its home country. Because of its commanding position in Germany's economy it is inextricably tied to virtually all German political, social and cultural institutions, including unions.

Writing for Fortune Magazine, Geoffrey Smith and Roger Parloff point out that, "VW's misbehavior did not come out of nowhere."[5] The company is run with an air of entitlement because it has a large umbilical cord to every level of German government. That means perennial political protection from Germany's top politicians. Chancellor Angela Merkel has repeatedly gone to bat to weaken emission standards for Volkswagen's benefit, and virtually make a mockery of the way that tail pipe emissions are inspected. During a visit to the California, Merkel had even tried to pressure the powerful CARB into relaxing their NOx standards.

The company has had numerous scandals in the past, including close brushes with the law, and has emerged with little consequence until Dieselgate. For example, top brass had advertising photos of "body fits" touched up to give the appearance of greater precision and higher quality workmanship than the cars actually had. To help keep labor on their side, management hired prostitutes for labor union representatives and all-expenses-paid-for luxury shopping trips to Paris for their wives, apparently in case the prostitute arrangements caused any domestic discord. That brilliant scheme included supporting the Brazilian mistress of the head of the labor union, Klaus Volkert, with 400,000 euros in spending money. Some German lawmakers were given salaries from Volkswagen for doing absolutely nothing that could qualify as work for the company.

From 1993 to 2002, Volkswagen's CEO fit perfectly into this paradigm. Smith and Parloff describe Ferdinand Piëch, as a "brilliant engineer and a ruthless, terrifying manager,"[5] who permeated the company culture with a nearly "above the law" condescension, and a deep abiding contempt for anything that stood in the way of success as he defined it. Piëch's approach to company management was certainly passed on to Martin Winterkorn. Under Winterkorn, Volkswagen's goal was to become the biggest car company in the world. And not just in sales, but in "profitability, innovation, customer satisfaction, everything."[5] And central to that goal was to triple U.S. sales by 2018. Their strategy to realize their ambitions was a collective stroke of genius--"Clean Diesel." Given the impetus for and importance of clean diesel to the company's goals, it is inconceivable that knowledge of the defeat devices didn't make its way to the top of the company. However, given the culture of the company, no one, including the engineers, had the courage to admit failure.

After the revelation of Dieselgate, numerous other car makers have started to squirm with the spotlight turning on them. BMW, Fiat Chrysler Automobiles (FCA), Daimler (maker of Mercedes), PSA (maker of Peugeots, Citroëns), and Renault-Nissan have all been targeted by regulatory authorities in Europe for suspicious emission profiles. In the US, the Justice Department has filed suit against FCA for placing their own defeat devices on diesel Jeep Grand Cherokees, and Dodge Ram pick-up trucks.

The scandal can best be understood in the context of the differences between America and Europe in several important ways. Diesel was more available and cheaper in Europe, which was all the more important in a part of the world not blessed with oil. That bias was magnified by important tax subsidies that prioritized the climate imperative of CO_2 emissions rather than the contributors to smog. This emphasis dramatically increased the number of diesel passenger cars in Europe, reaching 50 percent of new vehicles sold in the European Union by 2014.[5]

A UK consulting firm tested more than 400 different European diesel cars and found that only five had emissions as low as their license required. In fact on average, "diesels were emitting four times the regulated maximum," and Volkswagens were not the worst.[5] In fact virtually all manufacturers have produced diesel cars that emit far more pollution than is revealed in lab tests. Using a wide variety of technical "tricks" and strategies, they are all guilty of manipulating pollution-control devices, even if not as egregious as Volkswagen.

It is fascinating that Volkswagen has insisted that its non-American customers have not been harmed simply because limits on NOx are more lax in foreign countries. Despite literally thousands of studies to the contrary, VW unequivocally denies that nitrogen oxides are a health threat. That claim is as much scientific and factual absurdity as our current President's claim to be a "stable genius" (see chapter 5).

Researchers have looked at the amount of NOx emitted from diesel engines on the road compared to measurements in the lab, the heart of the Volkswagen scandal. They estimate that 38,000 thousand people die every year worldwide because of those illegal emissions. The countries hardest hit, by virtue of the number of diesel vehicles in their inventories are in Europe, China, and India.[6]

One of the researchers, Ray Minjares, at the International Council on Clean Transportation (ICCT) in the US, said, "Manufacturers know how to make their cars clean and they are actively choosing not to."[7]

In fact the deaths caused could indeed be much higher because the researchers only estimated mortality directly related to early death known to be caused by diesel emissions' contribution to ozone and particulate matter. Nitrogen oxides have their own toxicity, however, and chronic exposure to the millions of tons of excess NOx from all these illegal diesel emissions could indeed be responsible for morbidity and mortality far in excess of these numbers. Prof. Roy Harrison, an environmental health expert at the University of Birmingham in the UK, authored research that indicates the mortality from these excess NOx emissions is closer to ten times that amount, i.e. 380,000 deaths per year.[8] Harrison said, "This rigorous study highlights the serious consequences which have resulted directly from the irresponsible actions of the motor manufacturers"[7] In other words, the automotive industry is knowingly, willingly, aggressively, and unapologetically causing the deaths of hundreds of thousands of people, despite having the technology and means to prevent those deaths. And they will continue to do so as long as they can get away with it.

If there is a silver lining to Dieselgate, it is this. A group has been formed, *Doctors against Diesel,*[9] based in the UK, to advocate for eliminating diesel engines for personal transportation. We should all hope someone listens.

Major European cities have been hit particularly hard by diesel pollution resulting from Dirty Dieselgate and wood burning stoves. London now sometimes has worse pollution than Beijing, China.

18

Psychopaths in the Board Room

"The greatest evil … is conceived and ordered (moved, seconded, carried, and minuted) in clean, carpeted, warmed and well-lighted offices, by quiet men with white collars and cut fingernails and smooth-shaven cheeks who do not need to raise their voices."

–C.S. Lewis, *The Screwtape Letters*

Psychopaths are thought to be responsible for about half of all serious crime.[21] Ted Bundy, the good looking, beguiling law student and homicidal maniac who killed at least 30 women in seven states and likely many more, was first incarcerated in my home town of Salt Lake City. He was as much of a psychopath as there ever was, as was Adam Lanza, the Sandy Hook murderer who slaughtered 26 children and adults. Bernie Madoff was a non-violent one. To no one's surprise prisons are filled with psychopaths. But so are the board rooms of Fortune 500 companies.

Famed University of Chicago economist Milton Friedman, in his sentinel book *Capitalism and Freedom*, argued, "there is one and only one social responsibility of business": to make money for its shareholders.[1] To make Friedman's argument palatable, or even approach rationality, requires the naive assumption that individuals who run corporations will do the right thing, individuals like Don Blankenship.

On April 5, 2010, an explosion in the Upper Big Branch mine in West Virginia originated 1,000 feet underground, shot through two and half miles of tunnel at a rate of 1,500 feet per second, and killed 29 miners, the deadliest US coal mine disaster in the last 40 years. It was the kind of mining accident that occurred in the 1900s, but was not supposed to happen in the 21st century. Investigators concluded the deadly accident was triggered by a spark that ignited excess concentrations of methane gas, further fueled by highly flammable coal dust, which had accumulated in dangerous concentrations due to grossly inadequate, and highly illegal, ventilation.[2]

Don Blankenship, the CEO of Massy Energy, once referred to as the "King of Coal," was indicted in 2014 for engineering a network of company practices that lead directly to the deadly explosion, placing profits ahead of safety. In 2015, he was convicted of just one misdemeanor for conspiring to violate federal mine regulations. Blankenship's conviction was the first time in US history that a CEO of a major company was convicted of a workplace safety crime that resulted in employee deaths. Assistant US Attorney Steve Ruby, who led the prosecution, said, "This was a coal mine and a company that was – it's not an exaggeration to say - run as a criminal enterprise. The men and women that we talked to who worked in this mine said that it was absolutely understood, it was expected that if you worked at that mine, you were going to break the law in order to produce as much coal as possible, as fast and as cheaply as possible."[3]

Some of the stories that they have to tell are horrifying. His workers described being forced to work without enough fresh air, being forced to work in water up to their necks, miles underground. Forced to work in areas where the roof and the walls of the mine were falling in around them. Massey Energy was earning more than $600,000 a day from the mine, and Blankenship was incentivized to maximize production because his pay was directly proportional to every pound of coal removed. Blankenship had taken home more than $80 million in the three years prior to the accident.

At Blankenship's trial, prosecutors showed that his business model was to "invariably press for more production even at mines that he knew were struggling to keep up with the safety laws . . .[Blankenship] chose to prioritize production and pay fines rather than take steps necessary to prevent the safety violations from continuing."[4]

In just the one year prior to the accident, federal inspectors had shut down portions of the mine 48 times for safety violations and had issued 557 safety citations. Blankenship was fully aware of every violation because his style was to micromanage everything that went on. He approved all hiring, pay distribution, and every expense down to the penny.

The prosecution had shown that Blankenship's subordinates had used a coded radio channel to stifle inspectors from the Mine Safety and Health Administration, and had a recorded conversation between them where they admitted, "if it weren't for MSHA, we'd blow ourselves up."[5]

Unrepentant, undaunted, and let's add unhinged, at the time of this writing, Blankenship is running for the US Senate. With his millions in coal money, he started running ads on West Virginia TV blaming the accident on the MSHA, mother nature, and his Democratic opponent, Sen. Joe Manchin. He called the accident an act of God. He left out an important word. It certainly was an act of a God-complex. Blankenship has never demonstrated a nanosecond of introspection, sorrow or guilt over his prioritizing balance sheets over the lives of his miners. United Mine Workers of America International President Cecil E. Roberts said "he was 'disgusted' by the ads,'" which he called a "desperate, low-life attempt to once again shift the blame for a decade of death, destruction and despair at Massey Energy while Blankenship was CEO."[6]

Roberts continued, "Don Blankenship, by creating a corporate culture that put production over safety, is responsible for these failures. It is appalling that he continues his despicable attempt to shift the blame from himself, each time ripping open the painful wounds the families of the victims will suffer forever. Although Don Blankenship may not have received the proper punishment in this world, those families can rest assured that he will receive it in the next."[6]

Don Blankenship

Unfortunately, Don Blankenship will join many other CEOs for whom justice will wait until a life-after-death tribunal. Blankenship served ten months in the proverbial country club jail of Taft Correctional Institute in Bakersfield, CA, where his biggest inconvenience was that he couldn't choose what to watch on TV. He then served a month in a half-way house followed by one month of home confinement. His sentenced ended in May, 2017. He served a bit more than twelve days for each of the 29 miners who had been knowingly, deliberately, and heartlessly allowed to die for his bank account. Still consumed by his slap-on-the-wrist punishment, upon his release, he urged President Trump to reject all those calling for enhanced criminal liability for mine supervisors.

What is a psychopath? Most psychopaths are male, and symptoms of the disorder usually manifest themselves by early adulthood, but can sometimes recede by middle age. On the short term, psychopaths are often very likeable, intelligent, excellent conversationalists, even charming. It is common for them to be the life of the party and the center of attention. Over time, character flaws become much more obvious. They do not feel guilt, remorse, or empathy, and do not accept responsibility for causing harm or hurt feeling to others. Their default behavior pattern is to always act in their own self-interest, with little to no restraint of conscience. A common phrase they might use is, "Someone deserved to be treated poorly." They are arrogant, self-absorbed, narcissistic, have delusional grandiosity about their intelligence, talents, importance to the world, and their potential for greatness. They feel they deserve an exalted station in life, and they judge and belittle others who are less exalted. They blame others for their failures and short comings. They believe that the rules that apply to everyone else don't apply to them. Sound like any presidents you might know?

In part because they believe they are above the rules, they are classically high risk takers. They are willing to lie, cheat, steal, and defraud others, and are often well organized and premeditated in "conning" others. They are prone to impulsive fits of anger, caring little about how that makes other people feel. Psychopaths are "master manipulators" of those around them--their emotions, their time, and their money. They are often chameleons, appearing superficially normal, but cunning in their attempts to control others. A psychopath will tell you exactly what you want to hear if you are in a position to further their ambitions.[7]

They are particularly good at detecting vulnerability in others. They willingly create chaos all around them because they know others have difficulty coping with it, and they play their family, friends, and associates against one another. Psychopaths may have the most pleasant of mental illnesses because they invariably feel good about themselves.

James Fallon, a neuroscientist at the University of California and author of *The Psychopath Inside: A Neuroscientist's Personal Journey into the Dark Side of the Brain* says, "Cognitive empathy is the ability to know what other people are feeling, and emotional empathy is the kind where you feel what they're feeling." Remember how Bill Clinton was famous for claiming to "feel your pain"? His ability to convince people that he did was perhaps the key to his political fortunes.[8]

Professor Robert Hare, a leading expert on psychopathy, a researcher in criminal psychology and co-author of the best seller, *Snakes in Suits: When Psychopaths Go to Work*, believes that the incidence of psychopathy is close to ten times greater among those who work on Wall Street than the general population, and that 4 percent of business leaders fit all the criteria for psychopathy. Others, like Dr. Christopher Bayer, a well-known therapist to Wall St. traders believes the rate is higher still, perhaps as high as 21 percent,[9] a figure comparable to what is found among prison inmates, and 21 times higher than the general population.

Hare is the creator of the PCL-R, a psychological assessment used to determine whether, and to what degree, someone is a psychopath. He says, "It stuns me, as much as it did when I started 40 years ago, that it is possible to have people who are so emotionally disconnected that they can function as if other people are objects to be manipulated and destroyed without any concern."[8]

Hare believes that "psychopaths are increasingly common in business because they're attracted to the pace and volatility of today's hypercompetitive workplaces, and because companies unwittingly nurture them."[10]

Honore de Balzac, the 19th century French novelist, wrote, "Behind every great fortunate is a crime." You might also add, "and a psychopath," with the possible exception of lottery winners. Psychopaths often appear normal, even charming, like Ted Bundy. Underneath, they lack conscience and empathy, making them manipulative, volatile and often (but by no means always) criminal.

The psychologist Keven Dutton in his book, *The Wisdom of Psychopaths*, notes society, especially Wall Street, admires and rewards many of the qualities of psychopaths--fearlessness, emotional sterility, supreme confidence, ruthlessness, lack of remorse, refusal to take responsibility, narcissism and delusions of grandeur. Who could argue that those characteristics virtually defined the Wall Street crowd responsible for blowing up the world's economy in 2008?

A 2005 British study compared the psychological profiles of 39 senior business executives at leading British companies with those of mental patients in the UK's Broadmoor Special Hospital for the criminally insane. The business leaders scored a clear "victory" in the three traits normally used to identify the emotional dysfunction inherent in Psychopathic Disorder: histrionic personality disorder, narcissistic personality disorder, and compulsive personality disorder. This might help us understand how once giant companies like Enron, WorldCom, and Polaroid were led by executives who lined their own pockets while torpedoing their own companies and leaving thousands of employees and retirees unemployed and penniless.

Psychopaths within organizations are often singled out for rapid promotion because of their polish, charm, and cool decisiveness. Because of these traits psychologists believe they are more likely to be found at the top of organizations rather than the bottom. Research by Babiak and Hare in the USA, and Board and Fritzon in the UK and in Australia, has shown that psychopaths are indeed to be found more often at senior levels of organizations than they are at junior levels.[19]

The positive spin on this research suggests that many of these traits aren't all bad; they can be an asset in career success and personal achievement. We promote psychopaths. We elect them, we cheer on their sports championships, they entertain us, we listen to their lectures, we invest with them, and we let them have scalpels to operate on us.

These traits are often necessary in leaders, paired with courage, bravery, a confidence they can achieve greatness, a willingness to make tough decisions, and an unwillingness to accept defeat. They may save our lives, on the operating table, from a flood, in a fire, or on a battle field. Adolph Hitler is inarguably the most infamous psychopath of modern times, but his demise could be credited to the psychopathic traits of another world leader, Winston Churchill. Several of the most celebrated surgeons I ever worked with were clearly psychopaths--

throwing instruments, belittling, intimidating and threatening colleagues and staff, demanding that the entire operating theatre revolve around them, but nonetheless skilled surgeons. In sublime irony, reviewing his own brain scan and comparing it to that of murderers, Fallon found that his own brain scan suggested he was a borderline psychopath.[11]

After the Supreme Court's "Citizen United" decision, Bernie Sanders and many others have called for a constitutional amendment to repeal the concept that corporations are entitled to personhood. He said, "This is an enormously important issue, and how it is resolved will determine, to a significant degree, the future of American democracy." What prompts that assessment is this observation---corporations function as gigantic, amorphous, often robotic psychopaths. Corporations are not imbued with a moral compass. In fact, they are duty bound to act in their own best financial interests, i.e. do whatever they can to make money, albeit theoretically within the law. But those laws are all too frequently crafted by corporations themselves, for their own self-interest, like the trade agreements, NAFTA and the Trans Pacific Partnership, which run rough shod over the public interest. Ethics and humanitarian considerations are not an inherent part of business models, even more true of large, international corporations. Furthermore, when institutional investors are involved, pressure to prioritize quarterly profits is persistent and intense.

California just passed a law requiring more women to be on the boards of directors of publicly held companies,[20] which is similar to requirements in many European countries. Ostensibly the reason is to promote more gender diversity in business leadership. But perhaps an unintended co-benefit is that women are less likely to be psychopaths, less likely to function in their business lives with toxic masculinity, which is the dominant atmosphere of so many board rooms.

David Niose, former president of the American Humanist Association, writes, "The pathological and narcissistic nature of corporate 'persons' is reason enough to deny them fundamental constitutional rights . . . but the fact that they also wield economic resources far in excess of those available to real persons, magnifies the need to restrain them."[12]

Certainly not every corporation is run by a cruel psychopath. But there is an abundance in board rooms of people showing at minimum, an indomitable ability to rationalize, compartmentalize, and marginalize their moral compass in the pursuit of profit.

In the aftermath of the collapse of the Soviet Union, Western technocrats helped facilitate the transfer of Russia's wealth into the hands of a few oligarchs. The US government also got involved. In March 1993, Bain & Co., a management consulting firm based in Boston, received a $3.9 million federal contract to advise Boris Yeltsin's administration on how to transform the state run Russian economy into one of free market capitalism and private businesses, one of which was the tobacco industry. In that role, Bain served as the facilitator for British American Tobacco (BAT) to enter the Russian market, exploiting strategies germane to Russia -- avoiding taxes, laundering money and bribing customs officials. Bain helped international tobacco corporations buy up Russian cigarette factories, and helped to significantly expand the smoking habit in the Russian population.

Bain's campaign of "Let's get Russia smoking Western cigarettes" was wildly successful, increasing the rate of smoking among Russian women by 300 percent in only a few years. The smoking rate among young Russians is now among the highest in the world – a public health catastrophe. An internal Bain memo from one of their consultants said, "The profit numbers . . . do not yet account for the lifetime value of young smokers. The 'get them early, keep them for life argument' is critical to the whole debate."[22] Thanks to Bain's strategic assistance and marketing analysis, BAT went from practically zero to 25 percent of the growing, lucrative Russian cigarette market.[22]

At the same time Bain also helped Philip Morris develop domestic strategies to help them fight the rising antagonistic public sentiment and social stigma against tobacco, and growing attention from federal regulators. That strategy included campaigns centered on protecting "smokers' rights" and preventing job losses from increasing cigarette taxes.

Big Tobacco became Bain's financial savior when it was still struggling to avoid insolvency, playing a large role in resuscitating their otherwise moribund client list. Bain went on to work for BAT in Eastern Europe and Asia, with the aspiration of making BAT the international king of Big Tobacco. And the person who steered the Bain ship to all these lucrative tobacco ports, was none other than high profile, iconic Mormon leader and politician, former Republican nominee for President, new Senator from Utah, and the hope of many people to bring sanity back to the Trump-controlled Republican Party, Mitt Romney.

According to an extensive investigative report [22] by Zach Carter and Jason Cherkis, drawing from the University of California at San Francisco's essential library of Legacy Tobacco Documents, at the very same time Mitt served as a Mormon Stake President (the lay spiritual leader of the Boston Mormon flock of over 4,000 parishioners) he was making himself rich promoting smoking and helping to ruin the health of millions of people, Russians and Americans alike.

To appreciate how jarring this is, one must first appreciate how abhorrent smoking is to Mormons. As a person who grew up immersed in the Mormon culture, served a Mormon mission just like Romney did, and held a significant supervisory position during my Mormon mission, I know well what Mormons think of smoking. While Mormons put a premium on "clean living" and exemplary "moral behavior," as Romney himself does, in the eyes of Mormons, nothing distinguishes them more from non-Mormons than their unique prohibition on smoking. It is considered not just a health proscription, but also carries moral overtones. Smoking would disqualify one from entering into one of the Mormon sacred temples, for example. Nothing more thoroughly symbolizes a person's departure from the standards and activities of the Mormon Church than cigarette smoking.

But regardless of his religion, Romney's willingness to help spread death and disease to millions of people in order to pocket millions himself, is a complete disconnect from what should be a universal moral platform. I would refer Mitt to the New Testament's story of Jesus's "temple tantrum," physically casting out the money changers from the temple (John 2:13-16). In a modern day cast, the money changers would be corporate CEOs, selling their souls in the "den of thieves."

A New York Times Magazine article by Roger Rosenblatt ran in March 1994, titled, "How Do Tobacco Executives Live With Themselves?"[13] At the time the tobacco industry was the iconic corporate purveyor of death and human suffering. Tobacco was killing more than 420,000 Americans a year, more than homicide, suicide, AIDS, auto accidents, alcohol and drug abuse combined. But the same question could and should be asked about executives of every other corporation and industry mentioned in this book and many that were omitted. After interviewing numerous tobacco executives for their thoughts on why they made their living selling a product that killed people, Rosenblatt offered this analysis.

"None of these executives think of themselves as morally bankrupt, and I do not think of them individually in that way, either. What often happens to people who work for a large, immensely successful company, however, is that they tend to adopt the values of the company, regardless of its product. Loyalty supersedes objectivity."[13] He might just as well have added loyalty supersedes rationality, empathy, integrity, and introspection.

Greed also supersedes objectivity, and feeds rationalization. Don Blankenship certainly wouldn't have taken out a gun and shot 29 people for $80 million. On second thought, for that much money, I'm not so sure. Rosenblatt continued:

> "How good, smart, decent individuals manage to contribute to a wicked enterprise is a question that has been applied to murderous governments as well as to industries. The best answer, which isn't particularly satisfying, is that people in groups behave differently, and usually worse, than they do singly. In speaking with these Philip Morris executives, I felt the presence of the company within the person. In the end, I felt that I was speaking with more company than person, or perhaps to a person who could no longer distinguish between the two. In this situation, in which the company has effectively absorbed its employees in its moral universe."

If at this point, a reader might be thinking that in order to arrive at corporations that behave less like psychopaths, our first step should be to more carefully screen the psychological profiles of corporate CEOs. Other research however suggests that corporations don't necessarily pick the wrong people, those with psychopathic traits, but that CEOs become the wrong people once they are picked. The mantle of power itself becomes the affliction, not that the afflicted ascended to power.

You have heard the phrase, "Power corrupts, and absolute power corrupts absolutely." In fact there is hard evidence that being placed in a position of significant power causes actual brain damage. Dacher Keltner, a psychology professor at UC Berkeley, after engaging in research on the psychology of power for over twenty years concluded that people placed in positions of power behaved as if they had suffered a "traumatic brain injury--becoming more impulsive, less risk-aware, and, crucially, less adept at seeing things from other people's point of view."[14]

Sukhvinder Obhi, a neuroscientist at McMaster University, in Ontario, studies the issue from an anatomical perspective. Using a transcranial-magnetic-stimulation machine (the kind we all keep under our bed just in case a neuroscientist breaks into our house), Obhi observed that people in an experimental position of power, compared to the "unempowered" demonstrate impaired neuronal activity called "mirroring," a trait closely related to the ability to feeling empathy.[15]

Adam Galinsky (Professor of Management and Organizations at the Kellogg School of Management) and his co-authors Joe Magee (Assistant Professor of Management at New York University), M. Ena Inesi (Assistant Professor of Organizational Behavior at the London Business School), and Deborah H. Gruenfeld (Professor of Leadership and Organizational Behavior at Stanford University) write that possessing power impairs a person's capability of understanding the point of view, feelings, and the needs of others, including subordinates.[16] Through a series of experiments on human subjects they found that the more empowered a person felt, the more likely they were to assume that others viewed the world as they did, the more they oriented their behavior towards themselves, rather than to others, and the less accurate they were in reading the emotions of others.

Within any business or organization, people tend to mimic the behavior and even the body language of their superiors. As I write this, the personal behavior of Donald Trump is infiltrating the rank of file of the country, especially those in the Republican Party, even more especially, those who attend his rallies. Many members of his own party have excoriated Trump for exactly that. [23] Hate crimes are up, racist and anti-immigrant incidents and behavior are on the rise, and children at school are committing more acts of bullying, the First Lady's "Be Best" campaign notwithstanding. But the power dynamic is also corrosive because leaders stop considering their subordinates. For a leader to laugh when others laugh, and worry when his subordinates worry, helps nurture and mature those concordant feelings in the leader. But those in power tend to stop driving on this two way street, opting for a one way street, i.e. "their way or the highway."

A study from The Journal of Finance found that CEOs who had lived through the trauma of natural disasters that involved significant fatalities were less likely to engage in risky company management.[17] In stark contrast, CEOs who had lived through such events where there were no losses of life, were even more likely to engage in risky management.

Dr. David Owen, a British neurologist, describes the "Hubris Syndrome" as an acquired "disorder, appearing after the possession of power, particularly power which has been associated with overwhelming success, held for a period of years and with minimal constraint on the leader"--a perfect description of a large corporate CEO.[18]

I'm not a Star Trek fan by any means. I have little interest in sci-fi movies or novels. But in attempting to understand how seemingly functional, main stream, "normal" people can become necessary, integral parts of truly pathological corporations, I became fascinated by the "hive mindset" that permeates and defines the fictional alien antagonists of humanity, "The Borg." Granted the only motive of a corporation is profit, and the motivation of the Borg is quite different; an unemotional, mechanical perfection. But the "hive" mentality inherent in the Borg seems eerily similar to the rationalization that must go on in the minds of people who, crouching behind the shield of corporate balance sheets, are willing to commit what is nothing less than monumental crimes against humanity.

Death by Corporation is more than the age-old story of greed and lust for power. In today's world, with the global reach of corporations, and their technological capability for unprecedented, and even global destruction, both in the physical world and the cyber world, and in the intersection of the two, those unrestrained, pathological urges smoldering within just a few key individuals are hurtling us all toward consequences unimaginable, and a future irredeemable. The Borg's slogan was, "Resistance is Futile." I disagree, resistance is imperative.

19

Dismantling the Death Star

"Corporations are not legal 'persons' with constitutional rights and freedoms of their own, but legal fictions that we created and must therefore control."

—Kalle Lasn,
Culture Jam: How To Reverse America's Suicidal Consumer Binge—And Why We Must

The legal, political, and economic clout of large corporations allow them virtually unchallenged influence on public policy, almost regardless of which party is in control, but especially so when the Republican Party is in power. Monstrous, amoral, greedy and unaccountable corporations have become such a cliché that it has become the theme of movies like *Soylent Green*, *Robocop, Avatar*, *Ex Machina*, and even children's movies like *Wall-E*.

For people, like Tea Party devotees, overcome with nostalgia for a bygone, mythical era of unfettered free markets, small government, and minimal taxes, they ignore the other side of the equation of modern of society--the growth and dominance of corporations. If the three pillars of 20th century American society were, at one time, Big Business, Big Government, and Big Labor, the latter has disappeared to the point of becoming just an historical footnote. Without a government big enough to challenge and restrain corporations, and committed to the public interest, there is nothing to left to provide a balance of power. One could argue that the emergence of corporate lobbyists has become the dominant force in American politics, virtually controlling the legislative and regulatory processes. But prior to the 1970s, they hardly existed.

The person who did more than anyone else to return the keys of government back over to corporations a la the era of the robber barons, was a Richmond, Virginia lawyer, and eventual Supreme Court Justice, Lewis Powell.

Powell's memorandum to the US Chamber of Commerce, "Attack On American Free Enterprise System"[2] of August 23, 1971, is to corporate rule like the Magna Carta is to democracy and human rights. In the memorandum Powell claims that socialist trends were sweeping across America's campuses, and he particularly vilified Ralph Nader, as the single most important antagonist of American business. Powell cited a *Fortune* magazine article published just before his memorandum claiming that, "He [Nader] thinks...that a great many corporate executives belong in prison for defrauding the consumer, with shoddy merchandise, poisoning the food supply with chemical additives, and willfully manufacturing unsafe products that will maim or kill the buyer." And that Nader was referring not just to "fly-by-night hucksters, but the top management of blue-chip business." To the extent that Fortune's assessment of Nader was correct, Nader's analysis is even more true now than it was in 1971. Powell also singled out for attack one of the most important books of my undergraduate education, *The Greening of America* by Yale Professor Charles Reich.

Powell's memorandum was a "Braveheart" call to arms for the business community to marshal all their resources against these sinister forces that would "destroy the free enterprise system." He laid out a detailed plan to undermine sympathy for socialism on America's campuses, to protect free enterprise from "unfair" portrayal in the radio, print and TV media, to vigorously pursue redress in the courts, and to realign government with the interests of big business.

Powell's depiction of existential threat to corporate America became the spring board for corporations to begin engaging in the political arena in a way they had never done before. In the next few years they succeeded in killing "a major labor law reform, rolled back regulation, lowered their taxes, and helped to move public opinion in favor of less government intervention in the economy."[3]

But it wasn't until the late 1980s that, with the help of their lobbyists, corporations became fully entrenched in the regulatory and legislative machinery such that government became more business's handmaiden than their watch dog.

As touched on in the introduction, corporate power has been cyclical. From the founding of the country to the late 19th century, corporations were largely held in check.

President Grover Cleveland included this statement in his 1888 State of the Union Speech. "Corporations, which should be the carefully restrained creatures of the law and the servants of the people, are fast becoming the people's masters."[25] Then the doctrine of corporate personhood emerged, large corporate monopolies started using their muscle, and the government started falling in line.

Pressure from John D. Rockefeller and other industrial magnates of the time, then started paying off in state governments like New Jersey and Delaware. Those states were eager to be the home of high profile corporate charters (known as "charter mongering"), like Standard Oil, and they started crafting state laws to coddle corporations. These laws opened the doors for corporations to own stock in other companies, to share board members, to define themselves for any conceivable legal purpose, and eliminated expiration dates that essentially imbued corporations with immortality. Corporate trusts then embarked on their own "gold rush," moving to New Jersey and Delaware to exploit the new protectionism. While News Jersey later back-peddled, even today over 60 percent of the Fortune 500 companies, over half of publicly traded companies, and over a million businesses in total are still incorporated in Delaware because of this move in the early 1900s.[26]

There was progressive push back against laws protecting corporations during the Teddy Roosevelt era. But the election of Warren G. Harding, touting tax cuts and deregulation of business, ushered in the roaring 20s, which ultimately ended in the spectacular stock market crash in 1929. Deregulation of the banks and brokerage firms played a significant role in the arrival of the Great Depression.

Nonetheless, over the next 60 years virtually every state stuck their toes, and sometimes both legs, into a cesspool of corporate friendly legislation similar to Delaware, presumably in order to compete.

There is no time machine that will allow us to return to the 1860s or even the 1960s. But it is certainly instructive, and hopefully productive, for citizens to understand that corporate rule is a rather recent sociologic and political phenomenon.

Nonetheless, we have arrived at staggering new world order. As of 2016, of the 100 largest economies in the world, 69 are corporations and only 31 are entire countries.[4] By 2016, the ten largest corporations together had more revenue than the 180 poorest countries combined.[5]

Apple's cash on hand is greater than the GDP of two thirds of the world's countries.[6]

Corporate agendas drive the governing bodies of the world economy like the International Monetary Fund, World Bank, and the World Trade Organization. Corporate billionaires like Donald Trump, Charles and David Koch (Koch Industries and American Legislative Exchange Council--ALEC), Rupert Murdoch (Fox News), Peter Theil (PayPal), Carl Icahn (corporate raider), Wilbur Ross (finance and Commerce Secretary), Sheldon Adelson (casino magnate), and Robert Mercer (hedge fund) and many others have blurred, if not completely erased the lines between big business and government as never before. In many cases profits are maintained, and even created, not by innovation, creativity, and entrepreneurship, but by political muscle, intervention, and let's call campaign contributions for what they are--bribery.

The bribery yields a stunning pay off. Tax codes allow corporations and billionaires to shield between $21 and 31 trillion from taxation.[7] Between 2008 and 2015, the government committed $16 trillion to the bank bailouts,[8] while their hired guns in Congress and the White House simultaneously claim the government is perennially broke, and social programs, health care, and safety nets for the poor must be "reformed," the obvious euphemism for "butchered."

Aggressive multi-nationals often have a legal residence in one country, corporate management operating and living in another, park their financial assets in a yet another country, and staff and employees spread over many others. Foreign Policy magazine observes that, "Some of the largest American-born firms — GE, IBM, Microsoft, to name a few — collectively are holding trillions of dollars tax-free offshore by having revenues from overseas markets paid to holding companies incorporated in Switzerland, Luxembourg, the Cayman Islands, or Singapore."[1]

One wonders what Lewis Powell would think of *Forbes* magazine observing, "During the housing bubble, Wall Street was considered the heart and soul of free market capitalism, but when they were in danger of total collapse they fell on their knees as socialists, begging the government and tax payers to bail them out."[9]

Consistent with the theme, President Trump's "big, beautiful" tax cuts passed with great fanfare in Dec. 2017, parachuted even more money down upon corporations and their ultra-rich CEOs.

It was hardly a shock that corporations didn't invest their tax cut savings the way that the proponents claimed they would, by hiring more workers, raising wages, and company expansion. They did exactly what skeptics said they would do. The destination for those savings was to the immediate bottom line of the companies, and to stock buy backs, both of which made the primary beneficiaries, the stock holders, especially the CEOs whose compensation is usually directly tied to company stock value.[21] At the end of 2017, CEO pay had already increased 17.6 percent over the year before, and this before the huge tax cuts.[22] Meanwhile compensation for regular workers actually dropped by 0.2 percent. Furthermore, the tax cuts massively increased the federal deficit and debt according to *Forbes.*[2] Eighty-four percent of corporate stocks are owned by just 10 percent of the U.S. population.[24]

Al Gore was mocked mercilessly by the press for a claim he never really made that, "he had invented the internet." But lost in that banter, and lost in the minds of voters, was that it was government research and public funds that created the internet, but the profits from Google, Facebook, and Apple are all privatized. Corporate power is maintained by the virtually free capital of "quantitative easing," while students with loans cannot renegotiate their loans below 7 percent.

Corporations have been enshrined as dominant players in the game of elections, thanks to, among other things, the notorious "Citizens United" ruling. But just as important, it is through corporations that the world's billionaire class has amassed their fortunes in the first place, creating a grotesque, growing and self-perpetuating wealth inequality unequaled in human history. President Ronald Reagan, patron saint of the Republican Party, famously said, "But what I want to see above all is that this country remains a country where someone can always get rich."[20] That phrasing is important. Yes, a precious few "someones" are getting rich, but it's not you, not me, and hardly anyone else. Furthermore, the "above all" clearly reveals Reagan's value system--gaining material wealth for a fortunate few is America's greatest gift to the world. Not personal freedom or democracy, not defending the free world, not protection of human rights, not charity, generosity, morality or human decency—but material wealth.

The twin sister of that value system is Reagan's often expressed antipathy to government regulation of business. "The nine most terrifying words in the English language are; 'I'm from the government, and I'm here to help.'" That has long been perhaps the most beloved Reagan quote of rock hard conservatives. The not-so-subtle message is that government should be impotent in the face of corporate supremacy. And the pervasion of that attitude has been a godsend to industry, and the bane of the working class.

Thanks to helping creating the necessary political scaffolding, Reagan's dream for America has become all too real. The richest eight individuals in the world have a combined net worth of $426 billion, as much as the bottom half of the entire world's population.[10] Corporations were the tools by which a few dozen people have arrived at world economic control in the first place.

Soulless corporations take a wrecking ball to our macro-environment, contaminate our personal environment, and pursue the impossible and catastrophic mantra of infinite growth in a finite world. And still the billionaires perched on top are pushing for more corporate control of the world's economy with schemes to further privatize education, health care, electricity, water resources, border control and prisons.

In a downward spiral, countries, states and local governments bid against each other with tax breaks, environmental compromises, labor concessions, and infrastructure perks in a zero sum game to attract everything from steel mills to Google server centers.

Thirty years ago, 90 percent of the print, TV, and radio media was divided among 50 different companies. Presently, that number has dwindled down to six.[11] In fact, Facebook, Google, Twitter, and the right wing Sinclair Broadcast Group now basically control the first amendment and the flow of information. Corporate media broadcast propaganda with little to no supervision, restraint, or accountability. Corporate political power stifles competition, and has led to massive consolidation in almost every industry during the last 30 years.[12]

Ten companies now control virtually all processed food and consumer goods in the United States.[13] Bernie Sanders speaks often about banks that are "too big to fail," should be considered "too big to exist." Despite this obvious economic Achilles heel causing the 2008 worldwide financial crisis, the banking industry is more consolidated now than ever. The five largest banks, JPMorgan Chase, Bank of America, Wells Fargo, Citigroup and U.S. Bancorp, control nearly half of the world's banking assets.[14]

If it feels like our entire planet is under enemy occupation that's because essentially it is. As George Monbiot observes, "Do you wonder why those who want a kind and decent and just world, in which both human beings and other living creatures are protected, so often appear to be opposed by the entire political establishment? If so, you have encountered corporate power . . . The key political question of our age, by which you can judge the intent of all political parties, is what to do about corporate power."[15]

Corporate power is self-perpetuating in that it has sapped citizens of even imagining a world without free enterprise and market fundamentalism. In a world where every form of communication is monetized and advertised, we drift into a morass where we unconsciously assume that the only power we have is as consumers.

The election of Donald Trump is rightly condemned on many fronts, including, if not especially, his servile approach to business interests, especially his own. But to the extent that it has pushed the pause button on international trade agreements, it has given us a chance to regroup and re-examine whether we should continue to allow international businesses to assemble the "Corporate Death Star"---The TransPacific Partnership (TPP).

Let's review how the TPP evolved and how both political parties were at one time its rabid cheerleaders, including Hillary Clinton and the Obama Administration. There is no question that the "force is still strong" to resurrect the TPP, and this battle is far from over.

Beginning around 2010, while the public and the media slept, the TPP was being forged in secret. No information was made available to the press or the public, and only extremely limited access was allowed to a few members of Congress. But in 2012, a document was leaked to the watchdog group, Public Citizen,[17] revealing the U.S. position at the time and the reason for the secrecy. The contents are surreal and shocking, and prima facie evidence for how corporations have become the master puppeteers of government. The leaked document reveals that the trade agreement would subordinate domestic law and policy to a binding international governance system.[18]

Specifically, TPP would 1) severely limit regulation of foreign corporations operating within U.S. boundaries, giving them greater rights than domestic firms, 2) extend incentives for US firms to move investments and jobs to lower-wage countries, 3) establish an alternative legal system, creating "investor states" that give foreign corporations and investors new rights to circumvent US courts and laws, allowing them to sue the U.S. government before foreign tribunals and demand compensation for lost revenue due to US laws they claim undermine their TPP privileges or their investment "expectations." Most of the TPP had nothing to do with trade. It was drafted with the oversight of 600 representatives of multinational corporations, who essentially stuffed corporate Christmas stockings with whatever gifts they wanted, including a dismantling of environmental and public health protection, worker safety, and further off-shoring of what was once a domestic workforce.

Wall Street intends to use the provisions of the TPP to resume gorging themselves on risky financial products such as the toxic derivatives that led to the $183 billion bailout of AIG. TPP is part of Big Pharma's agenda to expand, strengthen and prolong monopoly protections, and constrain generic competition.

Eighty-four percent of the seafood consumed by Americans is imported, mostly from TPP nations, and most of it is farmed, which has been shown to contaminate the fish and other seafood with a menagerie of drugs and chemicals. In monitoring the safety of imported seafood, the FDA currently only looks for chemical residue in 0.1 percent of all imported fish [19](see Chapter 16), and the TPP would lead to even more imported and contaminated seafood.

Residents of the American West should be particularly alarmed. TPP would sanction plunder of our natural resources by foreign corporations allowed to bypass U.S. law. Disputes over Western land use for mining and timber, for example, would be settled by international tribunals. Even if you are oblivious to environmental concerns, you should be outraged at the total circumvention of national sovereignty. Foreign investors could bypass our legal framework, take any dispute to an international tribunal, and pursue compensation for being denied exploitation, if not exhaustion of our resources.

It gets worse. Those tribunals would be staffed by private-sector lawyers that rotate between acting as judges and as advocates for the corporations suing the governments. American taxpayers could be forced to pay those corporations virtually unlimited compensation for curtailing their profits via domestic laws that currently protect our air, land, and water. Furthermore, there would be no appeal of decisions made by these tribunals.

All of this sounds like an Alex Jones conspiracy hallucination. Oh, that we could bury this frightening tale so easily. Unknown to most Americans, predecessors of these tribunals already exist, having been established by the WTO. These kangaroo courts for corporations have already ordered governments to pay over $3.5 billion in investor-state cases under existing U.S. agreements. This includes payments over bans of toxic chemicals, land-use policies, forestry rules, and more. More than $14.7 billion remain in pending claims under U.S. agreements alone. Even when governments win, they often must pay for the tribunal's costs and legal fees, which average $8 million per case. Using these "investor-state" privileges, Chevron is trying to evade responsibility for an enormous oil spill in Ecuador; Phillip-Morris is circumventing Australia's cigarette labeling laws; Eli Lilly is attacking Canada's drug patent laws; European firms are fighting Egypt's post-revolution minimum wage increase and South Africa's post-Apartheid affirmative action law. The Canadian Cattlemen for Fair Trade sued the U.S. for a ban on imports of live Canadian cattle after the discovery of a case of mad cow disease in Canada. The TPP would allow more corporate attacks on domestic laws intended for our protection. The current cast of TPP characters includes twelve countries, but in fact, corporations have ambitions to extend the TPP to half the world's population.

TPP is much worse than trade agreements like NAFTA, which itself eviscerated middle-class jobs and wealth in the US. The clear winners are those who sit in the control towers of international corporations. The clear losers are everyone else. The US Chamber of Commerce can't get it signed fast enough, and politicians from both parties will once again run interference for them. While the TPP has been stalled in large part by the rhetoric of President Trump, it will no doubt rise again when the political winds change.

To the extent that corporations pool resources and expertise toward a common goal, they have been engines to the advancement of civilization. It's hard to imagine an affordable i-phone being created and sold by your next-door neighbor out of his garage, or skyscrapers or aircraft carriers (in case you think those are necessary) being financed or built by the neighborhood gardening club. But when that corporate goal becomes to make money only, and life's most basic necessities -- clean air, clean water, non-contaminated food, intact ecosystems, and a livable climate -- stand in the way of making that money, many of the most powerful corporations have indeed become a gang of Frankenstein monsters, turning on us with a zombie-like indifference, with diabolical schemes of profitability at our expense, even to the point of dragging us into the abyss of an apocalyptic, uninhabitable world.

International corporations thrive with impunity in regulatory gaps that exist because there is little to no cooperation between nations in holding them accountable. There are virtually no treaties to counter "Death Star" trade agreements that trample on global environments, local economies, and human rights.

There is however a ray of hope. There is a new United Nations initiative for a global treaty that binds transnational corporations to behavior and business practices that protect human rights, and elevates the rights of countries and people above the rights of corporations. This proposed UN treaty is the work of numerous countries primarily from the global south to create international corporate laws to hold transnational corporations to higher standards. In 2014, these countries, led by Ecuador, convinced the UN Human Rights Council (UNHRC) to pass a resolution establishing the need for such a treaty.

A treaty would make it more difficult for them to engage in human rights abuses, environmental destruction, dangerous business practices, or to play countries against one another in a race to the bottom, by standardizing how governments deal with multinationals.

Friends of the Earth (FOE) is the world's largest grassroots environmental network. They "challenge the current model of economic and corporate globalization, and promote solutions that will help to create environmentally sustainable and socially just societies."[16] Their name, however, should be changed to "Friends of Humanity" because the earth itself will survive corporations, but humanity may not. They are helping to spearhead the international adoption of this treaty.

Sen. Elizabeth Warren (D-Massachusetts) has proposed a novel approach to making corporations better global citizens. She has a bill, the Accountable Capitalism Act,[25] that would challenge the philosophy that corporations are people. More specifically, her act is based on the presumption that most people usually act to the general benefit of society at large. Changing the composition of corporate boards could make their decision making more humane.

The act would call for corporations with annual revenues of at least $1 billion to have 40 percent of board seats be elected by the company's workers. The companies would be required to get a new federal charter that would mandate corporate directors consider the impact of their decisions on workers, communities, and society at large, not just stockholders. Companies would be shielded from shareholder lawsuits that might claim the companies were acting contrary to their obligation to maximize stock value at all costs.

Thom Hartmann says it's time to bring back the "corporate death penality,"[27] meaning that when corporations behave as criminals in society (like virtually every company or industry mentioned in this book) they should be subject to dissolution—i.e., have their charters revoked.

The darling of progressive politics, 2019 edition, is Rep. Alexandria Ocasio-Cortez (D-New York). Within about five minutes of being sworn in, she started calling for a 60-70 percent marginal tax on income over $10 million and devoting the proceeds towards the transition to renewable energy. That, of course, set the hair on fire of everyone at Fox News. But Nobel Prizing winning economist Paul Krugman agrees with Ocasio-Cortez, even suggesting that number may be too low. Corporate executives make up about two-thirds of the top 1 percent of wage earners. In addition to obviously addressing the breathtaking wealth inequality of 21st century America, increasing the tax on the excessive incomes of corporate CEOs would reduce the incentive for rabid pursuit of corporate profits, and therefore their own salaries, at the expense of society at large. There is an inverse correlation between executive compensation and the performance of corporations in the arena of social responsibility. Corporations would benefit as well. A recent study showed an inverse correlation between CEO pay and future profitability and company operating performance,[28] and a positive correlation between the financial performance of a corporation and their socially responsible behavior.[23]

The 21st century is a world where Monsanto's deadly chemicals applied in Honduras threaten the health of people in Alabama, where recklessness, incompetence, and corruption at the Tokyo Electric Power Company (TEPCO) in a Fukushima nuclear plant can contaminate the entire Pacific Ocean with radioactivity, where Big Macs eaten in North Carolina contribute to deadly heat waves in Afghanistan, where greed festering on Wall Street snuffs out the lives of babies in Africa. Reigning in corporations is a global imperative of the same magnitude as preventing all-out nuclear war. It will require the same degree of international cooperation and consent.

Slavery to corporate interests did not happen overnight, and emancipation from corporations will likewise take time. But it was a political process, not an immutable act of God that expanded the rights of corporations in the first place, while failing to expand their responsibilities or accountability to society.

Beyond international treaties, voters in developed democracies must rise above the recent alarming nationalistic trends in Europe and America that pander to racist, ethnic, and immigrant prejudices, and exalt narrow, short term, autocratic self-interest. These attitudes must be replaced by the realization that we are all first and foremost citizens of the world, co-inhabitants of planet that is becoming ever more a closed system, and increasingly a crowded life boat carrying billions of other human beings. It is a life boat that is rapidly taking on water, and we cannot allow it to be navigated or sabotaged by corporate monsters, and our dwindling rations to be divided up by psychopaths.

Supreme Court Justice Lewis Powell, wrote the "Magna Carta" for corporate rule in the second half of the 20th century

"But what I want to see above all is that this country remains a country where someone can always get rich."
---President Ronald Reagan

Milton Friedman, Nobel Prize winning economist, authored the sentinel book *Capitalism and Freedom*, arguing, "there is one and only one social responsibility of business:" to make money for its shareholders.

Sen. Elizabeth Warren, sponsor of the Accountable Capitalism Act offering a ray of hope to end global corporate rule.

ABOUT THE AUTHOR

Dr. Brian Moench is a physician in private practice and a former faculty member of Harvard Medical School. His environmental expertise includes a former adjunct faculty position at the University of Utah Honors College teaching public health and the environment.

He was nominated by President Obama and was a finalist for the award, Champion of Change, for his work on public health and the climate crisis in 2013. He is the president and founder of Utah Physicians for a Healthy Environment, (UPHE). Since 2007, UPHE has become the largest civic organization of medical professionals in the Western U.S. who are actively engaged in many environmental battles throughout the country and Western Canada.

Dr. Moench has appeared on national television, on MSNBC and Fox Business Chanel to discuss the relationship between neonatal death and pollution from fracking. He has been quoted in numerous national newspapers, appears regularly on local television and radio programs in Utah discussing the health consequences of air pollution, and is well recognized by the Utah public as an expert on air pollution and health. He gives dozens of lectures a year on numerous issues related to environmental contamination. He has testified as an expert witness in environmental lawsuits.

On a lighter note, Dr. Moench grew up the child of extensive brain washing and corruption by *MAD Magazine*. Still under its formative influence as an adult, he became the creator and founder of a humorous greeting card company, In Your Face Cards, which sold 35 million cards over 11 years. CNN did a personal profile on him in the late 1990s highlighting his idiosyncratic mix of careers as a physician, humorous cartoonist, and creator of In Your Face Cards.

Dr. Moench is a thorn in the side of state and federal government, industrial polluters, oil and gas drillers, wood burners, climate deniers, purveyors of "fake news" and perpetrators of corporate malfeasance.

Dr. Moench speaking at the "Clean Air, No Excuses" protest rally in Salt Lake City, Jan. 2014--the largest public pollution protest in the United States in the last several decades.

References

Introduction

1.http://www.nytimes.com/2012/12/06/world/asia/3-walmart-suppliers-made-goods-in-bangladeshi-factory-where-112-died-in-fire.html?_r=1&
2. Corporate Laws Committee, ABA Business Law Section, Benefit Corporation White Pa- per, 68 Bus. Law. 1083, 1083–87 (2013)
3. https://newint.org/features/2002/07/05/history
4. https://newint.org/features/2002/07/05/history
5. https://hbr.org/2010/04/what-the-founding-fathers-real.html
6. http://reclaimdemocracy.org/corporate-accountability-history-corporations-us/
7. https://www.opensecrets.org/lobby/index.php
8. http://articles.washingtonpost.com/2011-08-11/politics/35270239_1_romney-supporters-mitt-romney-private-sector-experience

Chapter One

1. https://www.freep.com/story/news/local/michigan/flint-water-crisis/2017/06/14/flint-water-crisis-charges/397425001/
2. https://www.aap.org/en-us/about-the-aap/aap-press-room/pages/AAP-Statement-CDC-Revised-Lead-Exposure-Guidelines.aspx
3. http://www.epa.gov/history/topics/perspect/lead.html
4. http://www.corrosion-doctors.org/Elements-Toxic/Lead-history.htm
5. http://www.dartmouth.edu/~toxmetal/toxic-metals/more-metals/lead-history.html
6. http://journals.sagepub.com/doi/pdf/10.2190/NS7.4.p
7. https://www.thenation.com/article/secret-history-lead/
8. http://www.damninteresting.com/the-ethyl-poisoned-earth/
9. Very low lead exposures and children's neurodevelopment. Bellinger, David C. 2, April 2008, Current Opinion in Pediatrics, Vol. 20, pp. 172–177.
10. http://pic.plover.com/Nevin/Nevin2007.pdf
11. Needleman HL, Gunnoe C, Leviton A, Reed R, Peresie H, Maher C, et al. Deficits in psychologic and classroom performance of chil- dren with elevated dentine lead levels [published erratum appears in N Engl J Med 1994;331:616-7]. New Engl J Med 1979;300:689- 95.
12 http://historyoflead.tumblr.com/page/3

13. http://www.telegraph.co.uk/earth/energy/9800019/British-company-selling-toxic-lead-fuel-to-poor-countries.html
14. http://www.sfo.gov.uk/press-room/latest-press-releases/press-releases-2014/four-sentenced-for-role-in-innospec-corruption.aspx
15.http://www.unep.org/transport/pcfv/PDF/Hatfield_Global_Benefits_Unleaded.pdf
16. http://www.ncbi.nlm.nih.gov/pmc/articles/PMC3230438/
17. http://www.scientificamerican.com/article/lead-in-aviation-fuel/
18. http://www.thune.senate.gov/public/index.cfm/press-releases?ID=392dfa4e-b5a1-4f85-9de5-427ca34fce12
19. http://www.bloomberg.com/news/2013-09-02/lead-paint-s-tragic-comeback.html
20.http://www.ipen.org/ipenweb/documents/work%20documents/global_paintstudy.pdf
21. http://www.huffingtonpost.com/2013/03/25/lead-paint-exports-pesticides_n_2949694.html
22. Gottesfeld P, Kuepouo G, Tetsopgang S, Durand K. Lead concentrations and labeling of new paint in Cameroon. J Occup Environ Hyg. 2013;10(5):243-9. doi: 10.1080/15459624.2013.768934.
23. http://www.aap.org/en-us/about-the-aap/aap-press-room/pages/AAP-Statement-CDC-Revised-Lead-Exposure-Guidelines.aspx
24. http://www.cdc.gov/biomonitoring/Lead_FactSheet.html
25. Lucchini, RG, S Zoni, S Guazzetti, E Bontempi, S Micheletti, K Broberg, G Parrinello and DR Smith. 2012. Inverse association of intellectual function with very low blood lead but not with manganese exposure in Italian adolescents. Environ Res. 2012 Oct;118:65-71. doi: 10.1016/j.envres.2012.08.003. Epub 2012 Aug 24
26. Miranda ML, Kim D, Overstreet Galeano MA, Paul C, Hull A, Morgan SP. The relationship between early childhood blood lead levels and performance on end of grade tests. Environ Health Perspect. 2007;115:1242–1247.
27. http://www.theatlantic.com/features/archive/2014/03/the-toxins-that-threaten-our-brains/284466/
28. http://ephtracking.cdc.gov/showLeadPoisoningEnv.action
29. https://www.thelancet.com/journals/lanpub/article/PIIS2468-2667(18)30025-2/fulltext
30.https://www.cdc.gov/tobacco/data_statistics/fact_sheets/fast_facts/index.htm
31. https://archive.epa.gov/epa/aboutepa/epa-takes-final-step-phaseout-leaded-gasoline.html
32.https://video.search.yahoo.com/search/video;_ylt=AwrTcYS464ZWujoAJUaJzbkF?p=On%20the%20WaterfrontMarlon%20Brando&fr=aaplw&

fr2=p%3As%2Cv%3Ai%2Cm%3Apivot#id=5&vid=646fe9747c695aba6e a579e18b52964b&action=view
33. http://www.bbc.com/news/magazine-27067615
34.http://www.biologicaldiversity.org/campaigns/get_the_lead_out/pdfs/health/Bellinger_2008b.pdf
35. https://undark.org/article/lead-ammunition-bullets-hunting-copper/
36. https://www.reuters.com/article/us-usa-interior-zinke/new-interior-head-lifts-lead-ammunition-ban-in-nod-to-hunters-idUSKBN16930Z
37. https://www.npr.org/sections/health-shots/2017/05/10/527648768/lead-dust-from-firearms-can-pose-a-silent-health-risk
38. National Research Council. 2013. Potential Health Risks to DOD Firing-Range Personnel from Recurrent Lead Exposure. Washington, DC: The National Academies Press. https://doi.org/10.17226/18249.
39. https://www.ncbi.nlm.nih.gov/pmc/articles/PMC5379568/
40. http://www.oregonlive.com/environment/index.ssf/page visitors_soldiers_children_exposed_to_lead_dust_in_armories.html
41. https://www.nature.com/articles/d41586-018-02481-5

Chapter Two

1. https://www.ncbi.nlm.nih.gov/pmc/articles/PMC3817964/
2. Goldberg, J. M. and Falcone, T., 1999. 'Effect of diethylstilbestrol on reproductive function', Fertil. Steril. Vol. 72, No 1, pp. 1–7.
3. Newbold, R. R., Hanson, R. B., Jefferson, W. N., Bullock, B. C., Haseman, J. and McLachlan, J. A., 1998. 'Increased tumors but uncompromised fertility in the female descendants of mice exposed developmentally to diethylstilbestrol', Carcinogenesis Vol. 19, No 9, pp. 1655–1663.
4. Miller, M., 1999. 'DES research heats up again after breast cancer finding', J. Nat. Cancer Inst. Vol. 91, No 16, pp. 1361–1363.
5. Noller, K. L. and Fish, C. R., 1974. 'Diethylstilbestrol usage: Its interesting past, important present and questionable future', Medical Clinics of North America Vol. 58, No 4, pp. 793–810.
6. Aschbacher, P. W., 1976. 'Diethylstilbestrol metabolism in food-producing animals', J. Toxicol. Environ. Health Suppl. Vol. 1, pp. 45– 59.
7. Shimkin, M. B. and Grady, H. G., 1941. 'Toxic and carcinogenic effects of stilbestrol in strain C3H male mice', J. Nat. Cancer Inst. Vol. 4, p. 55.
8. Gardner, W. U., 1959. 'Carcinoma of the uterine cervix and upper vagina: induction under experimental conditions in mice', Ann. NY Acad. Sci. Vol. 75, pp. 543–564.
9. Dunn, T. B. and Green, A. W., 1963. 'Cysts of the epididymis, cancer of the cervix, granular cell myoblastoma, and other lesions after estrogen

injection in newborn mice', J. Nat. Cancer Inst. Vol. 31, No 2, pp. 425–455.
10. https://www.ncbi.nlm.nih.gov/pmc/articles/PMC3817964/#R1510
11. Bergkvist L, et al. The risk of breast cancer after estrogen and estrogen-progestin replacement. N Engl J Med. 1989 Aug 3;321(5):293-7.
12. Rosenberg L, Boggs D, Wise L, Adams-Campbell L, Palmer J. Oral Contraceptive Use and Estrogen/Progesterone Receptor–Negative Breast Cancer among African American Women. Cancer Epidemiol Biomarkers Prev August 2010 19; 2073. Published OnlineFirst July 20, 2010; doi: 10.1158/1055-9965.EPI-10-0428
13. https://www.onfaith.co/onfaith/2010/05/05/the-pills-50th-anniversary-science-reason-and-womens-rights/3902
14. http://www.lifesitenews.com/news/surgeon-birth-control-pill-a-molotov-cocktail-for-breast-cancer/
15. https://www.nejm.org/doi/full/10.1056/NEJMoa1700732
16. http://www.bloomberg.com/news/2011-12-05/bayer-withheld-yasmin-clot-risk-data-from-u-s-ex-agency-head-tells-court.html
17. http://noprescriptionnecessary.wordpress.com/2011/12/08/yaz-the-sad-and-sorry-saga-of-americas-best-selling-oral-contraceptive/
18. http://www.bloomberg.com/news/2011-12-05/bayer-withheld-yasmin-clot-risk-data-from-u-s-ex-agency-head-tells-court.html
19. http://msmagazine.com/blog/2012/02/09/just-how-safe-is-yaz-women-need-to-know/
20. http://www.lawyersandsettlements.com/lawsuit/yasmin-side-effects-yaz-blood.html#.UpwD_pE2VBU
21. Samuel Epstein, Criminal Indifference of the FDA to Cancer Prevention
22. https://jamanetwork.com/journals/jama/fullarticle/2653735

Chapter Three

1. https://www.ncbi.nlm.nih.gov/pmc/articles/PMC1122833/
2. Green C, Bassett K, Foerster V, Kazanjian A. Bone mineral density testing: does the evidence support its selective use in well women? Vancouver, BC: British Columbia Office of Health Technology Assessment; 1997.
3. Pinnock C, Stapleton A, Marshall V. Erectile dysfunction in the community: a prevalence study. Med J Aust. 1999;171:353–357. [PubMed]
4. Melody Petersen's Our Daily Meds: How the Pharmaceutical Companies Transformed Themselves into Slick Marketing Machines and Hooked the Nation on Prescription Drugs, pp. 224.
5. http://healthimpactnews.com/2012/pharmaceutical-companies-causing-106000-deaths-a-year/
6. http://www.cdc.gov/nchs/data/databriefs/db42.htm

References

7.http://online.wsj.com/article/SB10001424052970203731004576046073896475588.html?mod=WSJ_hpp_editorsPicks_1
8. http://abcnews.go.com/US/study-shows-foster-children-high-rates-prescription-psychiatric/story?id=15058380#.UdQdTM3B_JH
9. Gardiner Harris and Benedict Carey, "Researchers Fail to Reveal Full
10.http://www.alternet.org/story/148907/15_dangerous_drugs_big_pharma_shoves_down_our_throats
11. ht
12.https://www.goodrx.com/actiq?gclid=CjwKCAjwuITNBRBFEiwA9N9YEG_jl_rV496bxL1DIqMY17T0aHxbgw7J-YXcQ88JV__-PfXX7rYktBoCbVw↵Barboza D. Tearing down the facade of 'Vitamins Inc.' 10 Oct 1999. New York Times.
www.nytimes.com/1999/10/10/business/tearing-down-the-facade-of-vitamins-inc.html
QAvD_BwE&hide_online_pharmacies=false&show_pet_friendly_pharmacies=true
13. http://www.nytimes.com/1999/10/10/business/tearing-down-the-facade-of-vitamins-inc.html
14. https://www.nytimes.com/interactive/2017/06/05/upshot/opioid-epidemic-drug-overdose-deaths-are-rising-faster-than-ever.html
15.http://www.cdc.gov/features/vitalsigns/painkilleroverdoses/index.html
16. http://bostonreview.net/archives/BR35.3/angell.php
17. http://www.vanityfair.com/politics/features/2011/01/deadly-medicine-201101
18. Irving Kirsch et al., "The Emperor's New Drugs: An Analysis of Antidepressant Medication Data Submitted to the US Food and Drug Administration," Prevention & Treatment, July 15, 2002.
19. Lisa Cosgrove et al., "Financial Ties Between DSM-IV Panel Members and the Pharmaceutical Industry," Psychotherapy and Psychosomatics, Vol. 75, No.
3 (2006).
20. David Tuller, "Seeking a Fuller Picture of Statins," The New York Times, July 20, 2004.
21. http://www.guardian.co.uk/books/2012/oct/17/bad-pharma-ben-goldacre-review
22. http://www.time.com/time/health/article/0,8599,1883449,00.html
23. https://www.csmonitor.com/Business/Latest-News-Wires/2013/0829/Tylenol-warnings-Don-t-take-too-much
24. http://www.propublica.org/article/tylenol-mcneil-fda-use-only-as-directed
25.http://online.wsj.com/article/SB10001424052748704779704574554071807123380.html
26. http://www.drugwatch.com/vioxx/recall/

27. http://www.kentucky.com/news/local/crime/article44396436.html
28. https://www.samhsa.gov/data/sites/default/files/NSDUH-FRR1-2014/NSDUH-FRR1-2014.pdf
29. http://www.motherjones.com/politics/2013/08/meth-pseudoephedrine-big-pharma-lobby/
30. http://www.motherjones.com/politics/2013/08/meth-pseudoephedrine-big-pharma-lobby
31. https://www.apnews.com/3d257452c24a410f98e8e5a4d9d448a7
32. https://www.alternet.org/news-amp-politics/who-profits-opioid-crisis-meet-secretive-sackler-family-making-billions-oxycontin
33. https://www.cdc.gov/drugoverdose/epidemic/index.html
34. https://www.cdc.gov/mmwr/volumes/65/rr/rr6501e1.htm
35. https://www.apnews.com/3d257452c24a410f98e8e5a4d9d448a7
36. https://www.alternet.org/news-amp-politics/who-profits-opioid-crisis-meet-secretive-sackler-family-making-billions-oxycontin
37. https://www.newyorker.com/business/currency/who-is-responsible-for-the-pain-pill-epidemic
38. http://www.latimes.com/projects/la-me-oxycontin-full-coverage/
39.http://archive.jsonline.com/watchdog/watchdogreports/119130114.html/
40. http://www.purduepharma.com/news-media/2017/03/fda-approves-symproic-naldemedine-once-daily-tablets-c-ii-for-the-treatment-of-opioid-induced-constipation-in-adults-with-chronic-non-cancer-pain/
41. https://www.washingtonpost.com/graphics/2017/investigations/dea-drug-industry-congress/?utm_term=.c2eec2a53d79
42. https://www.cbsnews.com/news/ex-dea-agent-opioid-crisis-fueled-by-drug-industry-and-congress/
43. http://www.latimes.com/projects/la-me-oxycontin-part3/
44.https://www.cdc.gov/mmwr/volumes/66/wr/mm6610a1.htm?s_cid=mm6610a1_w
45. https://www.washingtonpost.com/national/health-science/oxycontin-maker-purdue-pharma-to-stop-promoting-the-drug-to-doctors/2018/02/10/c59be118-0ea7-11e8-95a5-c396801049ef_story.html?utm_term=.3cf3ec870af3
46. http://www.scientificamerican.com/article.cfm?id=pharmaceuticals-in-the-water
47. http://www.the-scientist.com/?articles.view/articleNo/43615/title/Drugging-the-Environment/
48. http://usatoday30.usatoday.com/news/nation/2008-03-10-drugs-tap-water_N.htm
49. https://www.publicintegrity.org/2018/02/12/21567/opioid-makers-paid-millions-advocacy-groups-promoted-their-painkillers-

amid?utm_source=email&utm_campaign=watchdog&utm_medium=publici-email&utm_source=Watchdog&utm_campaign=8beeec9ef4-EMAIL_CAMPAIGN_2017_12_13&utm_medium=email&utm_term=0_ffd1d0160d-8beeec9ef4-102622149&mc_cid=8beeec9ef4&mc_eid=973036fe07
50. https://www.ncbi.nlm.nih.gov/pmc/articles/PMC4107906/
51. Marshall J, Aldhous P. Swallowing the Best Advice? [last visited July 1, 2013];New Scientist. 2006 192:18–22. available at < http://www.newscientist.com/article/mg19225755.100-patient-groups-special-swallowing-the-best-advice.html>.
52. https://www.huffingtonpost.com/entry/patient-advocacy-groups-take-in-millions-from-drugmakers_us_5ac7c355e4b01e0b61b762a2
53.https://www.sciencedirect.com/science/article/pii/S014765131530109
54. https://pubs.acs.org/doi/ipdf/10.1021/acs.est.7b02912
55.https://www.sciencedirect.com/science/article/pii/S187853521630033
56. https://www.bmj.com/content/345/bmj.e4348
57. https://www.yahoo.com/news/malady-mongers-drug-companies-sell-treatments-inventing-diseases-100040360.html
58.https://www.nuedexta.com/utm_source=VerticalHealth&utm_medium=display&utm_term=Stroke&utm_content=300x600&utm_campaign=PHM#take-the-pba-quiz
59. https://www.explorejournal.com/article/S1550-8307(06)00348-X/pdf
60. https://www.ncbi.nlm.nih.gov/pmc/articles/PMC1420696/
61. https://www.ncbi.nlm.nih.gov/pmc/articles/PMC1369125/
62. https://www.consumerreports.org/prescription-drugs/too-many-meds-americas-love-affair-with-prescription-medication/
63. https://www.ncbi.nlm.nih.gov/pmc/articles/PMC3864987/
64. Hohl CM, Dankoff J, Colacone A, et al. Polypharmacy, adverse drug-related events, and potential adverse drug interactions in elderly patients presenting to an emergency department. Ann Emerg Med. 2001;38:666–71. [PubMed]
65. https://www.investopedia.com/investing/which-industry-spends-most-lobbying-antm-so/
66. https://www.scientificamerican.com/article/most-experimental-drugs-are-tested-offshore-raising-concerns-about-data/
67. https://www.bmj.com/content/346/bmj.f707
68. Drug Companies and Doctors: A story of Corruption." NY Review of Books, Jan. 15, 2009.
69. http://bostonreview.net/archives/BR35.3/angell.php
70. https://www.corporatecrimereporter.com/thunepstein051005.htm
71. McKay Jenkins, My Quest to Survive in a Toxic World. Penguin Random House

Chapter Four

1. https://www.nature.com/articles/s41559-016-0051
2. http://www.thelancet.com/journals/lancet/article/PIIS0140-6736(17)32345-0/fulltext
3. Gale RW, WL Cranor, DA Alvarez, JN Huckins, JD Petty and GL Robertson. Semivolatile organic compounds in residential air along the Arizona – Mexico border. Environmental Science and Technology doi: 10.1021/es803482u.
4. Bernard D. Goldstein, M.D., Howard J. Osofsky, M.D., Ph.D., and Maureen Y. Lichtveld, M.D., M.P.H.The Gulf Oil Spill N Engl J Med 2011; 364:1334-1348April 7, 2011
5. https://www.vox.com/a/explain-food-america
6. http://www.npr.org/2017/06/16/533255590/alarming-number-of-americans-believe-chocolate-milk-comes-from-brown-cows
7. https://www.scientificamerican.com/article/chemical-tainted-food/
8. Bouatra S, Aziat F, Mandal R, Guo AC, Wilson MR, et al. (2013) The Human Urine Metabolome. PLoS ONE 8(9): e73076. doi:10.1371/journal.pone.0073076
9. http://www.scientificamerican.com/article.cfm?id=newborn-babies-chemicals-exposure-bpa
10. Woodruff TJ, et al. Environmental Chemicals in Pregnant Women in the US: NHANES 2003-2004. Environmental Health Perspectives, 2011; DOI: 10.1289/ehp.1002727
11. Vogt R, Bennett D, Cassady D, Frost J, Ritz B, Hertz-Picciotto I. Cancer and non-cancer health effects from food contaminant exposures for children and adults in California: a risk assessment. Environ Health. 2012 Nov 9;11:83. doi: 10.1186/1476-069X-11-83.
12. http://articles.latimes.com/2007/may/25/nation/na-fetuses25
13. The Standing Committee of European Doctors – which brings together the continent's top physicians' bodies, including the Bristish Medical Association stated, "Chemical pollution represents a serious threat to children, and to Man's survival."
14.https://academic.oup.com/humrep/article/28/2/462/599181/Decline-in-semen-concentration-and-morphology-in-a
15. Jensen TK, Jorgensen N, Punab M, et al. Association of in utero exposure to maternal smoking with reduced semen quality and testis size in adulthood: a cross-sectional study of 1,770 young men from the general population in five European countries. Am J Epidemiol. 2004;159(1):49–49

16. https://www.scientificamerican.com/article/chemical-tainted-food/
17. http://www.chicagotribune.com/news/watchdog/flames/ct-met-flames-regulators-20120510,0,6880244,full.story
18. http://bangordailynews.com/2012/11/28/health/toxic-flame-retardants-found-in-couches-in-maine-and-u-s/?ref=inline
19. http://www.chicagotribune.com/news/watchdog/flames/ct-met-flames-regulators-20120510,0,6880244,full.story
20. http://www.ewg.org/research/flame-retardants-2014
21. Daniels R, et al. Mortality and cancer incidence in a pooled cohort of US firefighters from San Francisco, Chicago and Philadelphia (1950–2009). Occup Environ Med oemed-2013-101662
22. Kang D, Davis L, Hunt P, Kriebel . Cancer Incidence Among Male MassachusettsFirefighters, 1987–2003. AMERICAN JOURNAL OF INDUSTRIAL MEDICINE 51:329–335 (2008)
23. Stapleton HM, et al. Novel and High Volume Use Flame Retardants in US Couches Reflective of the 2005 PentaBDE Phase Out. Environ. Sci. Technol., 2012, 46 (24), pp 13432–13439
DOI: 10.1021/es303471d
24. http://www.huffingtonpost.com/2014/06/24/flame-retardants-toxic-_n_5525973.html
25. https://www.nature.com/articles/d41586-018-02481-5
26. http://www.truth-out.org/news/item/43517-the-precautionary-principle-asks-how-much-harm-is-avoidable-rather-than-how-much-harm-is-acceptable
27.http://precaution.org/lib/ruckelshaus_risk_in_a_free_society_(risk_assessment).1984.pdf
28. http://precaution.org/lib/rehn420.pdf
29. https://www.nap.edu/read/1802/chapter/1
30. http://www.truth-out.org/news/item/43517-the-precautionary-principle-asks-how-much-harm-is-avoidable-rather-than-how-much-harm-is-acceptable
31. Gordon L, Joo JE, Powell JE, Ollikainen M, Novakovic B, Li X, Andronikos R, Cruickshank MN, Conneely KN, Smith AK, Alisch RS, Morley R, Visscher PM, Craig JM, Saffery R. Neonatal DNA methylation profile in human twins is specified by a complex interplay between intrauterine environmental and genetic factors, subject to tissue-specific influence. Genome Res, July 16, 2012 DOI: 10.1101/gr.136598.111
32. https://www.ncbi.nlm.nih.gov/pmc/articles/PMC2726844/
33. http://www.businessinsider.com/epa-only-restricts-9-chemicals-2016-2
34. https://www.nbcnews.com/news/us-news/epa-eases-path-new-chemicals-raising-fears-health-hazards-n838201
35. http://www.businessinsider.com/epa-only-restricts-9-chemicals-2016-

2#nitrites-mixed-with-p6-mixed-mono-and-diamides-of-an-organic-acidpp7-triethanolamine-salts-of-a-substituted-organic-acidpp8-triethanolanime-salt-of-tricarboxylic-acidpp9-tricarboxylic-acidp-6
36. https://silentspring.org/media/environmental-health-news-presidents-cancer-panel
37.http://www.pulitzer.org/files/finalists/2013/chictrib2013/chictrib01.pdf
38.https://www.cdc.gov/exposurereport/pdf/FourthReport_UpdatedTables_Volume2_Mar2018.pdf
39. http://press.endocrine.org/doi/10.1210/jc.2015-1841
40. Macdonald RW, Harner T, Fyfe J. Recent climate change in the Arctic and its impact on contaminant pathways and interpretation of temporal trend data. Sci Total Environ. 2005;342:5–86
41. Ritter R, Scheringer M, MacLeod M, Hungerbuhler K. Assessment of nonoccupational exposure to DDT in the tropics and the north: relevance of uptake via inhalation from indoor residual spraying. Environ Health Perspect. 2011;119:707–712
42. https://news.nationalgeographic.com/2015/06/15616-breast-cancer-ddt-pesticide-environment/#close
43. Layton LUS. Cites fears on chemical in plastics. Washington Post. April 16, 2008:A1
44. Richter CA, Birnbaum LS, Farabollini F, et al. In vivo effects of bisphenol A in laboratory rodent studies. Reprod Toxicol. 2007;24(2):199–224
45. https://www.telegraph.co.uk/news/health/3660141/Men-under-threat-from-gender-bending-chemicals.html
46. https://www.scientificamerican.com/article/bpa-free-plastic-containers-may-be-just-as-hazardous/
47. http://www.pnas.org/content/early/2015/01/07/1417731112
48. https://www.ncbi.nlm.nih.gov/pmc/articles/PMC3222987/
49. Barnard A. Group blasts Bush nominee for industry-tied research. Boston Globe. March 13, 2001:A2
50. Gray GM, Cohen JT, Cunha G, et al. Weight of the evidence evaluation of low-dose reproductive and developmental effects of bisphenol A. Hum Ecol Risk Assess. 2004;10:875–921
51. vom Saal FS, Hughes C. An extensive new literature concerning low-dose effects of bisphenol A shows the need for a new risk assessment. Environ Health Perspect. 2005;113(8):926–933
52. vom Saal FS, Akingbemi BT, Belcher SM, et al. Chapel Hill bisphenol A expert panel consensus statement: integration of mechanisms, effects in animals and potential to impact human health at current levels of exposure. Reprod Toxicol. 2007;24(2):131–138
53. http://www.chemtrust.org/wp-content/uploads/chemtrust-toxicsoup-

mar-18.pdf
54. https://www.ncbi.nlm.nih.gov/pmc/articles/PMC2072821/
55. https://www.ewg.org/research/poisoned-legacy/lab-accident-global-pollutant#.WzEUGC2ZOA8
56. https://www.nofluoride.com/3MSecrets.cfm
57. https://theintercept.com/2018/07/31/3m-pfas-minnesota-pfoa-pfos/
58. https://www.mprnews.org/story/2018/03/12/3m-professor-giesy-influence-pfcs
59. https://www.cbc.ca/news/canada/saskatoon/u-of-s-professor-denies-suppressing-toxic-pollution-research-for-3m-1.4554634
60. https://theintercept.com/2015/08/11/dupont-chemistry-deception/
61. https://www.ewg.org/research/poisoned-legacy/lab-accident-global-pollutant#.WzEUGC2ZOA8
62. https://www.treehugger.com/environmental-policy/what-you-need-know-about-pfoa-and-pfos-chemicals-behind-pruitts-recent-epa-scandal.html
63. https://theintercept.com/2015/08/17/teflon-toxin-case-against-dupont/
64. https://ehjournal.biomedcentral.com/articles/10.1186/s12940-018-0405-y
65. https://www.politico.com/story/2018/05/14/emails-white-house-interfered-with-science-study-536950
66. https://ehjournal.biomedcentral.com/articles/10.1186/s12940-018-0405-y
67. https://theintercept.com/2016/03/03/new-teflon-toxin-causes-cancer-in-lab-animals/
68. http://www.urinemetabolome.ca
69. http://greensciencepolicy.org/wp-content/uploads/2014/03/Sac-Bee-Op-Ed.pdf
70.https://www.durabilityanddesign.com/news/?fuseaction=view&id=11563

Chapter Five

1. http://www.history.com/news/the-killer-fog-that-blanketed-london-60-years-ago
2. Bell, M.L.; Davis, D.L. & Fletcher, T. (2004). "A Retrospective Assessment of Mortality from the London Smog Episode of 1952: The Role of Influenza and Pollution". Environ Health Perspect. 112 (1; January): 6–8. doi:10.1289/ehp.6539
3. Bharadwaj P, et al. Early Life Exposure to the Great Smog of 1952 and the Development of Asthma. Am J Respir Crit Care Med. First published online 08 Jul 2016 as DOI: 10.1164/rccm.201603-0451OC
4. https://www.washingtonpost.com/news/energy-

environment/wp/2017/10/19/pollution-kills-9-million-people-each-year-new-study-finds/?utm_term=.9f99b9aa1e42
5. D. Helmig, C. R. Thompson, J. Evans, P. Boylan, J. Hueber, and J.-H. Park
Highly Elevated Atmospheric Levels of Volatile Organic Compounds in the Uintah Basin, Utah. Environ. Sci. Technol. 2014, 48, 4707−4715
dx.doi.org/10.1021/es405046r |
6. http://www.newsweek.com/2014/05/30/utah-boom-town-spike-infant-deaths-raises-questions-251605.html
7. http://www.newsweek.com/after-midwife-sounds-alarm-utah-confirms-spike-infant-deaths-oil-and-gas-315915
8. http://beta.latimes.com/nation/la-na-utah-baby-deaths-20150111-story.html
9. https://www.denverpost.com/2014/10/25/dead-babies-near-oil-drilling-sites-raise-questions-for-researchers/
10. https://www.youtube.com/watch?v=wbrwEOcscvI
11. https://www.rollingstone.com/culture/features/fracking-whats-killing-the-babies-of-vernal-utah-20150622
12. http://billmoyers.com/2015/01/12/utah-oil-boomtown-hostile-midwifes-concern-skyrocketing-infant-deaths/
13. http://archive.sltrib.com/story.php?ref=/sltrib/news/57914660-78/basin-birth-deaths-department.html.csp
14. https://www.penguinrandomhouse.com/books/531803/an-enemy-of-the-people-by-arthur-miller/9780140481402/
15. http://uphe.org/air-pollution-health/cardiovascular-system/
16. Janet Currie, Michael Greenstone, Katherine Meckel. Hydraulic
17. Muzhe Yang, Rhea A. Bhatta, Shin-Yi Chou and Cheng-I Hsieh. The Impact of Prenatal Exposure to Power Plant Emissions on Birth Weight: Evidence from a Pennsylvania Power Plant Located Upwind of New Jersey. Journal of Policy Analysis and Management, 4 APR 2017 DOI: 10.1002/pam.21989
18. http://www.sciencedirect.com/science/article/pii/S0013935111001484
19. https://magazine.byu.edu/article/clearing-the-air/
20. https://magazine.byu.edu/article/clearing-the-air/
21. http://www.thelancet.com/journals/lancet/article/PIIS0140-6736(17)32345-0/fulltext
22. http://www.who.int/news-room/detail/02-05-2018-9-out-of-10-people-worldwide-breathe-polluted-air-but-more-countries-are-taking-action
23. http://www.sciencedirect.com/science/article/pii/S1352231013004548
24. Mendola P, Ha S, Pollack AZ, Zhu Y, Seeni I, Kim SS, Sherman S, Liu D. Chronic and Acute Ozone Exposure in the Week Prior to Delivery Is Associated with the Risk of Stillbirth.Int J Environ Res Public Health. 2017

Jul 6;14(7). pii: E731. doi: 10.3390/ijerph14070731.
25. Mendola P, Ha S, Pollack AZ, Zhu Y, Seeni I, Kim SS, Sherman S, Liu D. Chronic and Acute Ozone Exposure in the Week Prior to Delivery Is Associated with the Risk of Stillbirth.Int J Environ Res Public Health. 2017 Jul 6;14(7). pii: E731. doi: 10.3390/ijerph14070731.
26. http://www.merchantsofdoubt.org
27. Di Q, et al. Air Pollution and Mortality in the Medicare Population. New England Journal of Medicine, 2017; 376 (26): 2513 DOI: 10.1056/NEJMoa1702747
28. Pope CA III, Burnett RT, Krewski D, et al. Cardiovascular mortality and expo- sure to airborne fine particulate matter and cigarette smoke: shape of the exposure-response relationship. Circulation 2009;120:941-8.
29. Humbert S, et al. Intake fraction for particulate matter: recommendations for life cycle impact assessment. Environ Sci Technol. 2011 Jun 1;45(11):4808-16. doi: 10.1021/es103563z. Epub 2011 May 12.
30. Peters, A. et al. (2006) Translocation and potential neurological effects of fine and ultrafine particles a critical update. Part. Fibre Toxicol. 3, 13
31. Saenen ND, Bove H, Steuwe C, Roeffaers MBJ, Provost EB, Lefebvre W, Vanpoucke C, Ameloot M, Nawrot TS. Children's Urinary Environmental Carbon Load: A Novel Marker Reflecting Residential Ambient Air Pollution Exposure? Am J Respir Crit Care Med [online ahead of print] 07 July 2017; www.atsjournals.org/doi/abs/10.1164/rccm.201704-0797OC

32. Miller MR, et al. Inhaled Nanoparticles Accumulate at Sites of Vascular Disease. ACS Nano. 2017 Apr 26. doi: 10.1021/acsnano.6b08551. [Epub ahead of print]
33. Urch B, Silverman F, Corey P, Brook J, Lukic K, Rajagopalan S, Brook R. Acute Blood Pressure Responses in Healthy Adults During Controlled Air Pollution Exposures. Environ Health Perspect. 2005 August; 113(8): 1052–1055.
34. Brook RD, Shin HH, Bard RL, Burnett RT, Vette A, Croghan C, et al. 2011. Exploration of the Rapid Effects of Personal Fine Particulate Matter Exposure on Arterial Hemodynamics and Vascular Function during the Same Day. Environ Health Perspect 119:688-694. doi:10.1289/ehp.1002107
35. Mills NL, Miller MR, Lucking AJ, Beveridge J, Flint L, Boere AJ, Fokkens PH, Boon NA, Sandstrom T, Blomberg A, Duffin R, Donaldson K, Hadoke PW, Cassee FR, Newby DE. Combustion-derived nanoparticulate induces the adverse vascular effects of diesel exhaust inhalation. European Heart Journal, July 13, 2011 DOI:10.1093/eurheartj/ehr195
36. Szyszkowicz M, Rowe BH, Brook RD. Even Low Levels of Ambient Air Pollutants Are Associated With Increased Emergency Department

Visits for Hypertension. Can J Cardiol. 2011 Sep 23. [Epub ahead of print]
37. Baccarelli A, Barretta F, Dou C, Zhang X, McCracken JP, Diaz A, Bertazzi PA, Schwartz J, Wang S, Hou L. Effects of Particulate Air Pollution on Blood Pressure in a Highly Exposed Population in Beijing, China: A repeated-measure study. Environ Health. 2011 Dec 21;10(1):108. [Epub ahead of print]
38. Schwartz J, Alexeeff SE, Mordukhovich I, Gryparis A, Vokonas P, Suh H, Coull BA. Association between long-term exposure to traffic particles and blood pressure in the Veterans Administration Normative Aging Study. Occup Environ Med. 2012 Mar 1. [Epub ahead of print]
39. Cosselman KE, Krishnan RM, Oron AP, Jansen K, Peretz A, Sullivan JH, Larson TV, Kaufman JD. Blood Pressure Response to Controlled Diesel Exhaust Exposure in Human Subjects. Hypertension. 2012 Mar 19. [Epub ahead of print]
40. Bilenko N, Rossem LV, Brunekreef B, Beelen R, Eeftens M, Hoek G, Houthuijs D, de Jongste JC, Kempen EV, Koppelman GH, Meliefste K, Oldenwening M, Smit HA, Wijga AH, Gehring U. Traffic-related air pollution and noise and children's blood pressure: results from the PIAMA birth cohort study. Eur J Prev Cardiol. 2013 Sep 18. [Epub ahead of print]
41. Mehta AJ, Zanobetti A, Koutrakis P, Mittleman MA, Sparrow D, Vokonas P, Schwartz J. Associations Between Short-term Changes in Air Pollution and Correlates of Arterial Stiffness: The Normative Aging Study, 2007-2011. Am J Epidemiol. 2013 Nov 13. [Epub ahead of print]
42. Zhao X, Sun Z, Ruan Y, Yan J, Mukherjee B, Yang F, Duan F, Sun L, Liang R, Lian H, Zhang S, Fang Q, Gu D, Brook JR, Sun Q, Brook RD, Rajagopalan S, Fan Z. Personal Black Carbon Exposure Influences Ambulatory Blood Pressure: Air Pollution and Cardiometabolic Disease (AIRCMD-China) Study. Hypertension. 2014 Jan 13. [Epub ahead of print]
43. Pieters N, et al. Blood Pressure and Same-Day Exposure to Air Pollution at School: Associations with Nano-Sized to Coarse PM in Children. Environ Health Perspect; DOI:10.1289/ehp.1408121
44. van Rossem L, et al. Prenatal Air Pollution Exposure and Newborn Blood Pressure. Environ Health Perspect; DOI:10.1289/ehp.1307419
45. Kaufman J, et al. Association between air pollution and coronary artery calcification within six metropolitan areas in the USA (the Multi-Ethnic Study of Atherosclerosis and Air Pollution): a longitudinal cohort study. Lancet. 2016 Aug 13;388(10045):696-704. doi: 10.1016/S0140-6736(16)00378-0. Epub 2016 May 24.
46. Adar SD, Klein R, Klein BE, Szpiro AA, Cotch MF, Wong TY, O'Neill MS, Shrager S, Barr RG, Siscovick DS, Daviglus ML, Sampson PD, Kaufman JD. Air Pollution and the Microvasculature: A Cross-Sectional Assessment of In Vivo Retinal Images in the Population-Based Multi-Ethnic Study of Atherosclerosis (MESA). PLoS Med. 2010 Nov

30;7(11):e1000372
47. Bauer M, Moebus S, Möhlenkamp S, Dragano N, Nonnemacher M, Fuchsluger M, Kessler C, Jakobs H, Memmesheimer M, Erbel R, Jöckel KH, Hoffmann B; HNR Study Investigative Group. Urban Particulate Matter Air Pollution Is Associated With Subclinical Atherosclerosis Results From the HNR (Heinz Nixdorf Recall) Study. J Am Coll Cardiol. 2010 Nov 23;56(22):1803-1808.
48. Breton CV, Wang X, Mack WJ, Berhane K, Lopez M, Islam TS, Feng M, Lurmann F, McConnell R, Hodis HN, Künzli N, Avol E. Childhood Air Pollutant Exposure and Carotid Artery Intima-Media Thickness in Young Adults. Circulation. 2012 Aug 15. [Epub ahead of print]
49. Wang T, Wang L, Moreno-Vinasco L, Lang GD, Siegler JH, Mathew B, Usatyuk PV, Samet JM, Geyh AS, Breysse PN, Natarajan V, Garcia JG Particulate matter air pollution disrupts endothelial cell barrier via calpain-mediated tight junction protein degradation. Part Fibre Toxicol. 2012 Aug 29;9(1):35. [Epub ahead of print]
50. Adar SD, Sheppard L, Vedal S, Polak JF, Sampson PD, et al. (2013) Fine Particulate Air Pollution and the Progression of Carotid Intima-Medial Thickness: A Prospective Cohort Study from the Multi-Ethnic Study of Atherosclerosis and Air Pollution. PLoS Med 10(4): e1001430. doi:10.1371/journal.pmed.1001430
51. Louwies T, Panis L, Kicinski M, De Boever P, Nawrot TS. Retinal Microvascular Responses to Short-Term Changes in Particulate Air Pollution in Healthy Adults. Environ Health Perspect. 2013 Jun 18. [Epub ahead of print]
52. Su TC, Hwang JJ, Shen YC, Chan CC. Carotid Intima-Media Thickness and Long-Term Exposure to Traffic-Related Air Pollution in Middle-Aged Residents of Taiwan: A Cross-Sectional Study. Environ Health Perspect. 2015 Mar 20. [Epub ahead of print]
53. Pope CA, Bhatnagar A, McCracken J, Abplanalp WT, Conklin DJ, O'Toole TE. Exposure to Fine Particulate Air Pollution Is Associated with Endothelial Injury and Systemic Inflammation. Circ Res. 2016 Oct 25. pii: CIRCRESAHA.116.309279.
54. Pope CA, Bhatnagar A, McCracken J, Abplanalp WT, Conklin DJ, O'Toole TE. Exposure to Fine Particulate Air Pollution Is Associated with Endothelial Injury and Systemic Inflammation. Circ Res. 2016 Oct 25. pii: CIRCRESAHA.116.309279.
55. Li W, et al. Short-Term Exposure to Ambient Air Pollution and Biomarkers of Systemic Inflammation: The Framingham Heart Study. Arterioscler Thromb Vasc Biol. 2017 Jul 27. pii: ATVBAHA.117.309799. doi: 10.1161/ATVBAHA.117.309799. [Epub ahead of print]
56. Chen R, et al. Cardiopulmonary benefits of reducing indoor particles of outdoor origin: a randomized, double-blind crossover trial of air

purifiers. J Am Coll Cardiol. 2015 Jun 2;65(21):2279-87. doi: 10.1016/j.jacc.2015.03.553.
57. Bind MA, et al. Quantile Regression Analysis of the Distributional Effects of Air Pollution on Blood Pressure, Heart Rate Variability, Blood Lipids, and Biomarkers of Inflammation in Elderly American Men: The Normative Aging Study. Environ Health Perspect. 2016 Mar 11. [Epub ahead of print]
58. Bell G, et al. Association of Air Pollution Exposures With High-Density Lipoprotein Cholesterol and Particle Number: The Multi-Ethnic Study of Atherosclerosis. Arteriosclerosis, Thrombosis, and Vascular Biology. 2017;37:976-982
59. Yin F, Lawal A, Ricks J, Fox JR, Larson T, Navab M, Fogelman AM, Rosenfeld ME, Araujo JA. Diesel Exhaust Induces Systemic Lipid Peroxidation and Development of Dysfunctional Pro-Oxidant and Proinflammatory High-Density Lipoprotein. Arterioscler Thromb Vasc Biol. 2013 Apr 4. [Epub ahead of print]
60. Wu XM, et al. Association between gaseous air pollutants and inflammatory, hemostatic and lipid markers in a cohort of midlife women. Environ Int. 2017 Jul 18;107:131-139. doi: 10.1016/j.envint.2017.07.004. [Epub ahead of print]
61. Di Q, et al. Air Pollution and Mortality in the Medicare Population. New England Journal of Medicine, 2017; 376 (26): 2513 DOI: 10.1056/NEJMoa1702747
62. Lepeule J, Laden F, Dockery D, Schwartz J. Chronic Exposure to Fine Particles and Mortality: An Extended Follow-Up of the Harvard Six Cities Study from 1974 to 2009. Environ Health Perspect. 2012 Mar 28. [Epub ahead of print]
63. Thurston GD, Burnett RT, Turner MC, Shi Y, Krewski D, Lall R, Ito K, Jerrett M, Gapstur SM, Diver WR, Pope CA 3rd. Ischemic Heart Disease Mortality and Long-Term Exposure to Source-Related Components of U.S. Fine Particle Air Pollution. Environ Health Perspect. 2015 Dec 2. [Epub ahead of print]
64. Andersen MS, et al. Co-benefits of climate mitigation: Counting statistical lives or life-years? Ecological Indicators, 2017; 79: 11 DOI: 10.1016/j.ecolind.2017.03.051
65. Hansell A, et al. Historic air pollution exposure and long-term mortality risks in England and Wales: prospective longitudinal cohort study. Thorax 2015;0:1–9. doi:10.1136/thoraxjnl-2015-207111
66. Peters A, et al "Times spent in traffic and the onset of myocardial infarction" AHA Meeting 2009.
67. Vreeland H, et al. Oxidative potential of PM 2.5 during Atlanta rush hour: Measurements of in-vehicle dithiothreitol (DTT) activity. Atmospheric Environment, 2017; 165: 169 DOI:

10.1016/j.atmosenv.2017.06.044
68. Wellenius G, et al. Ambient Air Pollution and the Risk of Acute Ischemic Stroke. Arch Intern Med. 2012;172(3):229-234
69. Davoodi G, Sharif AY, Kazemisaeid A, Sadeghian S, Farahani AV, Sheikhvatan M, Pashang M. Comparison of heart rate variability and cardiac arrhythmias in polluted and clean air episodes in healthy individuals. Environ Health Prev Med. 2010 Jul;15(4): 217-21. Epub 2010 Jan 22.
70. Van Hee VC, Szpiro AA, Prineas R, Neyer J, Watson K, Siscovick D, Park SK, Kaufman JD. Association of Long-term Air Pollution With Ventricular Conduction and Repolarization Abnormalities. Epidemiology. 2011 Sep 13. [Epub ahead of print]
71. Weichenthal S, Kulka R, Dubeau A, Martin C, Wang D, Dales R 2011. Traffic- Related Air Pollution and Acute Changes in Heart Rate Variability and Respiratory Function in Urban Cyclists. Environ Health Perspect 119:1373-1378. http://dx.doi.org/ 10.1289/ehp.1003321
72. Jia X, Hao Y, Guo X. Ultrafine carbon black disturbs heart rate variability in mice. Toxicol Lett. 2012 Apr 15. [Epub ahead of print]
73. Pieters N, Plusquin M, Cox B, Kicinski M, Vangronsveld J, Nawrot TS. An epidemiological appraisal of the association between heart rate variability and particulate air pollution: a meta-analysis. Heart. 2012 May 23. [Epub ahead of print]

74. Carll AP, et al. Inhaled ambient-level traffic-derived particulates decrease cardiac vagal influence and baroreflexes and increase arrhythmia in a rat model of metabolic syndrome. Part Fibre Toxicol. 2017 May 25;14(1):16. doi: 10.1186/s12989-017-0196-2.
75. Vora R, Zareba W, Utell MJ, Pietropaoli AP, Chalupa D, Little EL, Oakes D, Bausch J, Wiltshire J, Frampton MW. Inhalation of ultrafine carbon particles alters heart rate and heart rate variability in people with type 2 diabetes. Part Fibre Toxicol. 2014 Jul 16;11(1):31. [Epub ahead of print]
76. Rich DQ, Mittleman MA, Link MS, Schwartz J, Luttmann-Gibson H, Catalano PJ, Speizer FE, Gold 77. DR, Dockery DW. Increased Risk of Paroxysmal Atrial Fibrillation Episodes Associated with Acute Increases in Ambient Air Pollution. Environ Health Perspect. 2006; 114:120-123.
78. Song X, Liu Y, Hu Y, Zhao X, Tian J, Ding G, Wang S Short-Term Exposure to Air Pollution and Cardiac Arrhythmia: A Meta-Analysis and Systematic Review. Int J Environ Res Public Health. 2016 Jun 28;13(7). pii: E642.
79. Mordukhovich I, Kloog I, Coull B, Koutrakis P, Vokonas P, Schwartz J. Association between Particulate Air Pollution and QT Interval Duration in an Elderly Cohort. Epidemiology. 2015 Nov 24. [Epub ahead of print]
80. Baccarelli A, Martinelli A, Zanobetti A, et al. Exposure to particulate air

pollution and risk of deep vein thrombosis. Arch Intern Med. 2008; 168:920-927
81. Jacobs L, Emmerechts J, Mathieu C, Hoylaerts MF, Fierens F, Hoet PH, Nemery B, Nawrot TS. Air pollution related prothrombotic changes in persons with diabetes. Environ Health Perspect. 2010 Feb;118(2):191-6.
82. Emmerechts J, Jacobs L, Van Kerckhoven S, Loyen S, Mathieu C, Fierens F, Nemery B, Nawrot TS, Hoylaerts MF. Air Pollution-Associated Procoagulant Changes: role of Circulating Microvesicles. J Thromb Haemost. 2011 Nov 8. doi: 10.1111/j. 1538-7836.2011.04557.x. [Epub ahead of print]
83. Kloog I, Zanobetti A, Nordio F, Coull BA, Baccarelli AA, Schwartz J. Effects of airborne fine particles (PM2.5) on Deep Vein Thrombosis Admissions in North Eastern United States. J Thromb Haemost. 2015 Feb 12. doi: 10.1111/jth.12873. [Epub ahead of print]
84. Martinelli N, Girelli D, Cigolini D, Sandri M, Ricci G, Rocca G, Olivieri O. Access rate to the emergency department for venous thromboembolism in relationship with coarse and fine particulate matter air pollution. PLoS One. 2012;7(4):e34831. Epub 2012 Apr 11.
85. Leary PJ, Barr RG, Bluemke DA, Hough CL, Kaufman JD, Szpiro AA, Kawut SM, Van Hee VC. The relationship of roadway proximity and NOx with right ventricular structure and function: the MESA-Right Ventricle and MESA-Air studies. Am J Respir Crit Care Med 2013;187(conf abstr):A3976
86. Cakmak S, Dales R, Leech J, Liu L. The influence of air pollution on cardiovascular and pulmonary function and exercise capacity: Canadian Health Measures Survey (CHMS). Environ Res. 2011 Nov;111(8):1309-12. doi: 10.1016/j.envres.2011.09.016. Epub 2011 Oct 13.
87. Cutrufello PT, Rundell KW, Smoliga JM, Stylianides GA. Inhaled whole exhaust and its effect on exercise performance and vascular function. Inhal Toxicol. 2011 Sep;23(11):658-67. doi:10.3109/08958378.2011.604106. Epub 2011 Aug 25.
88. Mentz RJ, O'Brien EC. Air Pollution in Patients With Heart Failure: Lessons From a Mechanistic Pilot Study of a Filter Intervention. JACC Heart Fail. 2016 Jan;4(1):65-7. doi: 10.1016/j.jchf.2015.11.008.
89. Wu Z, He EY, Scott GI, Ren J. α,β-Unsaturated aldehyde pollutant acrolein suppresses cardiomyocyte contractile function: Role of TRPV1 and oxidative stress. Environ Toxicol. 2013 Dec 23. doi: 10.1002/tox.21941. [Epub ahead of print]
90. Shah ASV, Langrish JP, Nair H, et al. Global association of air pollution and heart failure: a systematic review and meta-analysis. The Lancet. Published online July 10 2013
91. Domínguez-Rodríguez A, Abreu-Afonso J, Rodríguez S, Juárez-Prera RA, Arroyo- Ucar E, Jiménez-Sosa A, González Y, Abreu-González P, Avanzas P. Comparative Study of Ambient Air Particles in Patients

Hospitalized for Heart Failure and Acute Coronary Syndrome. Rev Esp Cardiol. 2011 Aug;64(8):661-6. Epub 2011 Jun 8.
92. Gorr MW , et al. Early life exposure to air pollution induces adult cardiac dysfunction. American Journal of Physiology - Heart and Circulatory PhysiologyPublished 1 November 2014Vol. 307no. 9, H1353-H1360DOI: 10.1152/ ajpheart.00526.2014
93. Andersen Z, Hvidberg M, Jensen S, et al. Chronic Obstructive Pulmonary Disease and Long-Term Exposure to Traffic-Related Air Pollution: A Cohort Study. Am. J. Respir. Crit. Care Med. 2010, doi:10.1164/rccm.201006-0937OC. Published ahead of print on September 24, 2010
94. Spira-Cohen A, Chen L-C, Kendall M, Lall R, Thurston GD 2011. Personal Exposures to Traffic-Related Air Pollution and Acute Respiratory Health Among Bronx School Children with Asthma. Environ Health Perspect :-. doi:10.1289/ehp.1002653
95. Nawrot TS, Vos R, Jacobs L, Verleden SE, Wauters S, Mertens V, Dooms C, Hoet PH, Van Raemdonck DE, Faes C, Dupont LJ, Nemery B, Verleden GM, Vanaudenaerde BM. The impact of traffic air pollution on bronchiolitis obliterans syndrome and mortality after lung transplantation. Thorax. 2011 Mar 23. [Epub ahead of print]
96. Stern G, Latzin P, Röösli M, Fuchs O, Proietti E, Kuehni C, Frey U. A Prospective Study of the Impact of Air Pollution on Respiratory Symptoms and Infections in Infants. Am J Respir Crit Care Med. 2013 Apr 17. [Epub ahead of print]
97. Lepeule J, Litonjua AA, Coull B, Koutrakis P, Sparrow D, Vokonas PS, Schwartz J. Long-term Effects of Traffic Particles on Lung Function Decline in the Elderly. Am J Respir Crit Care Med. 2014 Jul 16. [Epub ahead of print]
98. Qiu H, Tian LW, Pun VC, Ho KF, Wong TW, Yu IT. Coarse particulate matter associated with increased risk of emergency hospital admissions for pneumonia in Hong Kong. Thorax. 2014 Aug 27. pii: thoraxjnl-2014-205429. doi: 10.1136/ thoraxjnl-2014-205429. [Epub ahead of print]
99. Darrow LA, Klein M, Flanders WD, Mulholland JA, Tolbert PE, Strickland MJ. Air Pollution and Acute Respiratory Infections Among Children 0-4 Years of Age: An 18-Year Time-Series Study. Am J Epidemiol. 2014 Oct 16. pii: kwu234. [Epub ahead of print]
100. Rice MB, Ljungman PL, Wilker EH, Dorans KS, Gold DR, Schwartz J, Koutrakis P, Washko GR, O'Connor GT, Mittleman MA. Long-Term Exposure to Traffic Emissions and Fine Particulate Matter and Lung Function Decline in the Framingham Heart Study. Am J Respir Crit Care Med. 2015 Jan 15. [Epub ahead of print]
101. Johannson KA, Balmes JR, Collard HR. Air pollution exposure: a

novel environmental risk factor for interstitial lung disease? Chest. 2015 Apr 1;147(4):1161-7. doi: 10.1378/chest.14-1299.
102. Psoter KJ, De Roos AJ, Wakefield J, Mayer JD, Rosenfeld M. Air pollution exposure is associated with MRSA acquisition in young U.S. children with cystic fibrosis. BMC Pulm Med. 2017 Jul 27;17(1):106. doi: 10.1186/s12890-017-0449-8.
103. Kim BJ, Seo JH, Jung YH, Kim HY, Kwon JW, Kim HB, Lee SY, Park KS, Yu J, Kim HC, Leem JH, Lee JY, Sakong J, Kim SY, Lee CG, Kang DM, Ha M, Hong YC, Kwon HJ, Hong SJ. Air pollution interacts with past episodes of bronchiolitis in the development of asthma. Allergy. 2013 Jan 25. doi: 10.1111/all.12104. [Epub ahead of print]
104. Pinkerton KE, Green FH, Saiki C, Vallyathan V, Plopper CG, Gopal V, Hung D, Bahne EB, Lin SS, Menache MG, Schenker MB. Distribution of particulate matter and tissue remodeling in the human lung. Environmental health perspectives 2000; 108: 1063-1069.
105. Li N, et al. Exposure to ambient particulate matter alters the microbial composition and induces immune changes in rat lung. Respir Res. 2017 Jul 25;18(1):143. doi: 10.1186/s12931-017-0626-6.
106. Karr CJ, Demers P, Koehoorn M, Lencar C, Tamburic L, Brauer M. Influence of Ambient Air Pollutant Sources on Clinical Encounters for Infant Bronchiolitis. Am. J. Respir. Crit. Care Med., Nov 2009; 180: 995 - 1001.
107. Strickland M, Darrow L, et al. Short Term Associations between Ambient Air Pollutants and Pediatric Asthma Emergency Department Visits. Am J Respir Crit Care Med Vol 182 pp307-316, 2010
108. Son J-Y, Bell ML, Lee J-T 2011. Survival Analysis of Long-Term Exposure to Different Sizes of Airborne Particulate Matter and Risk of Infant Mortality Using a Birth Cohort in Seoul, Korea. Environ Health Perspect 119:725-730. doi:10.1289/ehp. 1002364
109. Peel JL, Klein M, Flanders WD, Mulholland JA, Freed G, Tolbert PE 2011. Ambient Air Pollution and Apnea and Bradycardia in High-Risk Infants on Home Monitors. Environ Health Perspect 119:1321-1327. http://dx.doi.org/10.1289/ehp.1002739
110. Andersen ZJ, Bønnelykke K, Hvidberg M, Jensen SS, Ketzel M, Loft S, Sørensen M, Tjønneland A, Overvad K, Raaschou-Nielsen O. Long-term exposure to air pollution and asthma hospitalisations in older adults: a cohort study. Thorax. 2011 Sep 2. [Epub ahead of print]
111. Zhu R, Chen Y, Wu S, Deng F, Liu Y, Yao W. The Relationship between Particulate Matter (PM(10)) and Hospitalizations and Mortality Of Chronic Obstructive Pulmonary Disease: A Meta-Analysis. COPD. 2013 Jan 16. [Epub ahead of print]
112. Gan WQ, Fitzgerald JM, Carlsten C, Sadatsafavi M, Brauer M. Associations of Ambient Air Pollution with Chronic Obstructive

Pulmonary Disease Hospitalization and Mortality. Am J Respir Crit Care Med. 2013 Feb 7. [Epub ahead of print]
113. Farhat SC, Almeida MB, Silva-Filho LV, Farhat J, Rodrigues JC, Braga AL. Ozone is associated with an increased risk of respiratory exacerbations in cystic fibrosis patients. Chest. 2013 Mar 14. doi: 10.1378/chest.12-2414. [Epub ahead of print]
114. Johannson KA, Vittinghoff E, Lee K, Balmes JR, Ji W, Kaplan GG, Kim DS, Collard HR. Acute exacerbation of idiopathic pulmonary fibrosis associated with air pollution exposure. Eur Respir J. 2013 Oct 31. [Epub ahead of print]
115. Li MH, Fan LC, Mao B, Yang JW, Choi AM, Cao WJ, Xu JF. Short Term Exposure to Ambient Fine Particulate Matter (PM2.5) Increases Hospitalizations and Mortality of Chronic Obstructive Pulmonary Disease: A Systematic Review and Meta-Analysis. Chest. 2015 Jun 25. doi: 10.1378/chest.15-0513.
116. Ware LB, et al. Long-Term Ozone Exposure Increases the Risk of Developing the Acute Respiratory Distress Syndrome. Am J Respir Crit Care Med. 2015 Dec 17. [Epub ahead of print]
117. Schelegle, E., Morales, C., Walby, W., et al, 6.6 Hour Inhalation of Ozone Concentrations from 60 to 87 Parts per Billion in Healthy Humans. American Journal of Respiratory and Critical Care Medicine Vol 180. pp 265-272, (2009).
118. Thaller, E., Petronella, S., Hochman, D. et al. Moderate increases in Ambient PM 2.5 and Ozone Are Associated With Lung Function Decreases in Beach Lifeguards. Journal of Occupational and Environmental Medicine. 50(2):202-211, Feb. 2008.
119. Steinvil A, Fireman E, Kordova-Biezuner L, Cohen M, Shapira I, Berliner S, Rogowski O. Environmental air pollution has decremental effects on pulmonary function test parameters up to one week after exposure. Am J Med Sci. 2009 Oct; 338(4):273-9.
120. Chang YK, Wu CC, Lee LT, Lin RS, Yu YH, Chen YC. The short-term effects of air pollution on adolescent lung function in Taiwan. Chemosphere. 2011 Dec 19. [Epub ahead of print]
121. Panis L, Provost EB, Cox B, Louwies T, Laeremans M, Standaert A, Dons E, Holmstock L, Nawrot T, De Boever P. Short-term air pollution exposure decreases lung function: a repeated measures study in healthy adults. Environ Health. 2017 Jun 14;16(1):60. doi: 10.1186/s12940-017-0271-z.
122. Jerrett M, Burnett R, Pope CA, et al. Long-Term Ozone Exposure and Mortality. NEJM. Vol. 360:1085-1095. March 12, 2009 Num 11.
123. Bell ML, Peng RD, Dominici F. The exposure-response curve for ozone and risk of mortality and the adequacy of current ozone regulations. Environ Health Perspect 2006;114:532-536.

124. Faustini A, Stafoggia M, Berti G, Bisanti L, Chiusolo M, Cernigliaro A, Mallone S, Primerano R, Scarnato C, Simonato L, Vigotti MA, Forastiere F; The relationship between ambient particulate matter and respiratory mortality: a multi-city study in Italy. Eur Respir J. 2011 Jan 13. [Epub ahead of print]
125. Gauderman WJ, Gilliland GF, Vora H, et al. Association between Air Pollution and Lung Function Growth in Southern California Children: results from a second cohort. Am J Respir Crit Care Med 2002;166:76-84.
126. Gauderman WJ, Gilliland GF, Vora H, et al. The effect of air pollution on lung development from 10 to 18 years of age. NEJM 2004;351:1057-67.
127. Gauderman WJ, et al. Association of Improved Air Quality with Lung Development in Children. N Engl J Med 2015; 372:905-913March 5, 2015DOI: 10.1056/ NEJMoa1414123
128. Jedrychowski WA, Perera FP, Maugeri U, Mroz E, Klimaszewska-Rembiasz M, Flak E, Edwards S, Spengler JD. Effect of prenatal exposure to fine particulate matter on ventilatory lung function of preschool children of non-smoking mothers. Paediatr Perinat Epidemiol. 2010 Sep;24(5):492-
129. Saravia J, You D, Thevenot P, Lee GI, Shrestha B, Lomnicki S, Cormier SA. Early-life exposure to combustion-derived particulate matter causes pulmonary immunosuppression. Mucosal Immunol. 2013 Oct 30. doi: 10.1038/mi.2013.88. [Epub ahead of print]
130. Jedrychowski WA, Perera FP, Maugeri U, Majewska R, Mroz E, Flak E, Camann D, Sowa A, Jacek R. Long term effects of prenatal and postnatal airborne PAH exposures on ventilatory lung function of non-asthmatic preadolescent children. Prospective birth cohort study in Krakow. Sci Total Environ. 2014 Oct 6;502C:502-509. doi: 10.1016/ j.scitotenv.2014.09.051. [Epub ahead of print]
131. Yang S, et al. Prenatal Particulate Matter/Tobacco Smoke Increases Infants' Respiratory Infections: COCOA Study. Allergy Asthma Immunol Res. 2015 Nov;7(6):573-82. doi: 10.4168/aair.2015.7.6.573. Epub 2015 Jun 25.
132. Morales E, et al. Intrauterine and early postnatal exposure to outdoor air pollution and lung function at preschool age. Thorax, 2014; DOI: 10.1136/thoraxjnl-2014-205413
133. Veras MM, et al. Before the first breath: prenatal exposures to air pollution and lung development.Cell Tissue Res. 2016 Oct 10. [Epub ahead of print]
134. Vaiserman A. Early-life exposure to endocrine disrupting chemicals and later-life health outcomes: an epigenetic bridge? Aging Dis 2014;5(6):419–29.
135. Wick P, Malek A, Manser P, Meili D, Maeder-Althaus X, Diener L, et al. 2010. Barrier capacity of human placenta for nanosized materials.

Environ Health Perspect 118:432-436.
136. Rudge CV, et al. The placenta as a barrier for toxic and essential elements in paired maternal and cord blood samples of South African delivering women. J Environ Monit. 2009 Jul;11(7):1322-30. doi: 10.1039/b903805a. Epub 2009 Jun 3.
137. http://www.ewg.org/research/body-burden-pollution-newborns#.WXU3FVKZNSw
138. Zhang A, Hu H, Sánchez BN, Ettinger AS, Park SK, Cantonwine D, et al. 2011. Association between Prenatal Lead Exposure and Blood Pressure in Female Offspring. Environ Health Perspect :-. http://dx.doi.org/10.1289/ehp.1103736
139. van den Hooven EH, de Kluizenaar Y, Pierik FH, Hofman A, van Ratingen SW, Zandveld PY, et al. 2012. Chronic Air Pollution Exposure during Pregnancy and Maternal and Fetal C-reactive Protein Levels. The Generation R Study. Environ Health Perspect :-. http://dx.doi.org/10.1289/ehp.1104345
140. Weldy CS, Y Liu, YC Chang, IO Medvedev, JR Fox, TV Larson, WM Chien, MT Chin. In utero and early life exposure to diesel exhaust air pollution increases adult susceptibility to heart failure in mice. Particle and Fibre Toxicology. 2013. http://bit.ly/ 18znRIR
141. Baïz N, et al. Maternal exposure to air pollution before and during pregnancy related to changes in newborn's cord blood lymphocyte subpopulations. The EDEN study cohort. BMC Pregnancy Childbirth. 2011 Nov 2;11:87. doi: 10.1186/1471-2393-11-87.
142. Bruner-Tran, KL and KG Osteen. 2010. Developmental exposure to TCDD reduces fertility and negatively affects pregnancy outcomes across multiple generations. Reproductive Toxicology http://dx.doi.org/10.1016/j.reprotox.2010.10.003
143. Manikkam M, Guerrero-Bosagna C, Tracey R, Haque MM, Skinner MK (2012) Transgenerational Actions of Environmental Compounds on Reproductive Disease and Identification of Epigenetic Biomarkers of Ancestral Exposures. PLoS ONE 7(2): e31901. doi:10.1371/journal.pone.0031901
144. Gregory, DJ, et al. Transgenerational transmission of asthma risk after exposure to environmental particles during pregnancy.. American Journal of Physiology - Lung Cellular and Molecular Physiology, 2017; ajplung.00035.2017 DOI: 10.1152/ajplung.00035.2017
145. Tillett T Potential Mechanism for PM10 Effects on Birth Outcomes: In Utero Exposure Linked to Mitochondrial DNA Damage. Environ Health Perspect 120:a363- a363. http://dx.doi.org/10.1289/ehp.120-a363b
146. Perera F, Tang W, Herbstman J. Relation of DNA Methylation of 5-CpG Island of ACSL3 to Transplacental Exposure to Airborne PAH and Childhood Asthma.. PloS ONE. Feb. 16, 2009.

147. Bocskay K, Tang D, Orjuela M, et al. Chromosomal Aberrations in Cord Blood Are Associated with Prenatal Exposure to Carcinogenic Polycyclic Aromatic Hydrocarbons. Cancer Epidem Biomarkers and Prev. Vol. 14, 506-511, Feb 2005
148. Perera F, Tang D, Tu Y, Biomarkers in Maternal and Newborn Blood Indicate Heightened Fetal Susceptibility to Procarcinogenic DNA Damage. Environ Health Persp Vol 112 Number 10 July 2004
149. Pilsner JR, Hu H, Ettinger A, Sanchez BN, et al. Influence of prenatal lead exposure on genomic methaylation of cord blood DNA. Environ Health Persp, April 2009
150. Baccarelli A. Breathe deeply into your genes!: genetic variants and air pollution effects.., Am J Respir Crit Care Med. 2009 Mar 15;179(6):431-2.
151. Baccarelli A, Wright RO, Bollati V, Tarantini L, Litonjua AA, Suh HH, Zanobetti A, Sparrow D, Vokonas PS, Schwartz J. Rapid DNA methylation changes after exposure to traffic particles. Am J Respir Crit Care Med. 2009 Apr 1;179(7):523-4.
152. Pedersen M, Wichmann J, Autrup H, Dang DA, Decordier I, Hvidberg M, Bossi R, Jakobsen J, Loft S, Knudsen LE. Increased micronuclei and bulky DNA adducts in cord blood after maternal exposures to traffic-related air pollution. Environ Res. 2009 Nov; 109(8):1012-20. Epub 2009 Sep 23.
153. Rubesa J, Rybara R, Prinosilovaa P, Veznika Z, et al. Genetic polymorphisms influence the susceptibility of men to sperm DNA damage associated with exposure to air pollution. Mutation Research 683 (2010) 9–15.
154. Jurewicz J, et al. The relationship between exposure to air pollution and sperm disomy. Environ Mol Mutagen. 2015 Jan;56(1):50-9. doi: 10.1002/em.21883. Epub 2014 Jul 3.
155. Rubes J, Selevan S, Evenson D, Zudova D, Vozdova M, Zudova Z, Robbins W, Perreault S. Episodic air pollution is associated with increased DNA fragmentation in human sperm without other changes in semen quality. Human Reproduction Vol.20, No.10 pp. 2776–2783, 2005 doi:10.1093/humrep/dei122. Advance Access publication June 24, 2005.
156. Han X, Zhou N, Cui Z, Ma M, Li L, Cai M, et al. 2011. Association between Urinary Polycyclic Aromatic Hydrocarbon Metabolites and Sperm DNA Damage: A Population Study in Chongqing, China. Environ Health Perspect 119:652-657. doi:10.1289/ehp. 1002340
157. Slama R, Bottagisi S, Solansky I, Lepeule J, Giorgis-Allemand L, Sram R. Short-Term Impact of Atmospheric Pollution on Fecundability. Epidemiology. 2013 Sep 18. [Epub ahead of print]
158. Jurewicz J, Radwan M, Sobala W, Polanska K, Radwan P, Jakubowski L, Ulańska A, Hanke W. The relationship between exposure to air pollution and sperm disomy. Environ Mol Mutagen. 2015 Jan;56(1):50-9. doi:

10.1002/em.21883. Epub 2014 Jul 3.
159. Seamen N, et al. In Utero Fine Particle Air Pollution and Placental Expression of Genes in the Brain-Derived Neurotrophic Factor Signaling Pathway: An ENVIRONAGE Birth Cohort Study. Environ Health Perspect; DOI:10.1289/ehp.1408549
160. Seamen N, et al. In Utero Fine Particle Air Pollution and Placental Expression of Genes in the Brain-Derived Neurotrophic Factor Signaling Pathway: An ENVIRONAGE Birth Cohort Study. Environ Health Perspect; DOI:10.1289/ehp.1408549
161. Morales E, et al. Intrauterine and early postnatal exposure to outdoor air pollution and lung function at preschool age. Thorax, 2014; DOI: 10.1136/thoraxjnl-2014-205413
162. Duan H, et al. Long-term exposure to diesel engine exhaust induces primary DNA damage: a population-based study. Occup Environ Med. 2015 Oct 21. pii: oemed-2015-102919. doi: 10.1136/oemed-2015-102919. [Epub ahead of print]
163. Hou L, et al. Particulate Air Pollution Exposure and Expression of Viral and Human MicroRNAs in Blood: The Beijing Truck Driver Air Pollution Study. Environ Health Perspect; DOI:10.1289/ehp.1408519
164. Ding R, et al. Dose- and time- effect responses of DNA methylation and histone H3K9 acetylation changes induced by traffic-related air pollution. Sci Rep. 2017 Mar 3;7:43737. doi: 10.1038/srep437

165. Lai CH, et al. Exposure to fine particulate matter causes oxidative and methylated DNA damage in young adults: A longitudinal study. Sci Total Environ. 2017 Apr 23;598:289-296. doi: 10.1016/j.scitotenv.2017.04.079. [Epub ahead of print]
166. Goodson J, et al. In utero exposure to diesel exhaust particulates is associated with an altered cardiac transcriptional response to transverse aortic constriction and altered DNA methylation. The FASEB Journal, 2017; fj.201700032R DOI: 10.1096/fj.201700032R
167. Solaimani P, Saffari A, Sioutas C, Bondy SC, Campbell A. Exposure to ambient ultrafine particulate matter alters the expression of genes in primary human neurons. Neurotoxicology. 2016 Nov 13. pii: S0161-813X(16)30225-X. doi: 10.1016/j.neuro.2016.11.001. [Epub ahead of print]
168. Rosa MJ, et al. Identifying sensitive windows for prenatal particulate air pollution exposure and mitochondrial DNA content in cord blood. Environ Int. 2016 Nov 11. pii: S0160-4120(16)30741-3. doi: 10.1016/j.envint.2016.11.007. [Epub ahead of print]
169. Grevendonk L, et al. Mitochondrial oxidative DNA damage and exposure to particulate air pollution in mother-newborn pairs. Environmental Health 2016, 15:10
170. Janssen BG, Byun HM, Gyselaers W, Lefebvre W, Baccarelli AA,

Nawrot TS. Placental mitochondrial methylation and exposure to airborne particulate matter in the early life environment: An ENVIRONAGE birth cohort study. Epigenetics. 2015 May 21:0. [Epub ahead of print]
171. Rode L, et al. Peripheral Blood Leukocyte Telomere Length and Mortality Among 64 637 Individuals From the General Population. J Natl Cancer Inst (2015) 107 (6): djv074. DOI: https://doi.org/10.1093/jnci/djv074
172. Hou L, et al. Blood Telomere Length Attrition and Cancer Development in the Normative Aging Study Cohort. EBioMedicine. Volume 2, Issue 6, June 2015, Pages 591-596
173. Lin N, et al. Accumulative effects of indoor air pollution exposure on leukocyte telomere length among non-smokers. Environ Pollut. 2017 Apr 24;227:1-7. doi: 10.1016/j.envpol.2017.04.054. [Epub ahead of print]
174. Martens DS, Nawrot TS. Air Pollution Stress and the Aging Phenotype: The Telomere Connection. Curr Environ Health Rep. 2016 Jun 29. [Epub ahead of print]
175. Hou L, Wang S, Dou C, Zhang X, Yu Y, Zheng Y, Avula U, Hoxha M, Díaz A, McCracken J, Barretta F, Marinelli B, Bertazzi PA, Schwartz J, Baccarelli AA. Air pollution exposure and telomere length in highly exposed subjects in Beijing, China: A repeated-measure study. Environ Int. 2012 Aug 4;48C:71-77. [Epub ahead of print]
176. Lee E, et al. Traffic-Related Air Pollution and Telomere Length in Children and Adolescents Living in Fresno, CA. Journal of Occupational and Environmental Medicine, 2017; 59 (5): 446 DOI: 10.1097/JOM.0000000000000996
177. Bijnens E, Zeegers MP, Gielen M, Kicinski M, Hageman GJ, Pachen D, Derom C, Vlietinck R, Nawrot TS. Lower placental telomere length may be attributed to maternal residential traffic exposure; a twin study. Environ Int. 2015 Mar 7;79:1-7. doi: 10.1016/j.envint.2015.02.008. [Epub ahead of print]
178. Hettfleisch, K, et al. Short-Term Exposure to Urban Air Pollution and Influences on Placental Vascularization Indexes. Environ Health Perspect; DOI:10.1289/EHP300
179. Veras MM, Damaceno-Rodrigues NR, Caldini EG, Maciel Ribeiro AA, Mayhew TM, Saldiva PH, et al. 2008. Particulate urban air pollution affects the functional morphology of mouse placenta. Biol Reprod 79:578–584.
180. Rennie M, et al. Vessel tortuousity and reduced vascularization in the fetoplacental arterial tree after maternal exposure to polycyclic aromatic hydrocarbons. Am J Physiol Heart Circ Physiol 300: H675–H684, 2011. First published December 10, 2010; doi:10.1152/ajpheart.00510.2010.
181. Liu Y, et al. Effect of Fine Particulate Matter (PM2.5) on Rat Placenta Pathology and Perinatal Outcomes. Med Sci Monit. 2016 Sep 15;22:3274-

80.
182. Wylie B, et al. Placental Pathology Associated with Household Air Pollution in a Cohort of Pregnant Women from Dares Salaam, Tanzania. Environ Health Perspect; DOI:10.1289/EHP256
183. Martens DS, et al. Neonatal Cord Blood Oxylipins and Exposure to Particulate Matter in the Early-Life Environment: An ENVIRONAGE Birth Cohort Study. Environ Health Perspect. 2016 Nov 4. [Epub ahead of print]
184. Lee PC, Talbott EO, Roberts JM, Catov JM, Sharma RK, Ritz B. Particulate Air Pollution Exposure and C-reactive Protein During Early Pregnancy. Epidemiology. 2011 Apr 21. [Epub ahead of print]
185. Nachman R, et al. Intrauterine Inflammation and Maternal Exposure to Ambient PM2.5 during Preconception and Specific Periods of Pregnancy: The Boston Birth Cohort. Environ Health Perspect; DOI:10.1289/EHP243
186. Christopher S. Malley, Johan C.I. Kuylenstierna, Harry W. Vallack, Daven K. Henze, Hannah Blencowe, Mike R. Ashmore. Preterm birth associated with maternal fine particulate matter exposure: A global, regional and national assessment. Environment International, 2017; DOI: 10.1016/j.envint.2017.01.023
187. Currie J, et al. The 9/11 Dust Cloud and Pregnancy Outcomes: A Reconsideration
The Journal of Human Resources 51(4):805-831, DOI: 10.3368/jhr.51.4.0714-6533R
188. Li S, et al. Acute Impact of Hourly Ambient Air Pollution on Preterm Birth. Environ Health Perspect; DOI:10.1289/EHP200
189. Clemens T, Turner S, Dibben C. Maternal exposure to ambient air pollution and fetal growth in North-East Scotland: A population-based study using routine ultrasound scans. Environ Int. 2017 Jul 25;107:216-226. doi: 10.1016/j.envint.2017.07.018. [Epub ahead of print]
190. Kessler, R. et al, "Followup in Southern California: decreased birth weight following prenatal wildfire smoke exposure." Environ Health Perspect. 2012 Sept; 120(9): a362
191. Jedrychowski, W. et al, "Gender differences in fetal growth of newborns exposed prenatally to airborne fine particulate matter." Environ Health Perspect. 2009 May; 109(4): 447-456
192. Carvalho MA, et al. Associations of maternal personal exposure to air pollution on fetal weight and fetoplacental Doppler: a prospective cohort study. Reprod Toxicol. 2016 Apr 18. pii: S0890-6238(16)30063-6. doi: 10.1016/j.reprotox.2016.04.013. [Epub ahead of print
193. Erickson AC, et al. The reduction of birth weight by fine particulate matter and its modification by maternal and neighbourhood-level factors: a multilevel analysis in British Columbia, Canada. Environ Health. 2016 Apr

14;15(1):51. doi: 10.1186/s12940-016-0133-0.
194. Coker E, et al. Multi-pollutant exposure profiles associated with term low birth weight in Los Angeles County. Environ Int. 2016 Feb 15;91:1-13. doi: 10.1016/j.envint.2016.02.011. [Epub ahead of print]
195. Yongping H, et al. Geographic Variation in the Association between Ambient Fine Particulate Matter (PM2.5) and Term Low Birth Weight in the United States. Environ Health Perspect; DOI:10.1289/ehp.1408798
196. Bell ML, Belanger K, Ebisu K, Gent JF, Lee HJ, Koutrakis P, Leaderer BP. Prenatal Exposure to Fine Particulate Matter and Birth Weight: Variations by Particulate Constituents and Sources. Epidemiology. 2010 Aug 31. [Epub ahead of print]
197. Hyunok Choi H, et al. Fetal Window of Vulnerability to Airborne Polycyclic Aromatic Hydrocarbons on Proportional Intrauterine Growth Restriction. PLoS One. 2012; 7(4): e35464. Published online Apr 24, 2012. doi: 10.1371/journal.pone.0035464
198. Fleisch AF, Rifas-Shiman SL, Koutrakis P, Schwartz JD, Kloog I, Melly S, Coull BA, Zanobetti A, Gillman MW, Gold DR, Oken E. Prenatal Exposure to Traffic Pollution: Associations with Reduced Fetal Growth and Rapid Infant Weight Gain. Epidemiology. 2015 Jan;26(1):43-50.
199. Chen LY, and Ho C. Incense Burning during Pregnancy and Birth Weight and Head Circumference among Term Births: The Taiwan Birth Cohort Study. Environ Health Perspect; DOI:10.1289/ehp.1509922
200. Janssen BG, Saenen ND, Roels HA, Madhloum N, Gyselaers W, Lefebvre W, Penders J, Vanpoucke C, Vrijens K, Nawrot TS. 2017. Fetal thyroid function, birth weight, and in utero exposure to fine particle air pollution: a birth cohort study. Environ Health Perspect 125:699–705
201. Saenen N, et al. In Utero Fine Particle Air Pollution and Placental Expression of Genes in the Brain-Derived Neurotrophic Factor Signaling Pathway: An ENVIRONAGE Birth Cohort Study. Environ Health Perspect; DOI:10.1289/ehp.1408549
202. Wu, J. et al., "Association between local traffic-generated air pollution and preeclampsia and preterm delivery in the South Coast Air Basin of California." Environ Health Perspect. 2009 Nov; 117(11): 1773-79
203. Zhu Y, et al. Ambient air pollution and risk of gestational hypertension. Am J Epidemiol. 2017 May 4. doi: 10.1093/aje/kwx097. [Epub ahead of print]
204. Ibrahimou B, Salihu HM, Aliyu MH, Anozie C. Risk of Preeclampsia From Exposure to Particulate Matter (PM2.5) Speciation Chemicals During Pregnancy. J Occup Environ Med. 2014 Dec;56(12):1228-1234.
205. Dadvand P, Figueras F, Basagaña X, Beelen R, Martinez D, Cirach M, Schembari A, Hoek G, Brunekreef B, Nieuwenhuijsen MJ. Ambient air pollution and preeclampsia: a spatiotemporal analysis. Environ Health Perspect. 2013 Nov-Dec;121(11-12): 1365-71. doi: 10.1289/ehp.1206430.

Epub 2013 Sep 9.
206. Pedersen M, Stayner L, Slama R, Sørensen M, Figueras F, Nieuwenhuijsen MJ, Raaschou-Nielsen O, Dadvand P. Ambient Air Pollution and Pregnancy-Induced Hypertensive Disorders: A Systematic Review and Meta-Analysis. Hypertension. 2014 Jun 16. pii: HYPERTENSIONAHA.114.03545. [Epub ahead of print]
207. Xu X, Hu H, Ha S, Roth J, Air pollution Ambient air pollution and hypertensive disorder of pregnancy. J Epidemiol Community Health 2014;68:13-20 doi:10.1136/ jech-2013-202902
208. Jedrychowski WA, Perera FP, Maugeri U, Spengler J, Mroz E, Flak E, Stigter L, Majewska R, Kaim I, Sowa A, Jacek R. Prohypertensive Effect of Gestational Personal Exposure to Fine Particulate Matter. Prospective Cohort Study in Non-smoking and Non-obese Pregnant Women. Cardiovasc Toxicol. 2012 Feb 11. [Epub ahead of print]
209. van den Hooven EH, de Kluizenaar Y, Pierik FH, Hofman A, van Ratingen SW, Zandveld PY, Mackenbach JP, Steegers EA, Miedema HM, Jaddoe VW. Air Pollution, Blood Pressure, and the Risk of Hypertensive Complications During Pregnancy: The Generation R Study. Hypertension. 2011 Jan 10. [Epub ahead of print]
210. Dadvand P, et al. Air Pollution and Preterm Premature Rupture of Membranes: A Spatiotemporal Analysis. Am J Epidemiol January 15, 2014 179 (2) 200-207

211. Wallace ME, et al. Exposure to Ambient Air Pollution and Premature Rupture of Membranes.Am J Epidemiol. 2016 Jun 15;183(12):1114-21. doi: 10.1093/aje/kwv284. Epub 2016 May 17.
212. Hu H, Ha S, Henderson BH, Warner TD, Roth J, Kan H, Xu X. Association of Atmospheric Particulate Matter and Ozone with Gestational Diabetes Mellitus. Environ Health Perspect. 2015 Mar 20. [Epub ahead of print]
213. Fleisch A, et al. Air Pollution Exposure and Abnormal Glucose Tolerance during Pregnancy: The Project Viva Cohort. Environ Health Perspect; DOI:10.1289/ehp. 1307065
214. Dadvand P, Rankin J, Rushton S, Pless-Mulloli T. Ambient air pollution and congenital heart disease; a register-based study. Environ Res. 2011 Feb 15. [Epub ahead of print]
215. Ren A, Qiu X, Jin L, Ma J, Li Z, et al. Association of selected persistent organic pollutants in the placenta with the risk of neural tube defects. PNAS 2011 : 1105209108v1-201105209.
216. Vrijheid M, Martinez D, Manzanares S, Dadvand P, Schembari A, Rankin J, Nieuwenhuijsen M. Environ Health Perspect. 2010 Dec 3. [Epub ahead of print] Ambient Air Pollution and Risk of Congenital Anomalies: A Systematic Review and Meta-Analysis.

217. Ritz B, Yu F, Fruin S, Chapa G, Shaw G, Harris J. Ambient Air Pollution and Risk of Birth Defects in Southern California. Am. J. Epidemiol. (2002) 155(1): 17-25 doi: 10.1093/aje/155.1.17
218. Lupo, PJ, E Symanski, DK Waller, MA Canfield and LE Mitchellet. 2010. Maternal exposure to ambient levels of benzene and neural tube defect among offspring, Texas, 1999-2004. Environmental Health Perspectives http://dx.doi.org/10.1289/ehp.1002212.
219. Stingone, J. et al: "Maternal exposure to criteria air pollutants and congenital hear defects in offspring: results from the National Birth Defects Prevention Study." Environ Health Perspect. 2014 Aug; 122(8): 863-872
220. Farhi A, Boyko V, Almagor J, Benenson I, Segre E, Rudich Y, Stern E, Lerner-Geva L. The possible association between exposure to air pollution and the risk for congenital malformations. Environ Res. 2014 Sep 29;135C:173-180. doi: 10.1016/ j.envres.2014.08.024. [Epub ahead of print]
221. Liang Z, Wu L, Fan L, Zhao Q. Ambient air pollution and birth defects in Haikou city, Hainan province. BMC Pediatr. 2014 Nov 22;14(1):283. [Epub ahead of print]
222. Girguisa M, et al. Maternal exposure to traffic-related air pollution and birth defects in Massachusetts. Environmental Research. Volume 146, April 2016, Pages 1–9
223. Faiz, A. et al: "Ambient air pollution and the risk of stillbirth." Amer J of Epid. 2012 Aug 15; 176(4): 308-316
224. Enkhmaa D, Warburton N, Javzandulam B, Uyanga J, Khishigsuren Y, Lodoysamba S, Enkhtur S, Warburton D. Seasonal ambient air pollution correlates strongly with spontaneous abortion in Mongolia. BMC Pregnancy Childbirth. 2014 Apr 23;14(1):146. [Epub ahead of print]
225. DeFranco E, et al. Air Pollution and Stillbirth Risk: Exposure to Airborne Particulate Matter during Pregnancy Is Associated with Fetal Death. PLoS One. 2015 Mar 20;10(3):e0120594. doi: 10.1371/journal.pone.0120594.
226. Siddika N, Balogun HA, Amegah AK, Jaakkola JJ. Prenatal ambient air pollution exposure and the risk of stillbirth: systematic review and meta-analysis of the empirical evidence. Occup Environ Med. 2016 May 24. pii: oemed-2015-103086. doi: 10.1136/oemed-2015-103086. [Epub ahead of print] Review.
227. Mendola P, et al. Chronic and Acute Ozone Exposure in the Week Prior to Delivery Is Associated with the Risk of Stillbirth. Int J Environ Res Public Health. 2017 Jul 6;14(7). pii: E731. doi: 10.3390/ijerph14070731.
228. Green RS, Malig B, Windham GC, Fenster L, Ostro B, Swan S. Residential exposure to traffic and spontaneous abortion. Environ Health Perspect. 2009 Dec;117(12):1939-44. doi: 10.1289/ehp.0900943. Epub 2009 Aug 26.
229. Jedrychowski WA, Majewska R, Spengler JD, Camann D, Roen EL,

Perera FP. Prenatal exposure to fine particles and polycyclic aromatic hydrocarbons and birth outcomes: a two-pollutant approach. Int Arch Occup Environ Health. 2017 Feb 7. doi: 10.1007/s00420-016-1192-9. [Epub ahead of print]
230. Hyunok Choi H, et al. Fetal Window of Vulnerability to Airborne Polycyclic Aromatic Hydrocarbons on Proportional Intrauterine Growth Restriction. PLoS One. 2012; 7(4): e35464. Published online Apr 24, 2012. doi: 10.1371/journal.pone.0035464
231. Calderón-Garcidueñas L, Reed W, Maronpot RR, Henríquez-Roldán C, Delgado- Chavez R, Calderón-Garcidueñas A, Dragustinovis I, Franco-Lira M, Aragón-Flores M, Solt AC, Altenburg M, Torres-Jardón R, Swenberg JA. Brain inflammation and Alzheimer's-like pathology in individuals exposed to severe air pollution. Toxicol Pathol. 2004 Nov-Dec;32(6):650-8.
232. Calderon-Garciduenas, L. et al. (2003) DNA damage in nasal and brain tissues of canines exposed to air pollutants is associated with evidence of chronic brain inflammation and neurodegeneration. Toxicol. Pathol. 31, 524–538
233. Gackière F, Saliba L, Baude A, Bosler O, Strube C. Ozone inhalation activates stress-responsive regions of the central nervous system. J Neurochem. 2011 Apr 6. doi: 10.1111/j.1471-4159.2011.07267.x. [Epub ahead of print]
234. Levesque S, Surace MJ, McDonald J, Block ML. Air pollution and the brain: Subchronic diesel exhaust exposure causes neuroinflammation and elevates early markers of neurodegenerative disease. J Neuroinflammation. 2011 Aug 24;8(1):105. [Epub ahead of print]
235. Calderón-Garcidueñas L, et al. Air pollution, a rising environmental risk factor for cognition, neuroinflammation and neurodegeneration: The clinical impact on children and beyond. Rev Neurol (Paris). 2015 Dec 21. pii: S0035-3787(15)00923-6. doi: 10.1016/j.neurol.2015.10.008. [Epub ahead of print]
236. Brockmeyer S, D'Angiulli A. How air pollution alters brain development: the role of neuroinflammation. Transl Neurosci. 2016 Mar 21;7(1):24-30. doi: 10.1515/tnsci-2016-0005. eCollection 2016.
237. Hartz A, Bauer B, Block M, Diesel exhaust particles induce oxidative stress, proinflammatory signaling, and P-glycoprotein, up-regulation at the blood-brain barrier. The FASEB Journal 2008;22:2723-2733.
238. Calderon-Garciduenas L, Solt AC, et al. Long-term air pollution exposure is associated with neuroinflammation, an altered innate immune response, disruption of the blood-brain barrier, ultrafine particulate deposition, and accumulation of amyloid beta-42 and alpha-synuclein in children and young adults. Toxicol Pathol. 2008;36(2): 289-310. Epub 2008 Mar 18

239. Calderón-Garcidueñas L, et al. Air Pollution and Children: Neural and Tight Junction Antibodies and Combustion Metals, the Role of Barrier Breakdown and Brain Immunity in Neurodegeneration. Journal of Alzheimer's Disease, August 2014 DOI: 10.3233/JAD-141365
240. Wardlaw J, et al. What are White Matter Hyperintensities Made of? J Am Heart Assoc. 2015 Jun; 4(6): e001140. Published online 2015 Jun 23. doi: 10.1161/JAHA.114.001140
241. Calderón-Garcidueñas L, Mora-Tiscareño A, Styner M, Gómez-Garza G, Zhu H,
Torres-Jardón R. et al. White matter hyperintensities, systemic inflammation, brain growth, and cognitive functions in children exposed to air pollution. J. Alzheimers Dis. 2012;31:183–191
242. Koppenborg RP, Nederkoorn PJ, Geerlings MI, van den Berg E. Presence and progression of white matter hyperintensities and cognition: a meta-analysis. Neurology. 2014;82:2127–2138.
243. Calderon-Garciduenas L, Mora-Tiscareno A, Ontiveros E, et al. Air pollution, cognitive deficits and brain abnormalities: a pilot study with children and dogs. Brain Cogn. 2008 Nov;68(2):117-27. Epub 2008 Jun 11.
244. G. Oberdörster, Z. Sharp, V. Atudorei, A. Elder, R. Gelein, W. Kreyling and C. Cox. Translocation of Inhaled Ultrafine Particles to the Brain. Inhalation Toxicology 2004, Vol. 16, No. 6-7, Pages 437-445
245. Lewis J, Bench G, Myers O, Tinner B, Staines W, Barr E, Divine KK, Barrington W, Karlsson J (2005) Trigeminal uptake and clearance of inhaled manganese chloride in rats and mice. Neurotoxicology 26:113–23.
246. Maher, B, et al. Magnetite pollution nanoparticles in the human brain. PNAS 2016 ; published ahead of print September 6, 2016, doi:10.1073/pnas.1605941113
247. González-Maciel A, Reynoso-Robles R, Torres-Jardón R, Mukherjee PS, Calderón-Garcidueñas L. Combustion-Derived Nanoparticles in Key Brain Target Cells and Organelles in Young Urbanites: Culprit Hidden in Plain Sight in Alzheimer's Disease Development. J Alzheimers Dis. 2017 Jun 3. doi: 10.3233/JAD-170012. [Epub ahead of print]
248. Calderón-Garcidueñas L, et al. Hallmarks of Alzheimer disease are evolving relentlessly in Metropolitan Mexico City infants, children and young adults. APOE4 carriers have higher suicide risk and higher odds of reaching NFT stage V at ≤ 40 years of age. Environmental Research, 2018; 164: 475 DOI: 10.1016/j.envres.2018.03.023
249. Bolton JL, Huff NC, Smith SH, Mason SN, Foster WM, Auten RL, et al. 2013. Maternal stress and effects of prenatal air pollution on offspring mental health outcomes in mice. Environ Health Perspect 121:1075-1082
250. Peterson B, et al. Effects of Prenatal Exposure to Air Pollutants (Polycyclic Aromatic Hydrocarbons) on the Development of Brain White Matter, Cognition, and Behavior in Later Childhood. JAMA Psychiatry.

Published online March 25, 2015. doi:10.1001/jamapsychiatry.2015.57
251. Perera FP, Tang D, Wang S, Vishnevetsky J, Zhang B, Diaz D, Camann D, Rauh V. Prenatal Polycyclic Aromatic Hydrocarbon (PAH) Exposure and Child Behavior at age 6-7. Environ Health Perspect. 2012 Mar 22. [Epub ahead of print]
252. Calderón-Garcidueñas L, Engle R, Mora-Tiscareño A, Styner M, Gómez-Garza G, Zhu H, Jewells V, Torres-Jardón R, Romero L, Monroy-Acosta ME, Bryant C, González- González LO, Medina-Cortina H, D'Angiulli A. Exposure to severe urban air pollution influences cognitive outcomes, brain volume and systemic inflammation in clinically healthy children. Brain Cogn. 2011 Oct 25. [Epub ahead of print]
253. Perera, FP, L Zhigang, R Whyatt, L Hoepner, S Wang, D Camann and V Rauh. 2009. 2009. Prenatal airborne polycyclic aromatic hydrocarbon exposure and child IQ at age 5 years. Pediatrics doi: 10.1542/peds.2008-3506.
254. Edwards SC, Jedrychowski W, Butscher M, Camann D, Kieltyka A, Mroz E, et al. 2010. Prenatal Exposure to Airborne Polycyclic Aromatic Hydrocarbons and Children's Intelligence at Age 5 in a Prospective Cohort Study in Poland. Environ Health Perspect :-. doi:10.1289/ehp.0901070
255. Perera, FP, L Zhigang, R Whyatt, L Hoepner, S Wang, D Camann and V Rauh. 2009. 2009. Prenatal airborne polycyclic aromatic hydrocarbon exposure and child IQ at age 5 years. Pediatrics doi: 10.1542/peds.2008-3506.
256. Edwards SC, Jedrychowski W, Butscher M, Camann D, Kieltyka A, Mroz E, et al. 2010. Prenatal Exposure to Airborne Polycyclic Aromatic Hydrocarbons and Children's Intelligence at Age 5 in a Prospective Cohort Study in Poland. Environ Health Perspect :-. doi:10.1289/ehp.0901070
257. Zhou Z, Yuan T, Chen Y, Qu L, Rauh V, Zhang Y, Tang D, Perera F, Li T. Benefits of Reducing Prenatal Exposure to Coal-Burning Pollutants to Children's Neurodevelopment in China Research Article, published 14 Jul 2008 | doi:10.1289/ ehp.11480
258. Perera FP, Tang D, Wang S, Vishnevetsky J, Zhang B, Diaz D, Camann D, Rauh V. Prenatal Polycyclic Aromatic Hydrocarbon (PAH) Exposure and Child Behavior at age 6-7. Environ Health Perspect. 2012
259. Perera F, Weiland K, Neidell M, Wang S. Prenatal exposure to airborne polycyclic aromatic hydrocarbons and IQ: Estimated benefit of pollution reduction. J Public Health Policy. 2014 May 8. doi: 10.1057/jphp.2014.14. [Epub ahead of print]
260. Prenatal exposure to PM10 and NO2 and children's neurdevelopment from birth to 24 months of age: mothers and Children's Environmental Health (MOCEH) study. Sci Total Environ. 2014 May 15;481:439-45. doi: 10.1016/j.scitotenv. 2014.01.107. Epub 2014 Mar 12.
261. Jedrychowski WA, Perera FP, Camann D, Spengler J, Butscher M,

Mroz E, Majewska R, Flak E, Jacek R, Sowa A. Prenatal exposure to polycyclic aromatic hydrocarbons and cognitive dysfunction in children. Environ Sci Pollut Res Int. 2014 Sep 26. [Epub ahead of print]
262. Perera FP, Chang H-w, Tang D, Roen EL, Herbstman J, et al. (2014) Early-Life Exposure to Polycyclic Aromatic Hydrocarbons and ADHD Behavior Problems. PLoS ONE 9(11): e111670. doi:10.1371/journal.pone.0111670
263. Lertxundi A, et al. Exposure to fine particle matter, nitrogen dioxide and benzene during pregnancy and cognitive and psychomotor developments in children at 15months of age. Environ Int. 2015 Apr10;80:33-40. doi: 10.1016/j.envint.2015.03.007. [Epub ahead of print]
264. Vishnevetskya J, et al. Combined effects of prenatal polycyclic aromatic hydrocarbons and material hardship on child IQ. Neurotoxicology and Teratology. doi: 10.1016/j.ntt.2015.04.002. Available online 23 April 2015
265. Yorifuji T, Kashima S, Higa Diez M, Kado Y, Sanada S, Doi H. Prenatal Exposure to Traffic-related 266. Air Pollution and Child Behavioral Development Milestone Delays in Japan. Epidemiology. 2015 Aug 5. [Epub ahead of print]
267. Cooper L, Eskenazi B, Romero C, Balmes J, Smith KR. Neurodevelopmental performance among school age children in rural Guatemala is associated with prenatal and postnatal exposure to carbon monoxide, a marker for exposure to woodsmoke. Neurotoxicology. 2012 Mar;33(2):246-54. doi: 10.1016/j.neuro.2011.09.004. Epub 2011 Sep 24.
268. Allen J, et al. Early Postnatal Exposure to Ultrafine Particulate Matter Air Pollution: Persistent Ventriculomegaly, Neurochemical Disruption, and Glial Activation Preferentially in Male Mice. Environ Health Perspect; DOI:10.1289/ehp.1307984
269. Chen JC, et al. Ambient Air Pollution and Neurotoxicity on Brain Structure: Evidence From Women's Health Initiative Memory Study. ANN NEUROL 2015;78:466–476
270. Levesque S, Taetzsch T, Lull ME, Kodavanti U, Stadler K, Wagner A. et al. Diesel exhaust activates and primes microglia: air pollution, neuroinflammation and regulation of dopaminergic neurotoxicity. Environ. Health Perspect. 2011;119:1149–1155. [PMC free article] [PubMed]
271. Levesque S, Taetzsch T, Lull ME, Johnson JA, McGraw C, Block ML. The role of MAC1 in diesel exhaust particle-induced microglial activation and loss of dopaminergic neuron function. J. Neurochem. 2013;125:756–765. [PMC free article] [PubMed] Cacciottolo M, et al. Particulate air pollutants, APOE alleles and their contributions to cognitive impairment in older women and to amyloidogenesis in experimental models. Translational Psychiatry (2017) 7, e1022; doi:10.1038/tp.2016.280 Published online 31 January

272. Calderón-Garcidueñas L, Reed W, Maronpot RR, Henríquez-Roldán C, Delgado- Chavez R, Calderón-Garcidueñas A, Dragustinovis I, Franco-Lira M, Aragón-Flores M, Solt AC, Altenburg M, Torres-Jardón R, Swenberg JA. Brain inflammation and Alzheimer's-like pathology in individuals exposed to severe air pollution. Toxicol Pathol. 2004 Nov-Dec;32(6):650-8.
273. Calderón-Garcidueñas L, Kavanaugh M, Block M, D'Angiulli A, Delgado-Chávez R, Torres-Jardón R. et al. Neuroinflammation, Alzheimer's disease-associated pathology and down regulation of the prion-related protein in air pollution exposed children and young adults. J. Alzhemers Dis. 2012;28:93–107. [PubMed]
274. Fonken LK, Xu X, Weil ZM, Chen G, Sun Q, Rajagopalan S. et al. Air pollution impairs cognition, provokes depressive-like behaviors and alters hippocampal cytokine expression and morphology. Mol. Psychiatry. 2011;16:987–995. [PMC free article] [PubMed]
275. Rivas-Arancibia S, Guevara-Guzmán R, López-Vidal Y, Rodríguez-Martínez E, Zanardo-Gomes M, Angoa-Pérez M. et al. Oxidative stress caused by ozone exposure induces loss of brain repair in the hippocampus of adult rats. Toxicol.Sci. 2010;113:187–197. [PubMed]
276. Jung CR, Lin YT, Hwang BF. Ozone, particulate matter, and newly diagnosed Alzheimer's disease: a population-based cohort study in Taiwan. J Alzheimers Dis. 2015;44(2):573-84. doi: 10.3233/JAD-140855.
277. Cheng H, et al. Nanoscale Particulate Matter from Urban Traffic Rapidly Induces Oxidative Stress and Inflammation in Olfactory Epithelium with Concomitant Effects on Brain. Environ Health Perspect; DOI:10.1289/EHP134
278. Colicino E, et al. Telomere Length, Long-Term Black Carbon Exposure, and Cognitive Function in a Cohort of Older Men: The VA Normative Aging Study. Environ Health Perspect. 2016 Jun 3. [Epub ahead of print]
279. Kim KN, et al. Long-Term Fine Particulate Matter Exposure and Major Depressive Disorder in a Community-Based Urban Cohort. Environ Health Perspect. 2016 Apr 29. [Epub ahead of print]
280. Oudin, A, et al. Traffic-Related Air Pollution and Dementia Incidence in Northern Sweden: A Longitudinal Study. Environ Health Perspect; DOI:10.1289/ehp.1408322
281. Best EA, Juarez-Colunga E, James K, LeBlanc WG, Serdar B. Biomarkers of Exposure to Polycyclic Aromatic Hydrocarbons and Cognitive Function among Elderly in the United States (National Health and Nutrition Examination Survey: 2001-2002). PLoS One. 2016 Feb 5;11(2):e0147632. doi: 10.1371/journal.pone.0147632.
282. Laura A,, et al. Effects of particulate matter exposure on multiple sclerosis hospital admission in Lombardy region, Italy. Environ Res. 2015

Nov 25;145:68-73. doi: 10.1016/j.envres.2015.11.017. [Epub ahead of print]
283. Power MC, et al. The relation between past exposure to fine particulate air pollution and prevalent anxiety: observational cohort study. BMJ, 2015; 350 (mar23 11): h1111 DOI: 10.1136/bmj.h1111
284. Ailshire JA, Crimmins EM. Fine Particulate Matter Air Pollution and Cognitive Function Among Older US Adults. Am J Epidemiol. 2014 Jun 24. pii: kwu155. [Epub ahead of print]
285. Pun VC, et al. Association of Ambient Air Pollution with Depressive and Anxiety Symptoms in Older Adults: Results from the NSHAP Study. Environ Health Perspect; DOI:10.1289/EHP494
286. Lee H, Myung W, Kim DK, Kim SE, Kim CT, Kim H. Short-term air pollution exposure aggravates Parkinson's disease in a population-based cohort. Sci Rep. 2017 Mar 16;7:44741. doi: 10.1038/srep44741.
287. Porta D, Narduzzi S, Badaloni C, Bucci S, Cesaroni G, Colelli V, Davoli M, Sunyer J, Zirro E, Schwartz J, Forastiere F. Air pollution and cognitive development at age seven in a prospective Italian birth cohort. Epidemiology. 2015 Sep 30. [Epub ahead of print]
288. Mez J, et al. Clinicopathological Evaluation of Chronic Traumatic Encephalopathy in Players of American Football. JAMA. 2017;318(4):360-370. doi:10.1001/jama.2017.8334
289. Yang Y, Glenn AL, Raine A. Brain abnormalities in antisocial individuals: implications for the law.
Behav Sci Law. 2008;26(1):65-83. doi: 10.1002/bsl.788.

290. Huang F, et al. Particulate Matter and Hospital Admissions for Stroke in Beijing, China: Modification Effects by Ambient Temperature. J Am Heart Assoc. 2016 Jul 13;5(7). pii: e003437. doi: 10.1161/JAHA.116.003437.
291. Shah A, et al. Short term exposure to air pollution and stroke: systematic review and meta-analysis. BMJ 2015;350:h1295
292. Wellenius G, et al. Ambient Air Pollution and the Risk of Acute Ischemic Stroke. Arch Intern Med. 2012;172(3):229-234
293. Han M, et al. Association between hemorrhagic stroke occurrence and meteorological factors and pollutants. BMC Neurol. 2016 May 4;16(1):59. doi: 10.1186/s12883-016-0579-2.
294. Chiu HF, Yang CY. Short-term effects of fine particulate air pollution on ischemic stroke occurrence: a case-crossover study. J Toxicol Environ Health A. 2013;76(21): 1188-97. doi:10.1080/15287394.2013.842463.
295. Chen R, Zhang Y, Yang C, Zhao Z, Xu X, Kan H. Acute Effect of
296. Mateen F, Brook R. Air Pollution as an Emerging Global Risk Factor for Stroke JAMA 2011;305(12):1240-1241.doi:10.1001/jama.2011.352
297. Kettunen, J. et al. (2007) Associations of fine and ultrafine particulate air pollution with stroke mortality in an area of low air pollution levels.

Stroke 38, 918–922
298. Kim D, et al. The joint effect of air pollution exposure and copy number variation on risk for autism. Autism Res. 2017 Apr 27. doi: 10.1002/aur.1799. [Epub ahead of print]
299. Lam J, Sutton P, Kalkbrenner A, Windham G, Halladay A, Koustas E, Lawler C, Davidson L, Daniels N, Newschaffer C, Woodruff T. A Systematic Review and Meta-Analysis of Multiple Airborne Pollutants and Autism Spectrum Disorder. PLoS One. 2016 Sep 21;11(9):e0161851. doi: 10.1371/journal.pone.0161851.
300. Flores-Pajot MC, Ofner M, Do MT, Lavigne E, Villeneuve PJ. Childhood autism spectrum disorders and exposure to nitrogen dioxide, and particulate matter air pollution: A review and meta-analysis. Environ Res. 2016 Aug 25. pii: S0013-9351(16)30317-6. doi: 10.1016/j.envres.2016.07.030. [Epub ahead of print]
301. Allen JL, et al. Developmental Neurotoxicity of Inhaled Ambient Ultrafine Particle Air Pollution: Parallels with Neuropathological and Behavioral Features of Autism and Other Neurodevelopmental Disorders. Neurotoxicology. 2015 Dec 22. pii: S0161-813X(15)30048-6. doi: 10.1016/j.neuro.2015.12.014. [Epub ahead of print
302. Dickerson AS, et al. Autism spectrum disorder prevalence and proximity to industrial facilities releasing arsenic, lead or mercury. Sci Total Environ. 2015 Jul 25;536:245-251. doi: 10.1016/j.scitotenv.2015.07.024. [Epub ahead of print]
303. Talbott E, et al. Fine particulate matter and the risk of autism spectrum disorder. Environmental Research. Volume 140, July 2015, Pages 414–420
304. Kalkbrenner AE, Windham GC, Serre ML, Akita Y, Wang X, Hoffman K, Thayer BP, Daniels JL. Particulate Matter Exposure, Prenatal and Postnatal Windows of Susceptibility, and Autism Spectrum Disorders. Epidemiology. 2014 Oct 3. [Epub ahead of print]
305. von Ehrenstein OS, Aralis H, Cockburn M, Ritz B. In Utero Exposure to Toxic Air Pollutants and Risk of Childhood Autism. Epidemiology. 2014 Jul 21. [Epub ahead of print]
306. Volk HE, Kerin T, Lurmann F, Hertz-Picciotto I, McConnell R, Campbell DB. Autism spectrum disorder: interaction of air pollution with the MET receptor tyrosine kinase gene. Epidemiology. 2014 Jan;25(1):44-7. doi: 10.1097/EDE.0000000000000030.
307. Volk HE, Hertz-Picciotto I, Delwiche L, Lurmann F, McConnell R. Residential proximity to freeways and autism in the CHARGE study. Environmental Health Perspectives. 2011;119(6):873–877
308. Rzhetsky A, Bagley SC, Wang K, Lyttle CS, Cook EH Jr, et al. (2014) Environmental and State-Level Regulatory Factors Affect the Incidence of Autism and Intellectual Disability. PLoS Comput Biol 10(3): e1003518.

doi:10.1371/journal.pcbi. 1003518
309. Becerra T, Wilhelm M, Olsen J, Cockburn M, Ritz B. Ambient Air Pollution and Autism in Los Angeles County, California. Environ Health Perspect 121:380–386 (2013). http://dx.doi.org/10.1289/ehp.1205827 [Online 18 December 2012]
310. Jung CR, Lin YT, Hwang BF. Air Pollution and Newly Diagnostic Autism Spectrum Disorders: A Population-Based Cohort Study in Taiwan. PLoS One. 2013 Sep 25;8(9):e75510.
311. Windham GC, Zhang L, Gunier R, Croen LA, Grether JK. Autism spectrum disorders in relation to distribution of hazardous air pollutants in the San Francisco Bay area. Environmental Health Perspectives. 2006;114(9):1438–1444
312. Rossignol D A, Genuis S J, Frye R E. Environmental toxicants and autism spectrum disorders: a systematic review. Trans. Psychiatry. 2014;4(2):e360. [PMC free article] [PubMed]
313. Lyall K, Schmidt R J, Hertz-Picciotto I. Maternal lifestyle and environmental risk factors for autism spectrum disorders. Int. J. Epi. 2014;43:443–464. [PMC free article] [PubMed]
314. Raz R, Roberts AL, Lyall K, Hart JE, Just AC, Laden F. et al. Autism spectrum disorder and particulate matter air pollution before, during, and after pregnancy: a nested case-control analysis within the Nurses' Health Study II Cohort. Env. Health Perspect. 2015;123(3):264–270. [PMC free article] [PubMed]
315. http://archive.sltrib.com/article.php?id=53816934&itype=CMSID
316. Raaschou-Nielsen O, Andersen Z, Hvidberg M, Jensen SS, Ketzel M, Sørensen M, Loft S, Overvad K, Tjønneland A. Lung Cancer Incidence and Long-Term Exposure to Air Pollution from Traffic. Environ Health Perspect. 2011 Jan 12. [Epub ahead of print]
317. Pearson RL, Wachtel H, Ebi KL. Distance-weighted traffic density in proximity to a home is a risk factor for leukemia and other childhood cancers. J Air Waste Manag Assoc 50(2):175-180.
318. Weng HH, Tsai SS, Chen CC, Chiu HF, Wu TN, Yang CYJ. Childhood leukemia development and correlation with traffic air pollution in Taiwan using nitrogen dioxide as an air pollutant marker. Toxicol Environ Health A. 2008;71(7):434-8.
319. Langholz B, et al.; Traffic density and the risk of childhood leukemia in a Los Angeles case-control study. Ann. Epi. 2002 Oct; 12(7):482-7.
320. Raaschou-Nielsen O, Andersen ZJ, Hvidberg M, Jensen SS, Ketzel M, Sørensen M, Hansen J, Loft S, Overvad K, Tjønneland A. Air pollution from traffic and cancer incidence: a Danish cohort study. Environ Health. 2011 Jul 19;10:67. doi: 10.1186/1476-069X-10-67.
321. Chiu HF, Tsai SS, Chen PS, Liao YH, Liou SH, Wu TN, Yang CY. Traffic air pollution and risk of death from gastric cancer in taiwan: petrol

station density as an indicator of air pollutant exposure. J Toxicol Environ Health A. 2011 Sep 15;74(18): 1215-24.
322. Turner MC, Krewski D, Pope Iii CA, Chen Y, Gapstur SM, Thun MJ. Long-Term Ambient Fine Particulate Matter Air Pollution and Lung Cancer in a Large Cohort of Never Smokers. Am J Respir Crit Care Med. 2011 Oct 6. [Epub ahead of print] PubMed PMID: 21980033.
323. Hung LJ, Chan TF, Wu CH, Chiu HF, Yang CY. Traffic Air Pollution and Risk of Death from Ovarian Cancer in Taiwan: Fine Particulate Matter (PM(2.5)) as a Proxy Marker. J Toxicol Environ Health A. 2012 Feb 1;75(3):174-82.
324. Pedersen M, et al. Ambient air pollution and primary liver cancer incidence in four European cohorts within the ESCAPE project. Environ Res. 2017 Jan 17;154:226-233. doi: 10.1016/j.envres.2017.01.006. [Epub ahead of print]
325. Janitza AE, et al. Traffic-related air pollution and childhood acute leukemia in Oklahoma. Environmental Research. Volume 148, July 2016, Pages 102–111
326. Poulsen AH, et al. Air pollution from traffic and risk for brain tumors: a nationwide study in Denmark. Cancer Causes Control. 2016 Feb 18. [Epub ahead of print]
327. Spycher BD, et al. Childhood cancer and residential exposure to highways: a nationwide cohort study. Eur J Epidemiol. 2015 Nov 2. [Epub ahead of print]
328. Hoot, J. et al. Residential Proximity to Heavy-Traffic Roads, Benzene Exposure, and Childhood Leukemia—The GEOCAP Study, 2002–2007. American Journal of Epidemiology Volume 182, Issue 8Pp. 685-693
329. Ding N, Zhou N, Zhou M, Ren GM. Respiratory cancers and pollution. Eur Rev Med Pharmacol Sci. 2015 Jan;19(1):31-7.
330. Javier Pintos J, et al. Use of wood stoves and risk of cancers of the upper aero- digestive tract: a case-control study. Int. J. Epidemiol. (1998) 27 (6): 936-940 doi: 10.1093/ije/27.6.936
331. Scheurer ME, Danysh HE, Follen M, Lupo PJ. Association of traffic-related hazardous air pollutants and cervical dysplasia in an urban multiethnic population: a cross-sectional study. Environ Health. 2014 Jun 13;13(1):52. doi: 10.1186/1476-069X-13-52.
332. Bulka C, et al. Residence Proximity to Benzene Release Sites Is Associated With Increased Incidence of Non-Hodgkin Lymphoma. Cancer. Article first published online: 29 JUL 2013. DOI: 10.1002/cncr.28083
333. Parent ME, Goldberg MS, Crouse DL, Ross NA, Chen H, Valois MF, Liautaud A. Traffic-related air pollution and prostate cancer risk: a case-control study in Montreal, Canada. Occup Environ Med. 2013 Mar 26. [Epub ahead of print]
334. Sapkota A, Zaridze D, Szeszenia-Dabrowska N, Mates D, Fabiánová

E, Rudnai P ,. Indoor air pollution from solid fuels and risk of upper aerodigestive tract cancers in central and eastern Europe. Environ Res 2013;120:90–5.
335. Crouse DL, Goldberg MS, Ross NA, Chen H, Labrèche F 2010. Postmenopausal Breast Cancer Is Associated with Exposure to Traffic-Related Air Pollution in Montreal, Canada: A Case–Control Study. Environ Health Perspect 118:1578-1583. doi:10.1289/ ehp.1002221
336. Visani G, et al. Environmental nanoparticles are significantly over-expressed in acute myeloid leukemia. Leuk Res. 2016 Nov;50:50-56.
337. von Ehrenstein O, et al. In Utero and Early-Life Exposure to Ambient Air Toxics and Childhood Brain Tumors: A Population-Based Case–Control Study in California, USA Environ Health Perspect; DOI:10.1289/ehp.1408582
338. Heck JE, Wu J, Lombardi C, Qiu J, Meyers TJ, Wilhelm M, Cockburn M, Ritz B. Childhood cancer and traffic-related air pollution exposure in pregnancy and early life. Environ Health Perspect. 2013 Nov-Dec;121(11-12):1385-91. doi: 10.1289/ehp. 1306761. Epub 2013 Sep 9.
339. Ghosh JK, Heck JE, Cockburn M, Su J, Jerrett M, Ritz B. Prenatal Exposure to Traffic-related Air Pollution and Risk of Early Childhood Cancers. Am J Epidemiol. 2013 Aug 28. [Epub ahead of print]
340. Prenatal air pollution associated higher rates of retinoblastomas, ALL, and germ cell tumors. http://www.aacr.org/home/public--media/aacr-in-the-news.aspx?d=3062
341. Ho CK, Peng CY, Yang CY. Traffic air pollution and risk of death from bladder cancer in Taiwan using petrol station density as a pollutant indicator. J Toxicol Environ Health A. 2010;73(1):23-32. doi: 10.1080/15287390903248869.
342. Hu H, Dailey AB, Kan H, Xu X. The effect of atmospheric particulate matter on survival of breast cancer among US females. Breast Cancer Res Treat. 2013 Apr 17. [Epub ahead of print]
343. Hung LJ, Chan TF, Wu CH, Chiu HF, Yang CY. Traffic Air Pollution and Risk of Death from Ovarian Cancer in Taiwan: Fine Particulate Matter (PM(2.5)) as a Proxy Marker. J Toxicol Environ Health A. 2012 Feb 1;75(3):174-82.
344. Grant WB. Air pollution in relation to U.S. cancer mortality rates: an ecological study; likely role of carbonaceous aerosols and polycyclic aromatic hydrocarbons. Anticancer Res. 2009 Sep;29(9):3537-45.
345. Wong CM, et al. Cancer Mortality Risks from Long-term Exposure to Ambient Fine Particle. Cancer Epidemiol Biomarkers Prev; Published OnlineFirst April 29, 2016; doi 10.1158/1055-9965.EPI-15-0626
346. Mehta M, Chen LC, Gordon T, Rom W, Tang MS. Particulate matter inhibits DNA repair and enhances mutagenesis. Mutat Res 2008;657:116–21

347. Pearson J, Bachireddy C, Shyamprasad S, Goldfine A, Brownstein J. Association Between Fine Particulate Matter and Diabetes Prevalence in the U.S.Diabetes Care October 2010 33:2196-2201; published ahead of print July 13, 2010, doi:10.2337/ dc10-0698
348. Krämer U, Herder C, Sugiri D, Strassburger K, Schikowski T, Ranft U, et al. 2010. Traffic-related Air Pollution and Incident Type 2 Diabetes: Results from the SALIA Cohort Study. Environ Health Perspect :-. doi:10.1289/ehp.0901689
349. Coogan PF, White LF, Jerrett M, Brook RD, Su JG, Seto E, Burnett R, Palmer JR, Rosenberg L. Air Pollution and Incidence of Hypertension and Diabetes in African American Women Living in Los Angeles. Circulation. 2012 Jan 4. [Epub ahead of print]
350. Liu C, Ying Z, Harkema J, Sun Q, Rajagopalan S. Epidemiological and Experimental Links between Air Pollution and Type 2 Diabetes. Toxicol Pathol. 2012 Oct 26. [Epub ahead of print]
351. Park SK, Wang W. Ambient Air Pollution and Type 2 Diabetes: A Systematic Review of Epidemiologic Research. Curr Environ Health Rep. 2014 Sep 1;1(3): 275-286.
352. Li C, Fang D, Xu D, Wang B, Zhao S, Yan S, Wang Y. MECHANISMS IN ENDOCRINOLOGY: Main air pollutants and diabetes-associated mortality: a systematic review and meta-analysis. Eur J Endocrinol. 2014 Nov;171(5):R183-R190.
353. Park SK, Adar SD, O'Neill MS, Auchincloss AH, Szpiro A, Bertoni AG, Navas- Acien A, Kaufman JD, Diez-Roux AV. Long-Term Exposure to Air Pollution and Type 2 Diabetes Mellitus in a Multiethnic Cohort. Am J Epidemiol. 2015 Feb 17. pii: kwu280. [Epub ahead of print]
354. Malmqvist E. Maternal exposure to air pollution and type 1 diabetes - Accounting for genetic factors. Environ Res. 2015 Apr 13;140:268-274. doi: 10.1016/j.envres.2015.03.024. [Epub ahead of print]
355. Meo SA, et al. Effect of environmental air pollution on type 2 diabetes mellitus. Eur Rev Med Pharmacol Sci. 2015 Jan;19(1):123-8.
356. Xu X, Liu C, Xu Z, Tzan K, et al. Long-term Exposure to Ambient Fine Particulate Pollution Induces Insulin Resistance and Mitochondrial Alteration in Adipose Tissue. Toxicological Sciences Volume 124, Issue 1Pp. 88-98
357. Thiering E, Cyrys J, Kratzsch J, Meisinger C, Hoffmann B, Berdel D, von Berg A, Koletzko S, Bauer CP, Heinrich J. Long-term exposure to traffic-related air pollution and insulin resistance in children: results from the GINIplus and LISAplus birth cohorts Diabetologia, DOI 10.1007/s00125-013-2925-x
358. Vella RE, Pillon NJ, Zarrouki B, Croze ML, Koppe L, Guichardant M, Pesenti S, Chauvin MA, Rieusset J, Géloën A, Soulage CO. Ozone exposure triggers insulin resistance through muscle c-Jun N-terminal

Kinases (JNKs) activation. Diabetes. 2014 Oct 2. pii: DB_131181. [Epub ahead of print]
359. Madhloum N, et al. Cord plasma insulin and in utero exposure to ambient air pollution. Environ Int. 2017 May 22. pii: S0160-4120(16)30886-8. doi: 10.1016/j.envint.2017.05.012. [Epub ahead of print]
360. Alderete T, et al. Longitudinal Associations Between Ambient Air Pollution with Insulin Sensitivity, β-Cell Function, and Adiposity in Los Angeles Latino Children. Diabetes 2017 Jan; db161416. https://doi.org/10.2337/db16-1416
361. Wolf K, Popp A, Schneider A, Breitner S, Hampel R, Rathmann W, Herder C, Roden M, Koenig W, Meisinger C, Peters A; Association Between Long-Term Exposure to Air Pollution and Biomarkers Related to Insulin Resistance, Subclinical Inflammation and Adipokines. Diabetes. 2016 Sep 7. pii: db151567. [Epub ahead of print]
362. Makaji E, Raha S, Wade MG, Holloway AC. Effect of environmental contaminants on Beta cell function. Int J Toxicol. 2011 Aug;30(4):410-8. Epub 2011 Jun 24.
363. Khafaie MA, Salvi SS, Ojha A, Khafaie B, Gore SS, Yajnik CS. Systemic Inflammation (C-Reactive Protein) in Type 2 Diabetic Patients Is Associated With Ambient Air Pollution in Pune City, India. Diabetes Care. 2012 Nov 19. [Epub ahead of print]
364. Liu C, et al. Air Pollution–. Susceptibility to Inflammation and Insulin Resistance: Influence of CCR2 Pathways in Mice. Environ Health Perspect; DOI:10.1289/ehp. 1306841
365. Nemmar A, Al-Salam S, Beegam S, Yuvaraju P, Yasin J, Ali BH. Cell Physiol Biochem. 2014 Feb 11;33(2):413-422. [Epub ahead of print] Pancreatic Effects of Diesel Exhaust Particles in Mice with Type 1 Diabetes Mellitus
366. Teichert T, Vossoughi M, Vierkötter A, Sugiri D, Schikowski T, Schulte T, Roden M, Luckhaus C, Herder C, Krämer U. Association between traffic-related air pollution, subclinical inflammation and impaired glucose metabolism: results from the SALIA study. PLoS One. 2013 Dec 10;8(12):e83042. doi: 10.1371/journal.pone.0083042. eCollection 2013.
367. Haberzettl P, O'Toole TE, Bhatnagar A, Conklin DJ. Exposure to Fine Particulate Air Pollution Causes Vascular Insulin Resistance by Inducing Pulmonary Oxidative Stress. Environ Health Perspect. 2016 Apr 29. [Epub ahead of print]
368. Wolf K, Popp A, Schneider A, Breitner S, Hampel R, Rathmann W, Herder C, Roden M, Koenig W, Meisinger C, Peters A; Association Between Long-Term Exposure to Air Pollution and Biomarkers Related to Insulin Resistance, Subclinical Inflammation and Adipokines. Diabetes. 2016 Sep 7. pii: db151567. [Epub ahead of print]
369. Wei Y, et al. Chronic exposure to air pollution particles increases the

risk of obesity and metabolic syndrome: findings from a natural experiment in Beijing. FASEB J. 2016 Feb 18. pii: fj.201500142. [Epub ahead of print]
370. Rundle A, Hoepner L, Hassoun A, et al. Association of Childhood Obesity With Maternal Exposure to Ambient Air Polycyclic Aromatic Hydrocarbons During Pregnancy. Am. J. Epidemiol. online April 13, 2012 doi:10.1093/aje/kwr45
371. Bolton J, Smith S, Huff N, Gilmour MI, Foster WM, Auten R, and Bilbo S. Prenatal air pollution exposure induces neuroinflammation and predisposes offspring to weight gain in adulthood in a sex-specific manner FASEB J fj.12-210989; published ahead of print July 19, 2012, doi:10.1096/fj.12-210989
372. Salim S, et al. Air pollution effects on the gut microbiota. A link between exposure and inflammatory disease Gut Microbes. 2014 Mar 1; 5(2): 215–219. Published online 2013 Dec 20. doi: 10.4161/gmic.27251
373. Beamish LA, Osornio-Vargas AR, Wine E. Air pollution: An environmental factor contributing to intestinal disease. J Crohns Colitis. 2011 Aug;5(4):279-86. doi: 10.1016/ j.crohns.2011.02.017. Epub 2011 Mar 23.
374. Kaplan GG, Hubbard J, Korzenik J, Sands BE, Panaccione R, Ghosh S, Wheeler AJ, Villeneuve PJ. The inflammatory bowel diseases and ambient air pollution: a novel association. Am J Gastroenterol. 2010 Nov;105(11):2412-9. doi: 10.1038/ajg.2010.252. Epub 2010 Jun 29.
375. Ananthakrishnan AN, McGinley EL, Binion DG, Saeian K. Ambient air pollution correlates with hospitalizations for inflammatory bowel disease: an ecologic analysis. Inflamm Bowel Dis. 2011 May;17(5):1138-45. doi: 10.1002/ibd.21455. Epub 2010 Aug 30.
376. Kish L, Hotte N, Kaplan GG, Vincent R, Tso R, et al. (2013) Environmental Particulate Matter Induces Murine Intestinal Inflammatory Responses and Alters the Gut Microbiome. PLoS ONE 8(4): e62220. doi:10.1371/journal.pone.0062220
377. Kaplan G, Dixon E, Panaccione R, Fong A, Chen L, et al. Effect of ambient air pollution on the incidence of appendicitis. Published online ahead of print October 5, 2009 CMAJ 10.1503/cmaj.082068
378. Kaplan GG, Tanyingoh D, Dixon E, Johnson M, Wheeler AJ, Myers RP, Bertazzon S, Saini V, Madsen K, Ghosh S, et al. Ambient ozone concentrations and the risk of perforated and nonperforated appendicitis: a multicity case-crossover study. Environ Health Perspect. 2013;121:939–43. [PMC free article] [PubMed]
379. Litchfield I, et al. The Role of Air Pollution as a Determinant of Sudden Infant Death Syndrome: A Systematic Review and Meta-analysis. Epidemiology: January 2011 - Volume 22 - Issue 1 - pp S165-S166, doi: 10.1097/01.ede.0000392182.47995.3c
380. Scheers H, Mwalili SM, Faes C, Fierens F, Nemery B, Nawrot TS

2011. Does Air Pollution Trigger Infant Mortality in Western Europe? A Case-Crossover Study. Environ Health Perspect :-. doi:10.1289/ehp.1002913
381. Bernatsky, S, M Fournier, CA Pineau, AE Clarke, E Vinet and A Smargiassi. 2010. Associations between ambient fine particulate levels and disease activity in systemic lupus erythematosus (SLE). Environmental Health Perspectives http://dx.doi.org/ 10.1289/ehp.1002123.
382. Zeft AS, Prahala S, Lefevre S, et al. Juvenile idiopathic arthritis and exposure to fine particulate air pollution. Clin Exp Rheumotol 2009 Sep-Oct; 27(5):877-84
383. Zeft AS, et al. Systemic onset juvenile idiopathic arthritis and exposure to fine particulate air pollution. Clin Exp Rheumatol. 2016 Sep 1. [Epub ahead of print]
384. Zanobetti A, Redline S, Schwartz J, et al. Associations of PM 10 with sleep and sleep-disordered breathing in adults from seven US urban areas. Am J Respir Crit Care Med 2010 Sept 15;182(6):819-25
385. Lue S, Wellenius G, Wilker E, Mostofsky E, Mittleman M. Residential proximity to major roadways and renal function. J Epidemiol Community Health Published Online First: 13 May 2013 doi:10.1136/jech-2012-202307
386. Lu J, et al. Polluted Morality: Air Pollution Predicts Criminal Activity and Unethical Behavior. Psychological Science, 2018; 095679761773580 DOI: 10.1177/0956797617735807
387. Bowe B, et al. Particulate Matter Air Pollution and the Risk of Incident CKD and Progression to ESRD. J Am Soc Nephrol. 2017 Sep 21. pii: ASN.2017030253. doi: 10.1681/ASN.2017030253. [Epub ahead of print]
388. Slama R, Bottagisi S, Solansky I, Lepeule J, Giorgis-Allemand L, Sram R. Short- Term Impact of Atmospheric Pollution on Fecundability. Epidemiology. 2013 Sep 18. [Epub ahead of print]
389. Frutos V, González-Comadrán M, Solà I, Jacquemin B, Carreras R, Checa Vizcaíno MA. Impact of air pollution on fertility: a systematic review. Gynecol Endocrinol. 2014 Sep 12:1-7. [Epub ahead of print]
390. Vierkötter A, et al. Airborne Particle Exposure and Extrinsic Skin Aging. Journal of Investigative Dermatology. December 2010Volume 130, Issue 12, Pages 2719–2726
391. Huls A, et al. Traffic-Related Air Pollution Contributes to Development of Facial Lentigines: Further Epidemiological Evidence from Caucasians and Asians. Journal of Investigative Dermatology, May 2016Volume 136, Issue 5, Pages 1053–1056
392. Krutmann, J., Liu, W., Li, L., Pan, X., Crawford, M., Sore, G. et al. Pollution and skin: from epidemiological and mechanistic studies to clinical implications. J Dermatol Sci. 2014; 76: 163–168
393. Byron Lew, B. Mak Arvin. Happiness and air pollution: evidence from

14 European countries. International Journal of Green Economics, 2013 (in press)
394. Ho HC, et al. Spatiotemporal influence of temperature, air quality, and urban environment on cause-specific mortality during hazy days. Environment International, Volume 112, March 2018, Pages 10-22
395. Jia Z, et al. Exposure to Ambient Air Particles Increases the Risk of Mental Disorder: Findings from a Natural Experiment in Beijing. Int J Environ Res Public Health. 2018 Jan 19;15(1). pii: E160. doi: 10.3390/ijerph15010160.
396. Oudin A, et al. The association between daily concentrations of air pollution and visits to a psychiatric emergency unit: a case-crossover study.Environ Health. 2018 Jan 10;17(1):4. doi: 10.1186/s12940-017-0348-8.
397. Casas L, et al. Does air pollution trigger suicide? A case-crossover analysis of suicide deaths over the life span. European Journal of Epidemiology. November 2017, Volume 32, Issue 11, pp 973–981
398. Sunyer J, et al. Traffic-related Air Pollution and Attention in Primary School Children: Short-term Association. Epidemiology: March 2017 – Volume 28 – Issue 2 – p 181–189. doi: 10.1097/EDE.0000000000000603
399. Hackshaw A, et al. Low cigarette consumption and risk of coronary heart disease and stroke: meta-analysis of 141 cohort studies in 55 study reports. BMJ, 2018; j5855 DOI: 10.1136/bmj.j5855
400. Tanwar V, et al. Preconception Exposure to Fine Particulate Matter Leads to Cardiac Dysfunction in Adult Male Offspring. Journal of the American Heart Association, 2018; 7 (24) DOI: 10.1161/JAHA.118.010797

Chapter Six

1.https://www.theguardian.com/environment/2003/mar/04/usnews.climatechange
2.https://dge.carnegiescience.edu/labs/caldeiralab/Caldeira%20downloads/PSAC,%201965,%20Restoring%20the%20Quality%20of%20Our%20Environment.pdf
3. https://www.theguardian.com/environment/climate-consensus-97-per-cent/2015/nov/05/scientists-warned-the-president-about-global-warming-50-years-ago-today
4. https://www.skepticalscience.com/LBJ-climate-1965.html
5. http://www.abc.net.au/radionational/programs/scienceshow/naomi-oreskes---merchants-of-doubt/3012690
6. ^ Svante Arrhenius (1896). "On the Influence of Carbonic Acid in the Air upon the Temperature of the Earth". Publications of the Astronomical Society of the Pacific. 9: 14. Bibcode:1897PASP....9...14A. doi:10.1086/121158

7. https://www.climateliabilitynews.org/2018/04/05/climate-change-oil-companies-knew-shell-exxon/
8. http://www.climatefiles.com/denial-groups/heartland-institute/global-climate-coalition-draft-primer-for-ipcc-2nd-assessment/
9. https://www.ucsusa.org/global-warming/fight-misinformation/climate-deception-dossiers-fossil-fuel-industry-memos#.WvkHvy-ZO7o
10. https://blog.ucsusa.org/peter-frumhoff/global-warming-fact-co2-emissions-since-1988-764?_ga=2.266565894.985928187.1526269785-480492249.1425487403
11. http://www.washingtonpost.com/wp-dyn/content/article/2007/01/31/AR2007013101808.html
12.https://www.nytimes.com/2006/04/16/books/review/rentagenius.html
13. https://insideclimatenews.org/content/Exxon-The-Road-Not-Taken
14. https://insideclimatenews.org/news/22122015/exxon-mobil-oil-industry-peers-knew-about-climate-change-dangers-1970s-american-petroleum-institute-api-shell-chevron-texaco
15.http://content.usatoday.com/communities/sciencefair/post/2011/11/defense-science-panel-climate-a-national-security-threat/1#.ULKQQ4WGaKs
16. http://www.businessinsider.com/insurance-industry-trust-us-global-warming-is-real-2012-3
17. http://climatechange.worldbank.org/
18.http://www.unep.org/publications/ebooks/emissionsgapreport/chapter1.asp?c=1.3
19. http://www.ametsoc.org/policy/2012climatechange.html
20. https://www.newscientist.com/article/2168847-worst-case-climate-change-scenario-is-even-worse-than-we-thought/
21.https://agupubs.onlinelibrary.wiley.com/doi/abs/10.1002/2018PA003341
22. https://insideclimatenews.org/news/07052018/atlantic-ocean-circulation-slowing-climate-change-heat-temperature-rainfall-fish-why-you-should-care
23.https://data.globalchange.gov/assets/a9/5b/4c4bfa80f8b4cb189cae5a495f22/sap3-4-brochure.pdf
24. http://advances.sciencemag.org/content/2/10/e1600873
25. https://www.nbcnews.com/news/world/scope-great-barrier-reef-s-massive-coral-bleaching-alarms-scientists-n867521
26. https://www.insidescience.org/news/bringing-plight-coral-reefs-our-screens
27. https://www.thenation.com/article/alec-exposed-koch-connection/
28. https://www.facebook.com/Brutlive/videos/1840542786244232/
29. Patrick T. Brown, Ken Caldeira. Greater future global warming inferred

References

from Earth's recent energy budget. Nature, 2017; 552 (7683): 45 DOI: 10.1038/nature24672
30. https://www.sciencedaily.com/releases/2017/12/171206132220.htm
31. http://whatweknow.aaas.org/get-the-facts/
32. http://www.sciencemag.org/news/2010/07/critical-ocean-organisms-are-disappearing
33. http://news.mit.edu/2015/ocean-acidification-phytoplankton-0720
34. https://www.nytimes.com/2017/06/03/us/politics/republican-leaders-climate-change.html
35. https://www.greenpeace.org/usa/global-warming/climate-deniers/koch-industries/
36. http://noclimatetax.com/the-pledge/
37. https://www.thelancet.com/journals/lancet/article/PIIS0140-6736(09)60935-1/fulltext
38. https://medsocietiesforclimatehealth.org
39. http://www.earth.columbia.edu/articles/view/3343
40. http://www.pnas.org/content/113/42/11649?sid=96714617-f7a1-47b3-bda6-e6cff2f57958
41. https://phys.org/news/2017-01-giant-middle-east-storm-climate.html
42. http://www.pnas.org/content/106/37/15594.short
43. https://www.ars.usda.gov/southeast-area/raleigh-nc/plant-science-research/docs/climate-changeair-quality-laboratory/ozone-effects-on-plants/
44. https://www.theguardian.com/environment/2016/dec/01/climate-change-trigger-unimaginable-refugee-crisis-senior-military
45. https://www.psychologicalscience.org/observer/global-warming-and-violent-behavior#.WTqiMjOZOHr
46. https://www.ncbi.nlm.nih.gov/pmc/articles/PMC4446935/
47. https://www.esquire.com/news-politics/a36228/ballad-of-the-sad-climatologists-0815/
48. https://www.thestar.com/news/world/2016/02/28/climate-change-is-wreaking-havoc-on-our-mental-health-experts.html
49. https://thinkprogress.org/epa-chief-exaggerates-coal-job-growth-2ab69ed36b6/
50. https://www.voanews.com/a/air-pollution-linked-to-early-death/1739804.html
51. https://www.nature.com/articles/s41558-017-0013-9.epdf?referrer_access_token=cIvITR7fMhcwgE0bx8PB7NRgN0jAjWel9jnR3ZoTv0MPTCfJUE3ksFmZmzoQEYcQk-1mQqwS7BiPlUuAOmIcmWrf4Loxm_sqSthlI7wnuZT3tBaPiMLg_hIXEbcChC614jflyI-jR8pr0erVruunVyBRj5r6KtGAd6xIbEEBQmmx-DBYsgbv1xKaUHTGvXDforC0xAi3q2rlTm7L3n3BzRi1I3OszqqgWK244uO3GpEtebT5xE1EtBrdUV_H8A2-_1r40VNnfEIzlReWbcY_lP-

bw5fn7TUapKGDqyS2xrs%3D&tracking_referrer=www.technologyreview.com
52. Pascual M, et al. Malaria, resurgence in the East AFrican highlands: temperature trends revisited. Proc. Natl Acad Sci USA 2006 103:5829034
53. Schlenker W, et al. Nonlinear temperature effects indicate severe damages to U.S. crop yields under climate change. PNAS, vol. 106 no. 3715594–15598, doi: 10.1073/pnas.0906865106
54. https://www.nature.com/articles/s41467-018-02992-9
55.https://www.researchgate.net/publication/40724209_Managing_the_Health_Effects_of_Climate_Change
56. https://www.ncbi.nlm.nih.gov/pmc/articles/PMC3346787/
57. https://www.ecowatch.com/climate-change-tipping-point-2170296371.html
58. https://www.nap.edu/read/13111/chapter/6
59. https://www.scientificamerican.com/article/global-warming-beyond-the-co2/
60. Ad Hoc Study Group on Carbon Dioxide and Climate (1979). "Carbon Dioxide and Climate: A Scientific Assessment" (PDF). National Academy of Sciences. Archived from the original (PDF) on 2008-07-25.
61.https://www.usatoday.com/story/news/politics/onpolitics/2015/01/22/mitt-romney-climate-change/81555810/
62. http://thehill.com/blogs/blog-briefing-room/news/335834-romney-tweets-support-for-paris-climate-deal
63. https://www.theguardian.com/us-news/2017/jun/01/republican-senators-paris-climate-deal-energy-donations
64. https://www.theguardian.com/us-news/2016/mar/03/oil-and-gas-industry-has-pumped-millions-into-republican-campaigns
65. https://www.yahoo.com/news/republican-congressman-explains-sea-level-203325866.html
66. https://www.independent.co.uk/environment/republican-voters-third-humans-climate-change-democrats-usa-a8085691.html
67. https://www.independent.co.uk/news/world/americas/us-politics/republican-climate-change-god-will-take-care-if-real-tim-walberg-comments-global-warming-a7767346.html
68.https://scholar.princeton.edu/sites/default/files/mgilens/files/gilens_and_page_2014_-testing_theories_of_american_politics.doc.pdf
69. https://www.washingtonpost.com/news/monkey-cage/wp/2016/05/23/critics-challenge-our-portrait-of-americas-political-inequality-heres-5-ways-they-are-wrong/?utm_term=.f57bf88268a5
70. https://www.theatlantic.com/business/archive/2015/04/how-corporate-lobbyists-conquered-american-democracy/390822/
71. http://www.lung.org/about-us/blog/2017/07/how-climate-change-has-led-to-an-increase-in-valley-fever.html

72. https://www.sciencedaily.com/releases/2018/05/180516123655.htm
73. https://www.theverge.com/2017/8/9/16116198/climate-change-report-extreme-weather-co2-donald-trump
74. https://www.theguardian.com/environment/keep-it-in-the-ground-blog/2015/mar/25/what-numbers-tell-about-how-much-fossil-fuel-reserves-cant-burn
75. http://news.exxonmobil.com/press-release/exxonmobil-releases-reports-shareholders-managing-climate-risk
76.https://www.americanprogress.org/issues/green/news/2013/11/05/78807/big-oil-big-profits-big-tax-breaks/
77. https://insideclimatenews.org/news/22122015/exxon-mobil-oil-industry-peers-knew-about-climate-change-dangers-1970s-american-petroleum-institute-api-shell-chevron-texaco
78. https://insideclimatenews.org/news/01122015/documents-exxons-early-co2-position-senior-executives-engage-and-warming-forecast
79. https://www.ecowatch.com/climate-change-global-temperatures-noaa-2569919615.html
80. http://pubs.usgs.gov/fs/2012/3145/
81. http://pubs.usgs.gov/fs/2012/3145/
82. http://www.eia-international.org/explosion-of-super-greenhouse-gases-expected-over-next-decade
83. https://thehill.com/policy/energy-environment/407614-epa-to-abandon-restrictions-against-climate-change-linked-chemical
84. https://www.desmogblog.com/competitive-enterprise-institute
85. https://www.livescience.com/63334-coal-affecting-climate-century-ago.html
86. https://history.aip.org/climate/20ctrend.htm#L_M0465
87. http://nymag.com/intelligencer/2017/07/climate-change-earth-too-hot-for-humans.html
88. https://www.wired.com/story/north-carolina-fends-off-swarms-of-super-sized-mosquitoes/
89. https://endcoal.org/wp-content/uploads/2018/09/TsunamiWarningEnglish.pdf
90. https://www.theguardian.com/environment/2017/jun/21/top-global-banks-still-lend-billions-extract-fossil-fuels
91. http://www.ipcc.ch/report/sr15/
92. http://news.cornell.edu/stories/2017/06/rising-seas-could-result-2-billion-refugees-2100
93. https://www.theguardian.com/environment/2018/oct/08/global-warming-must-not-exceed-15c-warns-landmark-un-report
94. https://truthout.org/video/noam-chomsky-the-future-of-human-life-is-at-risk-thanks-to-climate-denial/
95. https://www.ran.org/wp-

content/uploads/2018/06/BankingonClimateChange2018_Japan_Summar y_FINAL_ENG_(1).pdf
96. https://www.env-health.org/wp-content/uploads/2018/06/hidden_price_tags_report.pdf
97. http://priceofoil.org/content/uploads/2017/10/OCI_US-Fossil-Fuel-Subs-2015-16_Final_Oct2017.pdf
98. https://www.nature.com/articles/s41560-017-0009-8
99. https://www.washingtonpost.com/news/federal-eye/wp/2015/12/16/unaccompanied-children-crossing-southern-border-in-greater-numbers-again-raising-fears-of-new-migrant-crisis/?utm_term=.a4e1850c59d4
100. https://reliefweb.int/disaster/dr-2014-000132-hnd
101. https://www.theguardian.com/world/2018/oct/30/migrant-caravan-causes-climate-change-central-america
102.https://journals.plos.org/plosone/article?id=10.1371/journal.pone.019 3570
103. https://www.wfp.org/news/news-release/new-study-examines-links-between-emigration-and-food-insecurity-dry-corridor-el-sa
104.https://www.nationalgeographic.com/environment/2018/10/drought-climate-change-force-guatemalans-migrate-to-us/
105.https://www.acs.org/content/acs/en/pressroom/presspacs/2018/acs -presspac-february-14-2018/coffee-threatened-by-climate-change-disease-pests.html
106. https://www.theguardian.com/world/2018/oct/30/migrant-caravan-causes-climate-change-central-america
107. https://www.aljazeera.com/news/2018/09/chronic-malnutrition-guatemalas-children-end-180911105550474.html
108.https://www.nationalgeographic.com/environment/2018/10/drought-climate-change-force-guatemalans-migrate-to-us/
109. https://www.globalcitizen.org/fr/content/2-billion-climate-change-refugees-2100/
110. https://thinkprogress.org/global-temperatures-have-been-above-average-for-406-months-in-a-row-5d32b5faba51/

Chapter Seven

1. Genetic engineering high nicotine plots. Bates no. 3001191/1212. Available at: http://legacy.library.ucsf.edu/tid/hpl41f00. Accessed June 18, 2004.
2. Genetic engineering of tobacco for nicotine biosynthesis. RJ Reynolds. Bates no. 507029142/9143. Available at: http://legacy.library.ucsf.edu/tid/jtq34d00. Accessed June 18, 2004.
3. Conkling MA. The molecular genetics of nicotine biosynthesis in

tobacco altering nicotine content through genetic engineering. October 1, 1997. Philip Morris. Bates no. 2063655463. Available at: http://legacy.library.ucsf.edu/tid/wmf67e00. Accessed June 18, 2004.
4. Lewan T. Brazil's secret: crazy tobacco. RJ Reynolds. Bates no. 522608883/8891. Available at: http://legacy.library.ucsf.edu/tid/xoc60d00. Accessed June 18, 2004.
5. Novotny TE, Zhao F. Consumption and production waste: another externality of tobacco use. Tob Control. 1999;8: 75–80. [PMC free article] [PubMed]
6. Skladanowski MA. Addition of nicotine extract to a blend. January 4, 1977. Lorillard. Bates no. 01305142/ 5145. Available at: http://legacy.library.ucsf.edu/tid/cvm61e00. Accessed June 18, 2004.
7. Brinkley A. Sam asked me to develop a process description for the high nicotine G-7 sheet and for the heat treated nicotine/extract G-7 sheet. September 27, 1990. RJ Reynolds. Bates no. 508902381/2385. Available at: http://legacy.library.ucsf.edu/tid/fmr83d00. Accessed June 18, 2004.
8. Amos Westmoreland 1991 objectives. RJ Reynolds. Bates no. 509304887/4888. Available at: http://legacy.library.ucsf.edu/tid/cor73d00. Accessed June 18, 2004.
9. KDN effluent/nicotine extract design. February 7, 1991. RJ Reynolds. Bates no. 512846285/6285. Available at: http://legacy.library.ucsf.edu/tid/osx23d00. Accessed June 18, 2004.
10. KDN effluent/nicotine extract design. February 26, 1991. RJ Reynolds. Bates no. 508029619/9619. Available at: http://legacy.library.ucsf.edu/tid/vfj13a00. Accessed June 18, 2004.
11. Flinchum G. German high nicotine extract. November 22, 1993. RJ Reynolds. Bates no. 510783956/3956. Available at: http://legacy.library.ucsf.edu/tid/bju53d00. Accessed June 18, 2004.
12. Additional investment and production cost calculations for the Winston-Salem cast sheet when a nicotine extract is added. December 8, 1993. RJ Reynolds. Bates no. 512596452/6454. Available at: http://legacy.library.ucsf.edu/tid/bfj33d00. Accessed June 18, 2004.
13. Charles JL, Davies B, DeNoble VJ, Horn JL, Mele PC. Behavioral pharmacology annual report. Philip Morris. Bates no. 2022144128/4211. Available at: http://legacy.library.ucsf.edu/tid/bly44e00. Accessed July 7, 2004.
14. DeNoble VJ, Harris CM, Horn J, Mele PC. Reinforcing activity of acetaldehyde [abstract]. Philip Morris. Bates no. 2071670753/0755. Available at: http://legacy.library.ucsf.edu/tid/can26c00. Accessed July 10, 2004.
15. Marjorie Jacobs, From the First to the Last Ash: The History, Economics & Hazards of Tobacco funded in 1993 by the Massachusetts Department of Public Health grant to The Cambridge Tobacco Education

Program, Cambridge Department of Human Service Programs.
16. Clark, Briggs & Cooke 2005, p. 1374
17. Prosecution Memorandum: Requesting a formal investigation by the United States Department Of Justice of the possible violation of federal criminal laws by named individuals and corporations in, or doing business with, the tobacco industry. Submitted by US Rep. Martin T Meehan to US Attorney General Janet Reno, 14 December 1994. http://www.stic.neu.edu/MN/6MMMEMO.HTM. Jones, Day, Reavis & Pogue. Fact Team Memorandum. 31 December 1985. Bates No.: 515873805-515873929.
18. C. White, Smoking and lung cancer: a landmark in the history of chronic disease epidemiology. Yale Journal of Biology and Medicine 1990 Jan-Feb; 63(1):29-46.
19. Teague C. Survey of cancer research. R. J. Reynolds Tobacco Co. 2 February 1953. Bates No. 501932947-501932968.
20. K M Cummings, C P Morley, A Hyland. Failed promises of the cigarette industry and its effect on consumer misperceptions about the health risks of smoking. Tob Control 2002;11:i110-i117 doi:10.1136/tc.11.suppl_1.i110.
21. Tucker, CA. (9/30/74). R. J. Reynolds. Presentation to the Board of Directors of RJR Industries
22. ^ 7 Cal. 4th 1057, 1073-74 (1994). R. J. Reynolds. Mangini v. R. J. Reynolds Tobacco Co.
23. http://www.youtube.com/watch?v=x4c_wI6kQyE.
24. http://senate.ucsf.edu/tobacco executives1994congress.html
25. http://www.philipmorrisusa.com/DisplayPageWithTopic.asp?ID=60.
26. Burns, D., et al. Designed for Addiction: How the Tobacco Industry Has Made Cigarettes More Addictive, More Attractive to Kids and Even More Deadly. available at: http://www.tobaccofreekids.org/content/what_we_do/industry_watch/product_manipulation/2014_06_19_DesignedforAddiction_web.pdf
27. Rabinoff, M, et al., "Pharmacological and Chemical Effects of Cigarette Additives," American Journal of Publich Health, Nov. 2007
28. Burns, D., et al. Designed for Addiction: How the Tobacco Industry Has Made Cigarettes More Addictive, More Attractive to Kids and Even More Deadly. available at: http://www.tobaccofreekids.org/content/what_we_do/industry_watch/product_manipulation/2014_06_19_DesignedforAddiction_web.pdf
29. http://www.theguardian.com/business/2013/sep/07/tobacco-philip-morris-millions-delay-eu-legislation
30. http://www.reuters.com/investigates/special-report/pmi-who-fctc/
31. http://www.who.int/bulletin/archives/78(7)902.pdf
32. https://www.npr.org/sections/health-

shots/2017/11/27/566014966/in-ads-tobacco-companies-admit-they-made-cigarettes-more-addictive
33. https://www.theguardian.com/world/2017/jul/11/how-big-tobacco-has-survived-death-and-taxes
34.https://psychnews.psychiatryonline.org/doi/full/10.1176/appi.pn.2018.1b2
35. https://ehp.niehs.nih.gov/ehp2175/
36.http://pediatrics.aappublications.org/content/early/2018/03/01/peds.2017-3557
37. https://www.theverge.com/2017/11/16/16658358/vape-lobby-vaping-health-risks-nicotine-big-tobacco-marketing
38.http://journals.plos.org/plosone/article?id=10.1371/journal.pone.0173055
39.https://ecigarettes.surgeongeneral.gov/documents/2016_sgr_full_report_non-508.pdf
40. http://www.tobaccoharmreduction.org/wpapers/006v1.pdf
41. https://www.pmi.com/smoke-free-products/iqos-our-tobacco-heating-system
42. https://www.ncbi.nlm.nih.gov/pmc/articles/PMC4532619/
43. https://abcnews.go.com/Lifestyle/wireStory/cigarette-sellers-turn-scholarships-promote-brands-55739990
44. https://usrtk.org/hall-of-shame/why-you-cant-trust-the-american-council-on-science-and-health/
45.https://apps.health.ny.gov/pubdoh/professionals/doctors/conduct/factions/FileDownloadAction.action?finalActionId=3432&fileName=lc116347.pdf&fileSeqNum=2
46. https://www.ncbi.nlm.nih.gov/pubmed/28880788
47. https://www.nap.edu/catalog/24952/public-health-consequences-of-e-cigarettes
48. https://www.vox.com/science-and-health/2018/1/23/16923070/nas-report-e-cigarettes-health-risks
49. https://www.reuters.com/article/us-smoking/smoking-could-kill-1-billion-this-century-who-idUSBKK25206020070702
50. https://ehp.niehs.nih.gov/ehp3451/
51.http://journals.plos.org/plosone/article?id=10.1371/journal.pone.0198047
52.https://www.ncbi.nlm.nih.gov/pmc/articles/PMC4363846/
53. https://abcnews.go.com/beta-story-container/Business/wireStory/altria-spending-128b-stake-vapor-company-juul-59928844
54. https://www.drugabuse.gov/related-topics/trends-statistics/monitoring-future
55. https://www.nytimes.com/2018/12/18/health/vaping-nicotine-

teenagers.html?fbclid=IwAR3x9jgbXnOUmkYZ_RoITFPsAFhSuGvChs7Hvzv3YHxZuU5gCRt7KrAwrXQ
56. https://cei.org/content/fear-profiteers

Chapter Eight

1. http://www.theguardian.com/environment/2013/oct/14/gm-crops-is-opposition-to-golden-rice-wicked
2. http://www.monbiot.com/2004/03/09/seeds-of-distraction/
3. http://www.spiegel.de/international/world/monsanto-papers-reveal-company-covered-up-cancer-concerns-a-1174233.html
4. "MSNBC," January 23, 2004. "Study Finds Link Between Agent Orange, Cancer." The Globe and Mail, June 12, 2008. "Last Ghost of the Vietnam War."
5. http://www.dailymail.co.uk/news/article-2401378/Agent-Orange-Vietnamese-children-suffering-effects-herbicide-sprayed-US-Army-40-years-ago.html
6. http://www.nytimes.com/1998/10/25/magazine/playing-god-in-the-garden.html?pagewanted=all&src=pm
7. David Schubert, "A different perspective on GM food," Nature Biotechnology, Vol. 20, 2002, p. 969
8. https://www.cancer.org/latest-news/study-finds-sharp-rise-in-colon-cancer-and-rectal-cancer-rates-among-young-adults.html
9. Netherwood et al, "Assessing the survival of transgenic plant DNA in the human gastrointestinal tract," Nature Biotechnology 22 (2004): 2.
10. Arisa A, Leblanc S. Maternal and fetal exposure to pesticides associated to genetically modified foods in Eastern Townships of Quebec, Canada Reproductive Toxicology Volume 31, Issue 4, May 2011, Pages 528-533doi:10.1016/j.reprotox.2011.02.004
11.http://www.ensser.org/fileadmin/user_upload/Eng_PR_3_No_Consensus_lv_01.pdf
12. Mezzomo BP, Miranda-Vilela AL, Freire IdS, Barbosa LCP, Portilho FA, et al. (2013) Hematotoxicity of Bacillus thuringiensis as Spore-crystal Strains Cry1Aa, Cry1Ab, Cry1Ac or Cry2Aa in Swiss Albino Mice. J Hematol Thromb Dis 1: 104. doi:10.4172/jhtd.1000104
13.http://www.epa.gov/scipoly/sap/meetings/2000/october/octoberfinal.pdf
14.Bernstein I, Bernstein J, Miller M, Tiewzieva S, Bernstein D, Lummus Z, Selgrade M, Doerfler D, and Seligy V "Immune responses in farm workers after exposure to Bacillus thuringiensis pesticides" 1999 Environ Health Perspect 107,575-82
15. Smith, JM. Genetic Roulette. Fairfield: Yes Books.2007. p.10
16. Finamore A, Roselli M, Britti S, et al. Intestinal and peripheral immune

response to MON 810 maize ingestion in weaning and old mice. J Agric. Food Chem. 2008; 56(23):11533-11539.

17. Malatesta M, Boraldi F, Annovi G, et al. A long-term study on female mice fed on a genetically modified soybean:effects on liver ageing. Histochem Cell Biol. 2008; 130:967-977.

18. Velimirov A, Binter C, Zentek J. Biological effects of transgenic maize NK603xMON810 fed in long term reproduction studies in mice. Report-Federal Ministry of Health, Family and Youth. 2008.

19. Ewen S, Pustzai A. Effects of diets containing genetically modified potatoes expressing Galanthus nivalis lectin on rat small intestine.Lancet. 354:1353-1354.

20. Kilic A, Aday M. A three generational study with genetically modified Bt corn in rats: biochemical and histopathological investigation. Food Chem. Toxicol. 2008; 46(3):1164-1170.

21. Kroghsbo S, Madsen C, Poulsen M, et al. Immunotoxicological studies of genetically modified rice expression PHA-E lectin or Bt toxin in Wistar rats. Toxicology. 2008; 245:24-34.

22. Chang FC, Simcik MF, Capel PD (2011) Occurrence and fate of the herbicide glyphosate and its degradate aminomethylphosphonic acid in the atmosphere. Environ Toxicol Chem 30: 548-555

23. http://www.ithaka-journal.net/druckversionen/e052012-herbicides-urine.pdf

24. https://www.organicconsumers.org/news/ucsf-presentation-reveals-glyphosate-contamination-people-across-america#close

25.http://www.mailtribune.com/apps/pbcs.dll/article?AID=/20140413/OPINION/404130320/-1/NEWSMAP

26. Paganelli A, Gnazzo V, Acosta H, Lopez SL, Carrasco A. Glyphosate-based herbicides produce teratogenic effects on vertebrates by impairing retinoic acid signaling. Chemical Research in Toxicology, 23 (2010), pp. 1586–1595

27. Antoniou, M., Robinson, C., & Fagan, J. (2012). Teratogenic effects of glyphosate-based herbicides: Divergence of regulatory decisions from scientific evidence. Journal of Environmental Analytical Toxicology S4:006. http://www.omicsonline.org/2161-0525/2161-0525-S4-006.php?aid=7453

28. Gasnier, C. et al, Glyphosate-based herbicides are toxic and endocrine disruptors in human cell lines. Toxicology doi:10.1016/j.tox.2009.06.00688.

29. Clair, É., Mesnage, R., Travert, C., Séralini, G-É. A glyphosate-based herbicide induces necrosis and apoptosis in mature rat testicular cells in vitro, and testosterone decrease at lower levels. Toxicology in Vitro 26 (2) 269-279 2012.

30. de Liz Oliveira Cavalli VL et al. Roundup® Disrupted Male Reproductive Functions By Triggering Calcium-Mediated Cell Death In Rat Testis And Sertoli Cells104.

31. Yousef, M.I. et al. Toxic effects of carbofuran and glyphosate on semen characteristics in rabbits. J Environ Sci Health B. 1995 Jul; 30(4):513-34
32. Eriksson M, Hardell L, Carlberg M, Akerman M. Pesticide exposure as risk factor for non- Hodgkin lymphoma including histopathological subgroup analysis100. Int J.Cancer. 2008, 123:1657-1663
33. Thongprakaisang S, Thiantanawat A, Rangkadilok N, Suriyo T, Satayavivad J. Glyphosate induces human breast cancer cells growth via estrogen receptors. Food Chem Toxicol. 2013, 59C: 129-136. doi: 10.1016/j.fct.2013.05.057101
34. Alavanja, M.C.R., Ross, M.K., Bonner, M.R. Increased Cancer Burden Among Pesticide Applicators and Others Due to Pesticide Exposure. CA Cancer J Clin 2013 63 (2): 120– 142
35. https://www.nature.com/articles/srep39328
36. Ya-xing Gui et al. Glyphosate induced cell death through apoptotic and autophagic mechanisms. Neurotoxicology and Teratology 2012, 34 (3): 344–349
37. Pezzoli G, Cereda E "Exposure to pesticides or solvents and risk of Parkinson disease" Neurology 2013; 80: 2035-2041.
38. Ross S, McManus IC, Harrison V, Mason O. Neurobehavioral problems following low-level exposure to organophosphate pesticides: a systematic and meta-analytic review. Critical Reviews in Toxicology, Ahead of Print : Pages 1-24 (doi: 10.3109/10408444.2012.738645)
39. Mañas F et al. Genotoxicity of AMPA, the environmental metabolite of glyphosate, assessed by the Comet assay and cytogenetic tests. Ecotoxicol Environ Saf. 2009 72 (3):834-7. doi: 10.1016/j.ecoenv.2008.09.01932
40. Simoniello, M.F. et al. DNA damage in workers occupationally exposed to pesticide mixtures. J Appl Toxicol. 2008, 28 (8): 957-65. doi: 10.1002/jat.1361
41. Benedetti D, et al. Genetic damage in soybean workers exposed to pesticides: Evaluation with the comet and buccal micronucleus cytome assays. Mutation Research/Genetic Toxicology and Environmental Mutagenesis 2013, 752 (1–2): 28–33
42. Samsel A, Seneff S. Review: Glyphosate's Suppression of Cytochrome P450 Enzymes and Amino Acid Biosynthesis by the Gut Microbiome: Pathways to Modern Diseases. Entropy 2013, 15(4), 1416-1463; doi:10.3390/e15041416
43. http://voiceofrussia.com/2010/04/16/6524765.html/
44. Ya-xing Gui et al. Glyphosate induced cell death through apoptotic and autophagic mechanisms. Neurotoxicology and Teratology 2012, 34 (3): 344–349
45. Mesnage, R.; Defarge, N.; Spiroux de Vendômois, J. and Séralini, G.-E. (2014, in press): Major pesticides are more toxic to human cells than their declared active principles.

http://www.hindawi.com/journals/bmri/aip/179691/
46. http://www.theguardian.com/world/2012/oct/14/kidney-disease-killing-sugar-cane-workers-central-america
47. Jayasumana C, et al. Glyphosate, Hard Water and Nephrotoxic Metals: Are They the Culprits Behind the Epidemic of Chronic Kidney Disease of Unknown Etiology in Sri Lanka? Int. J. Environ. Res. Public Health 2014, 11(2), 2125-2147; doi:10.3390/ijerph110202125
48. http:// www.gmwatch.org/latest-listing/1-news- items/10585-why-gm-crops-are-dangerous
49. Shehata, A.A. et al. The Effect of Glyphosate on Potential Pathogens and Beneficial Members of Poultry Microbiota In Vitro. Current Microbiology 2013, 66 (4): 350-358
50. Krüger, M. et al. Glyphosate suppresses the antagonistic effect of Enterococcus spp. on Clostridium botulinum. Anaerobe 2013, Feb 6
51. Clair, E. et al. Effects of Roundup® and Glyphosate on Three Food Microorganisms: Geotrichum candidum, Lactococcus lactis subsp. cremoris and Lactobacillus delbrueckii subsp. bulgaricus. Curr Microbiol. Epub 2012 Feb 24
52. http://www.organic-systems.org/journal/81/8106.pdf
53. Ewen SW, Pusztai A. Effect of diets containing genetically modified potatoes expressing Galanthus nivalis lectin on rat small intestine. Lancet. 1999 Oct 16;354(9187):1353-4.
54. Molodecky N, and Kaplan G. Environmental Risk Factors for Inflammatory Bowel Disease. Gastroenterol Hepatol (N Y). May 2010; 6(5): 339–346.
55. Murch S, et al. Autism, inflammatory bowel disease, and MMR vaccine. The Lancet, Volume 351, Issue 9106, Page 908, 21 March 1998.
56. Walker S, et al. Identification of Unique Gene Expression Profile in Children with Regressive Autism Spectrum Disorder (ASD) and Ileocolitis. March 08, 2013DOI: 10.1371/journal.pone.0058058
57. Chen B. et al. Abnormal gastrointestinal histopathology in children With autism spectrum disorders. J Pediatr Gastroenterol Nutr. February 2011.
58. Pardo C, et al. Immunity, neuroglia and neuroinflammation in autism. International Review of Psychiatry, December 2005; 17(6): 485–495
59. http://www.ksdk.com/story/news/health/2014/02/27/birth-defects-spike-yakima-valley/5861527/
60. Fritz Kreiss: News Report Sunday 17 March 2013
61. https://publichealth.wustl.edu/events/food-fear-find-facts-todays-culture-alarmism/
62. http://www.aaemonline.org/gmopressrelease.html
63. http://www.ithaka-journal.net/herbizide-im-urin?lang=en
64. https://jamanetwork.com/journals/jama/article-

abstract/2658306?redirect=true
65. http://www.thelancet.com/journals/lanonc/article/PIIS1470-2045%2815%2970134-8/abstract
66. http://documents.foodandwaterwatch.org/doc/MonsantoReport.pdf
67. Mella M and Rissler J (2004), Gone to Seed: Transgenic Contaminates in the Traditional Seed Supply, Union of Concerned Scientists
68. http://documents.foodandwaterwatch.org/doc/MonsantoReport.pdf
69.http://documents.foodandwaterwatch.org/doc/PublicResearchPrivateGain.pdf
70.http://documents.foodandwaterwatch.org/doc/PublicResearchPrivateGain.pdf
71. Lesser, L, et al. "Relationship between Funding Source and Conclusion among Nutrition-Related Scientific Articles." PLOS MEDICINE.
72. Diels, Johan. "Association of financial or professional conflict of interest to research outcomes on health risks or nutritional assessment studies of genetically modified products." Food Policy. November 22, 2010 at 200 to 201.
73. http://www.momsacrossamerica.com/glyphosate_testing_results
74.http://inthesetimes.com/features/monsanto_epa_glyphosate_roundup_investigation.html
75. Waltz, Emily. "Under Wraps." Nature Biotechnology. October 2009 at 882.
76. Public comment to FIFRA Scientific Advisory Panel. Document ID:EPA-HQ-OPP-2008-0836-0043. Available at http://www.regulations, gov/#!documentDetail;D=EPA-HQ-OPP-2008-0836-0043.
77.http://www.mailtribune.com/apps/pbcs.dll/article?AID=/20140413/OPINION/404130320/-1/NEWSMAP
78. http://www.businessinsider.com/gm-pig-study-is-deeply-flawed-2013-6#ixzz2xy2eSbtx
79. http://www.forbes.com/sites/henrymiller/2013/07/17/you-can-put-lipstick-on-a-pig-study-but-it-still-stinks/
80. http://www.forbes.com/sites/henrymiller/2013/10/23/junk-science-attacks-on-important-products-and-technologies-diminishes-us-all/
81. http://www.forbes.com/sites/jonentine/2013/10/24/michael-pollan-promotes-denialist-anti-gmo-junk-science-brags-he-manipulates-new-york-times-editors/2/
82. http://www.popularmechanics.com/science/health/what-can-we-do-about-junk-science-16674140-2
83. "The sinister sacking of the world's leading GM expert – and the trail that leads to Tony Blair and the White House." By Andrew Rowell. Daily Mail July 7, 2003.
84. Lancet, Editorial, May 22, p1769).
85. Daily Mail, UK, 13 Sept 1999. See also BBC news 7 sept 1999

86. Ignacio Chapela, interview by John Ross Feb. 2004
87. http://www.boulderweekly.com/article-12640-you-are-confused-cs.html
88. Zhang L1, Hou D, Chen X, Li D, Zhu L, Zhang Y, Li J, Bian Z, Liang X, Cai X, Yin Y, Wang C, Zhang T, Zhu D, Zhang D, Xu J, Chen Q, Ba Y, Liu J, Wang Q, Chen J, Wang J, Wang M, Zhang Q, Zhang J, Zen K, Zhang CY. Exogenous plant MIR168a specifically targets mammalian LDLRAP1: evidence of cross-kingdom regulation by microRNA. Cell Res. 2012 Jan;22(1):107-26. doi: 10.1038/cr.2011.158. Epub 2011 Sep 20
89. http://www.independentsciencenews.org/health/seralini-and-science-nk603-rat-study-roundup/
90. Gilles-Eric Séralinia, Emilie Claira, Robin Mesnagea, Steeve Gressa, Nicolas Defargea, Manuela Malatestab, Didier Hennequinc, Joël Spiroux de Vendômoisa Long term toxicity of a Roundup herbicide and a Roundup-tolerant genetically modified maize. Food and Chemical Toxicology Volume 50, Issue 11, November 2012, Pages 4221–4231
91. http://www.endsciencecensorship.org/en/page/Statement#signed-by
92. http://www.loe.org/shows/segments.html?programID=13-P13-00049&segmentID=2
93. http://www.ehn.org/monsanto-glyphosate-cancer-smear-campaign-2509710888.html
94. https://www.reuters.com/article/us-monsanto-herbicide/monsanto-seeks-retraction-for-report-linking-herbicide-to-cancer-idUSKBN0MK2GF20150324
95. http://www.ehn.org/monsanto-takes-on-world-health-organization-2509721283.html?utm_source=EHN&utm_campaign=b5fcab2cab-EMAIL_CAMPAIGN_2017_11_25&utm_medium=email&utm_term=0_8573f35474-b5fcab2cab-99026613
96. http://www.thenation.com/blog/176863/twenty-six-countries-ban-gmos-why-wont-us#
97. http://image.guardian.co.uk/sys-files/Environment/documents/2011/10/19/GMOEMPEROR.pdf
98. Catherine Badgley et al., Organic Agriculture and the Global Food Supply, Cambridge Journals, 9 June 2006, Introduction, doi:10.1017/S1742170507001640.
99. Olivier De Schutter, Food Commodities Speculation and Food Price Crises, issue brief, Geneva, Switzerland: United Nations, 2010, p. 1-2, http://www.srfood.org/images/stories/pdf/otherdocuments/20102309_briefing_note_02_en.pdf (accessed 18 January 2011).
100. http://www.mailtribune.com/apps/pbcs.dll/article?AID=/20140413/OPINION/404130320/-1/NEWSMA
101. William Neuman & Andrew Pollack, Farmers Cope with Round-Up

Resistance Weeds, New York Times, 4th May 2010.
102. https://www.nytimes.com/2017/11/01/business/soybeans-pesticide.html
103. http://www.organic-center.org/science.pest.php?action=view&report_id=159.
104. "GM Crops: Global socio- economic and environmental impacts 1996- 2009" Graham Brookes and Peter Barfoot. PG Economics Ltd. UK. 2011
105. "Benefits of Bt cotton elude farmers in China" GM Watch, http://www.gmwatch.org/ latest-listing/1-news-items/13089
106. "Who Benefits from GM Crops? Feed the Biotech Giants, Not the World's Poor." Friends of the Earth International, February 2009
107.http://www.mailtribune.com/apps/pbcs.dll/article?AID=/20140413/OPINION/404130320/-1/NEWSMAP
108. http://stlouis.cbslocal.com/2012/01/17/argentina-claims-monsanto-contractor-abused-farm-workers/
109. Are there Benefits from the Cultivation of Bt cotton? Review of Agrarian Studies Vol 1(1) January- June 2011. Madhura Swaminathan* and Vikas Rawal
110. http://www.news.cornell.edu/stories/1999/04/toxic-pollen-bt-corn-can-kill-monarch-butterflies
111. http://www.globalresearch.ca/the-seeds-of-suicide-how-monsanto-destroys-farming/5329947 The Indian state of Maharashtra eventually banned Bt cotton in 2012
112. http://naturalsociety.com/monsanto-abusing-illegal-workers-in-slave-like-conditions/
113. http://www.truth-out.org/archive/item/92751:war-over-monsanto-gets-ugly
114. http://www.truth-out.org/archive/item/92751:war-over-monsanto-gets-ugly
115.http://inthesetimes.com/features/monsanto_epa_glyphosate_roundup_investigation.html
116.http://seattletimes.com/html/opinion/2023357601_egvallianatosoped epa13xml.html?cmpid=2628
117. Bøhna T, et al. Compositional differences in soybeans on the market: Glyphosate accumulates in Roundup Ready GM soybeans. Food Chemistry. Volume 153, 15 June 2014, Pages 207–215
118. http://www.biointegrity.org
119. http://www.nytimes.com/2001/01/25/business/25FOOD.html
120. http://freedomoutpost.com/2013/06/5-million-farmers-locked-in-lawsuit-against-monsanto/#FSTU1EK5YVwB7K5D.99
121. http://www.youtube.com/watch?v=lPLfVfZGsls
122. http://rt.com/usa/monsanto-glyphosate-roundup-epa-483/

123. http://www.theecologist.org/News/news_analysis/2337631 extreme_levels_of_roundup_are_the_norm_in_gmo_soya.html
124. http://www.nydailynews.com/news/national/u-s-state-dept-helped-promote-monsanto-products-overseas-article-1.1343801
125. https://www.theguardian.com/environment/2017/oct/18/warning-of-ecological-armageddon-after-dramatic-plunge-in-insect-numbers
126.http://journals.plos.org/plosone/article?id=10.1371/journal.pone.0185809
127.https://web.archive.org/web/20170220012554/https://www.forbes.com/sites/henrymiller/2015/03/20/march-madness-from-the-united-nations/#4443f5672e93 127.
128. https://www.nytimes.com/2017/08/01/business/monsantos-sway-over-research-is-seen-in-disclosed-emails.html?mcubz=1&_r=0
129. http://www.newsweek.com/campaign-organic-food-deceitful-expensive-scam-785493
130. https://www.wsj.com/articles/gene-editing-is-here-and-desperate-patients-want-it-1507847260
131. https://ehp.niehs.nih.gov/ehp3127/
132. Sudo N, Chida Y, Aiba Y, Sonoda J, Oyama N, Yu X-N, et al. 2004. Postnatal microbial colonization programs the hypothalamic-pituitary-adrenal system for stress response in mice. J Physiol (Lond) 558(Pt 1):263–275, PMID: 15133062, 10.1113/jphysiol.2004.063388.
133. Clarke G, Grenham S, Scully P, Fitzgerald P, Moloney RD, Shanahan F, et al. 2013. The microbiome-gut-brain axis during early life regulates the hippocampal serotogenic system in a sex-dependent manner. Mol Psychiatry 18(6):666–673, PMID: 22688187, 10.1038/mp.2012.77.
134. Hsiao EY. 2014. Gastrointestinal issues in autism spectrum disorder. Harv Rev Psychiatry 22(2):104–111, PMID: 24614765, 10.1097/HRP.0000000000000029.
135. Kang DW, Adams JB, Gregory AC, Borody T, Chittick L, Fasano A, et al. 2017. Microbiota Transfer Therapy alters gut ecosystem and improves gastrointestinal autism symptoms: an open label study. Microbiome 5(1):10, PMID: 28122648, 10.1186/s40168-016-0225-7.
136. Roegge CS, Timofeeva OA, Seidler FJ, Slotkin TA, Levin ED. 2008. Developmental diazinon neurotoxicity in rats: later effects on emotional response. Brain Res Bull 75(1):166–172, PMID: 18158111, 10.1016/j.brainresbull.2007.08.008.
137. Slotkin TA, Ryde IT, Levin ED, Seidler FJ. 2008. Developmental neurotoxicity of low dose diazinon exposure of neonatal rats: effects on serotonin systems in adolescence and adulthood. Brain Res Bull 75(5):640–647, PMID: 18355640, 10.1016/j.brainresbull.2007.10.008.
138. Timofeeva OA, Roegge CS, Seidler F, Slotkin TA, Levin ED. 2008. Persistent cognitive alterations in rats after early postnatal exposure to low

doses of the organophosphate pesticide, diazinon. Neurotoxicol Teratol 30(1):38–45, PMID: 18096363, 10.1016/j.ntt.2007.10.002.
139. https://gmoanswers.com/ask/hi-does-senior-monsanto-scientist-dan-goldstein-still-maintain-if-ingested-glyphosate-excreted
140.http://www.centerforfoodsafety.org/files/monsantoexsum1142005.pdf
141. https://truthout.org/articles/monsanto-in-haiti/
142.https://experiencelife.com/article/frankenfood-genetically-modified-foods/
143. http://omicsonline.org/open-access/detection-of-glyphosate-residues-in-animals-and-humans-2161-0525.1000210.pdf
144. http://bookstore.teri.res.in/e_issue_text_1.php?oj_id=59§or=119
145.https://www.soilassociation.org/media/7202/glyphosate-and-soil-health-full-report.pdf
146.https://www.theguardian.com/environment/2017/oct/18/warning-of-ecological-armageddon-after-dramatic-plunge-in-insect-numbers
147. https://www.ncbi.nlm.nih.gov/pmc/articles/PMC5756058/
148.https://www.huffingtonpost.com/entry/insect-population-decline-extinction_us_5c611921e4b0f9e1b17f097d
149. Sánchez-Bayo F, et al. [a] Worldwide decline of the entomofauna: A review of its drivers. Biological Conservation. Volume 232, April 2019, Pages 8-27

Chapter Nine

1. http://www.upi.com/Science_News/2013/06/22/Oregon-bumblebee-die-off-surpasses-50000/UPI-41391371946087/
2. http://rspb.royalsocietypublishing.org/content/274/1608/303.full
3.http://www.un.org/apps/news/story.asp?NewsID=37731#.UdG2ps3B_JE
4. Mullin CA, Frazier M, Frazier JL, et al., 2010. High levels of miticides and agrochemicals in North American apiaries: implications for honey bee health.PLoS One. 5(3):e9754.
5. Pettis JS, Lichtenberg EM, Andree M, Stitzinger J, Rose R, et al. (2013) Crop Pollination Exposes Honey Bees to Pesticides Which Alters Their Susceptibility to the Gut Pathogen Nosema ceranae. PLoS ONE 8(7): e70182. doi:10.1371/journal.pone.0070182
6. http://unric.org/en/uk-a-ireland-news-archive/28303-more-than-honey-the-plight-of-the-honey-bee
7. http://science.sciencemag.org/content/356/6345/1395
8. Tsvetkov1 N, et al. Chronic exposure to neonicotinoids reduces honey bee health near corn crops. Science 30 Jun 2017:
Vol. 356, Issue 6345, pp. 1395-1397. DOI: 10.1126/science.aam7470

9. http://www.xerces.org/wp-content/uploads/2016/10/HowNeonicsCanKillBees_XercesSociety_Nov2016.pdf
10. Woodcock BA, et al. Country-specific effects of neonicotinoid pesticides on honey bees and wild bees. Science 30 Jun 2017: Vol. 356, Issue 6345, pp. 1393-1395
11. David Vogel, The Politics of Precaution, Princeton University Press, 2012
12.http://ec.europa.eu/food/animals/live_animals/bees/study_on_mortality/index_en.htm
13. http://www.cropscience.bayer.com/en/Media/Press-Releases/2014/Bayer-CropScience-opens-North-American-Bee-Care-C
14. http://events.unl.edu/acreage/2013/04/01/76879/
15. http://viacampesina.org/en/index.php/main-issues-mainmenu-27/biodiversity-and-genetic-resources-mainmenu-37/441-syngenta-murder-and-private-militias-in-brazil
16. http://www.bostonglobe.com/magazine/2013/06/22/the-harvard-scientist-linking-pesticides-honeybee-colony-collapse-disorder/nXvIA5I6IcxFRxEOc8tpFI/story.html
17. http://www.scientificamerican.com/article/robobee-project-building-flying-robots-insect-size/
18. Hallmayer J, Cleveland S, Torres A, et al. "Genetic Heritability and Shared Environmental Factors Among Twin Pairs With Autism," Arch Gen Psychiatry. 2011;68(11):1095-1102. doi:10.1001/archgenpsychiatry.2011.76.
19. James SJ, Slikker W, Melnyk S, New E, Pogribna M, Jernigan S. "Thimerosol Neurotoxicity is Associated with Glutathione Depletion: Protection with Glutathione Precursors," NeuroToxicology 26.(2005) 1-8.
20. Goodrich AJ, et al. Joint effects of prenatal air pollutant exposure and maternal folic acid supplementation on risk of autism spectrum disorder. Autism Res. 2017 Nov 9. doi: 10.1002/aur.1885. [Epub ahead of print]
21. Adams J, Baral M, Geis E, et al. "The Severity of Autism Is Associated with Toxic Metal Body Burden and Red Blood Cell Glutathione Levels," Journal of Toxicology Volume 2009.(2009), Article ID 532640, 7 pages. doi:10.1155/2009/532640.
22. Croen L, Grether J, Yoshida C, Odouli R, Hendrick V, "Antidepressant Use During Pregnancy and Childhood Autism Spectrum Disorders," Arch Gen Psychiatry. 2011;68(11):1104-1112. doi:10.1001/archgenpsychiatry.2011.73
23. Volk H, Hertz-Picciotto I, Delwiche L , Lurmann F, McConnell R. "Residential Proximity to Freeways and Autism in the CHARGE study," Environ Health Perspect. 2010 December 13. (Epub ahead of print.) PMID: 21156395.

24. Whyatt RM, Liu X, Rauh VA, Calafat AM, Just AC, Hoepner L, et al. 2011. "Maternal Prenatal Urinary Phthalate Metabolite Concentrations and Child Mental, Psychomotor and Behavioral Development at 3 Years of Age," Environ Health Perspect 120:290-295.
25. Kern J, Geier D, Adams J, Mehta J, Grannemann B, Geier M. "Toxicity biomarkers in autism spectrum disorder: A blinded study of urinary porphyrins," Pediatrics International. (2011) 53, 147–153 doi: 10.1111/j.1442-200X.2010.03196.x.
26. Miodovnik, A, SM Engel, C Zhu, X Ye, LV Soorya, MJ Silva, AM Calafat and MS Wolff. 2011. "Endocrine disruptors and childhood social impairment," Neurotoxicology.
27. Roberts, EM et al. "Maternal residence near agricultural pesticide applications and autism spectrum disorders among children in the California Central Valley," Environmental Health Perspectives. 115(10):1482-1489.
28. Henrik Viberg anders Fredriksson, Sonja Buratovic, Per Eriksson. "Dose-dependent behavioral disturbances after a single neonatal Bisphenol A dose," Toxicology, 2011; DOI: 10.1016/j.tox.2011.09.006.
29. Whyatt RM, Liu X, Rauh VA, Calafat AM, Just AC, Hoepner L, et al. 2011. "Maternal Prenatal Urinary Phthalate Metabolite Concentrations and Child Mental, Psychomotor and Behavioral Development at Age Three Years," Environ Health Perspect.
30. Holmes AS, Blaxill MF, Haley BE; "Reduced levels of mercury in first baby haircuts of autistic children," Int J Toxicol. 2003 Jul-Aug;22(4):277-85.
31. Allen J, Shanker G, Tan K, Aschner M. "The Consequences of Methylmercury Exposure on Interactive Functions between Astrocytes and Neurons," Neurotoxicology 23.(2002) 755-759
32. "Body Burden - The Pollution in Newborns," Environmental Working Group, 2005.
33. Woodruff TJ, Zota AR, Schwartz JM 2011. "Environmental Chemicals in Pregnant Women in the United States: NHANES 2003-2004," Environ Health Perspect 119:878-885.
34. M. Henry et al. "A common pesticide decreases foraging success and survival in honey bees," Science. doi: 10.1126/science.1215039.
35. P.R. Whitehorn et al. "Neonicotinoid pesticide reduces bumble bee colony growth and queen production," Science. doi: 10.1126/science.1215025.
36. C. Lu, K.M. Warchol and R.A. Callahan. "In situ replication of honey bee colony collapse disorder," Bulletin of Insectology, Vol. 65, June 2012
37. Lu C, Warchol K, Callahan R. Sub-lethal exposure to neonicotinoids impaired honey bees winterization before proceeding to colony collapse disorder. Bulletin of Insectology 67 (1): 125-130, 2014
38. http://www.theguardian.com/environment/2014/may/09/honeybees-

dying-insecticide-harvard-study
39. http://onlinelibrary.wiley.com/doi/10.1111/1365-2435.12292/full
40. http://www.ithaka-journal.net/herbizide-im-urin?lang=en
41. http://www.thelancet.com/journals/laneur/article/PIIS1474-4422%2813%2970278-3/abstract
42. http://www.scientificamerican.com/article/autism-rise-driven-by-environment/
43. http://www.momsacrossamerica.com/glyphosate_testing_results
44. gmoanswers.com/ask/given-glyphosate-lipid-soluble-and-knowing-its-really-only-ingested-humans-through-gm-foods-how
45. http://www.nejm.org/doi/full/10.1056/NEJMoa1307491
46. Rauh V, Arunajadai S, Horton M, Perera F, Hoepner L, Barr DB, et al. 2011. Seven-Year Neurodevelopmental Scores and Prenatal Exposure to Chlorpyrifos, a Common Agricultural Pesticide. Environ Health Perspect 119:1196-1201. http://dx.doi.org/10.1289/ehp.1003160
47. Engel S, et al. Prenatal Exposure to Organophosphates, Paraoxonase 1, and Cognitive Development in Childhood. Environmental Health Perspectives, 2011; DOI: 10.1289/ehp.1003183
48. Horton M, et al. Impact of Prenatal Exposure to Piperonyl Butoxide and Permethrin on 36-Month Neurodevelopment. Pediatrics 2011; 127:3 e699-e706; doi:10.1542/peds.2010-0133
49. Horton M, Kahn L, Perera F, Barr D, Rauh V. Does the home environment and the sex of the child modify the adverse effects of prenatal exposure to chlorpyrifos on child working memory? Neurotoxicology and Teratology, 2012; DOI: 10.1016/j.ntt.2012.07.004
50. Ross S, McManus IC, Harrison V, Mason O. Neurobehavioral problems following low-level exposure to organophosphate pesticides: a systematic and meta-analytic review. Critical Reviews in Toxicology, Ahead of Print : Pages 1-24 (doi: 10.3109/10408444.2012.738645)
51. Rauh V, et al. Brain anomalies in children exposed prenatally to a common organophosphate pesticide. PNAS 2012 109 (20) 7871-7876; published ahead of print April 30, 2012, doi:10.1073/pnas.1203396109
52. Oulhote Y, Bouchard M, Urinary Metabolites of Organophosphate and Pyrethroid Pesticides and Behavioral Problems in Canadian Children Environ Health Perspect; DOI:10.1289/ehp.1306667
53. Ostrea EM, et al. 2011. Fetal exposure to propoxur and abnormal child neurodevelopment at two years of age. Neurotoxicology http://dx.doi.org/10.1016/j.neuro.2011.11.006.
54. Greenop K, Peters S, Bailey H, et al. Exposure to pesticides and the risk of childhood brain tumors. Cancer Causes & Control. April 2013
55. Kimura-Kuroda J, Komuta Y, Kuroda Y, Hayashi M, Kawano H (2012) Nicotine-Like Effects of the Neonicotinoid Insecticides Acetamiprid and Imidacloprid on Cerebellar Neurons from Neonatal Rats. PLoS ONE

7(2): e32432. doi:10.1371/journal.pone.003243
56. Pezzoli G, Cereda E. "Exposure to pesticides or solvents and risk of Parkinson disease" Neurology 2013; 80: 2035-2041.
57. Ross S, McManus IC, Harrison V, Mason O. Neurobehavioral problems following low-level exposure to organophosphate pesticides: a systematic and meta-analytic review. Critical Reviews in Toxicology, Ahead of Print : Pages 1-24 (doi: 10.3109/10408444.2012.738645)
58. Jason R. Richardson, PhD1,2; Ananya Roy, ScD2; Stuart L. Shalat, ScD1,2; Richard T. von Stein, PhD2; Muhammad M. Hossain, PhD1,2; Brian Buckley, PhD2; Marla Gearing, PhD4; Allan I. Levey, MD, PhD3; Dwight C. German, PhD5 Elevated Serum Pesticide Levels and Risk for Alzheimer Disease JAMA Neurol. Published online January 27, 2014. doi:10.1001/jamaneurol.2013.6030
59. http://www.pnas.org/content/114/36/9653
60. http://www.sciencemag.org/news/2017/07/antisocial-bees-share-genetic-profile-people-autism
61.https://www.motherjones.com/politics/2013/10/american-council-science-health-leaked-documents-fundraising/

Chapter Ten

1. — Memo from a trustee of the Manville Trust, 1988
2. http://www.ewg.org/research/asbestos-think-again/industry-hid-dangers-decades
3. Fatal Deception: The Terrifying True Story of How Asbestos Is Killing America. Michael Bowker, pg. 171
4. https://static.ewg.org/reports/asbestos/documents/pdf/BX-091266.pdf?_ga=2.25908671.303125222.1530475214-521960154.1430338886
5. https://www.washingtonpost.com/archive/politics/1978/11/12/new-data-on-asbestos-indicate-cover-up-of-effects-on-workers/028209a4-fac9-4e8b-a24c-50a93985a35d/?noredirect=on&utm_term=.2c4a2a26ee03
6. https://www.washingtonpost.com/archive/politics/1978/11/12/new-data
7. The American Corporation Today, p. 346, Carl Kaysen
8. Source: Insurance industry memo 10/09/75
9. https://www.ncbi.nlm.nih.gov/pmc/articles/PMC5497111/
10. https://www.asbestos.com/exposure/secondary/
11. https://www.cleveland.com/court-justice/index.ssf/2014/01/tri-c_professor_with_lung_canc.html
12. https://insiderexclusive.com/3677-2/
13. http://www.asbestosnation.org/wp-content/uploads/2015/02/doc_1973_ATI_Full_EX3.gif

14. http://www.asbestosnation.org/wp-content/uploads/2015/02/doc_FMSI121076.gif
15. http://www.asbestosnation.org/wp-content/uploads/2015/02/doc_AusternEX2.gif
16. https://www.gpo.gov/fdsys/pkg/CFR-2011-title40-vol31/pdf/CFR-2011-title40-vol31-part763-subpartI.pdf
17. https://openjurist.org/947/f2d/1201/corrosion-proof-fittings-v-environmental-protection-agency
18. http://www.motherjones.com/environment/2000/05/libbys-deadly-grace
19. https://www.theguardian.com/world/2009/mar/08/usa-mining-libby-montana
20. http://www.motherjones.com/environment/2000/05/libbys-deadly-grace
21. https://www.motherjones.com/environment/2000/05/libbys-deadly-grace/4/
22. http://blogs.wsj.com/five-things/2014/02/03/5-takeaways-from-the-w-r-grace-bankruptcy/
23. https://blogs.wsj.com/briefly/2014/02/03/5-takeaways-from-the-w-r-grace-bankruptcy/
24. http://www.newstatesman.com/sci-tech/2014/03/killer-dust-why-asbestos-still-killing-people
25. https://www.washingtonpost.com/archive/politics/1978/11/12/new-data-on-asbestos-indicate-cover-up-of-effects-on-workers/028209a4-fac9-4e8b-a24c-50a93985a35d/?utm_term=.85f6564e8dcb
26. https://www.washingtonpost.com/archive/politics/1978/11/12/new-data-on-asbestos-indicate-cover-up-of-effects-on-workers/028209a4-fac9-4e8b-a24c-50a93985a35d/?utm_term=.172244a0550d
27.https://www.businesswire.com/news/home/20180416005884/en/ADAO-Announces-New-Findings-Show-Asbestos-Related-Deaths
28. https://www.rollingstone.com/politics/politics-news/trump-asbestos-707642/
29. http://www.mesothelioma-help-network.com/mesothelioma/articles/history_of_asbestos/index.html

Chapter Eleven

1. http://www.cowspiracy.com/facts/
2. https://www.alternet.org/food/university-oxford-has-disturbingly-cozy-connection-monsanto
3. http://time.com/3035872/sixth-great-extinction/
4. http://www.newscientist.com/article/mg20227024.400-rainforests-may-pump-winds-worldwide.html

5. https://www.scientificamerican.com/article/earth-talks-daily-destruction/
6. http://www.mcspotlight.org/media/reports/beyond.html
7. Kaimowitz et al. 2004. Hamburger Connection Fuels Amazon Destruction: Cattle ranching and deforestation in Brazil's Amazon. Centre for International Forestry Research (CIFOR). 10 pp.
8. http://www.nature.com/news/one-third-of-our-greenhouse-gas-emissions-come-from-agriculture-1.11708
9. http://www.scientificamerican.com/article/the-greenhouse-hamburger/
10. http://blueandgreentomorrow.com/2014/04/01/eat-less-meat-and-dairy-to-dodge-climate-disaster-urges-study/
11. http://www.scientificamerican.com/article/the-greenhouse-hamburger/
12. http://www.waterfootprint.org/
13. http://www.telegraph.co.uk/earth/8359076/US-farmers-fear-the-return-of-the-Dust-Bowl.html
14.http://www.washingtonpost.com/blogs/govbeat/wp/2014/07/24/study-colorado-river-basin-drying-up-faster-than-previously-thought/
15. http://www.theguardian.com/environment/2014/jul/21/giving-up-beef-reduce-carbon-footprint-more-than-cars
16. http://link.springer.com/article/10.1007%2Fs10584-014-1169-1
17. http://www.pnas.org/content/early/2014/07/17/1402183111
18. http://www.theguardian.com/environment/2014/jul/21/giving-up-beef-reduce-carbon-footprint-more-than-cars
19. http://shrinkthatfootprint.com/food-carbon-footprint-diet
20. http://www.businessinsider.com/the-35-companies-that-spent-1-billion-on-ads-in-2011-2012-11?op=1
21. http://www.pcrm.org/health/diets/vegdiets/how-can-i-get-enough-protein-the-protein-myth
22. http://news.harvard.edu/gazette/2003/03.13/09-kidney.html
23. Butler LM, Sinha R, Millikan RC, et al. Heterocyclic amines, meat intake, and association with colon cancer in a population-based study. Am J Epidemiol. 2003;157:434-445.
24. Sinha R. An epidemiologic approach to studying heterocyclic amines. Mutat Res. 2002;506:197.
25. Zheng W, Lee SA. Well-done meat intake, heterocyclic amine exposure, and cancer risk. Nutr and Cancer. 2009;61:437-446.
26. http://healthfree.com/nutritional_power_myth.html
27. Bernstein, A.M., et al., Major dietary protein sources and risk of coronary heart disease in women. Circulation, 2010. 122(9): p. 876-83.
28. Aune, D., G. Ursin, and M.B. Veierod, Meat consumption and the risk of type 2 diabetes: a systematic review and meta-analysis of cohort studies. Diabetologia, 2009. 52(11): p. 2277-87.

References

29. Pan, A., et al., Red meat consumption and mortality: results from 2 prospective cohort studies. Arch Intern Med, 2012. 172(7): p. 555-63.
30. http://www.health.harvard.edu/fhg/updates/Red-meat-and-colon-cancer.sht
31. Pan, A., et al., Red meat consumption and risk of type 2 diabetes: 3 cohorts of US adults and an updated meta-analysis. Am J Clin Nutr, 2011. 94(4): p. 1088-96.
32. http://www.columbia.edu/~lnp3/mydocs/ecology/cattle.htm
33.http://www.theecologist.org/News/news_analysis/2337631/extreme_levels_of_roundup_are_the_norm_in_gmo_soya.html
34. http://omicsonline.org/open-access/detection-of-glyphosate-residues-in-animals-and-humans-2161-0525.1000210.pdf
35. http://www.who.int/mediacentre/factsheets/fs225/en/
36. https://www.niehs.nih.gov/health/topics/agents/dioxins/index.cfm
37. http://www.nytimes.com/2014/03/09/opinion/sunday/the-fat-drug.html?_r=0
38. http://mbio.asm.org/content/3/5/e00190-12.full
39.http://news.google.com/newspapers?nid=1915&dat=19970707&id=jwchAAAAIBAJ&sjid=w3YFAAAAIBAJ&pg=4806,1159584
40. http://www.vibrancyuk.com/dangersmeat.html
41. Jeremy Rifkin, Beyond Beef, published 1992.
42. https://www.ncbi.nlm.nih.gov/pubmed/23261125
43.http://documents.foodandwaterwatch.org/doc/CarbonMonoxide_web.pdf
44.http://preventcancer.aicr.org/site/News2?page=NewsArticle&id=15642&news_iv_ctrl=0&abbr=pr_
45. http://organicconsumers.org/foodsafety/processedmeat050305.cfm
46. http://www.seacoastonline.com/articles/20130602-LIFE-306020316
47. https://www.youtube.com/watch?v=hXXrB3rz-xU
48. http://blog.sfgate.com/hleon/2011/11/04/the-mcrib-sandwich-and-a-yoga-mat-what-do-they-have-in-common/
49. http://consumerist.com/2014/08/13/pink-slime-plant-reopens-because-high-beef-prices-mean-theres-a-need-for-cheaper-ingredients/
50. http://www.nytimes.com/2012/04/05/opinion/kristof-arsenic-in-our-chicken.html?_r=2&nl=todaysheadlines&emc=edit_th_20120405&
51. https://www.epa.gov/ghgemissions/global-greenhouse-gas-emissions-data
52. https://www.ecowatch.com/oxford-university-monsanto-2506308904.html
53.http://www.washingtonpost.com/wp-dyn/content/article/2009/06/25/AR2009062503381.html
54.https://www.weforum.org/agenda/2018/11/deforestation-in-the-brazilian-amazon-reaches-decade-high

55. http://www.mightyearth.org/mysterymeat/
56. https://www.ucsusa.org/press/2016/burger-king-rbi-sustainability-framework-tropical-forests#.XGowLy2ZO7o
57.https://www.eco-business.com/opinion/a-most-unlikely-hope-how-the-companies-that-destroyed-the-worlds-forests-can-save-them/
58. https://www.smh.com.au/world/multinationals-carving-up-africa-for-food-20131229-301jk.html

Chapter Twelve

1. http://thehill.com/homenews/house/353714-scalise-shooting-fortified-view-on-gun-rights
2. https://everytownresearch.org/gun-violence-by-the-numbers/
3. https://www.vox.com/policy-and-politics/2015/12/8/9870240/gun-ownership-deaths-homicides
4. http://beta.latimes.com/science/sciencenow/la-sci-sn-united-states-mass-shooting-20150824-story.html
5. https://www.theguardian.com/us-news/2017/nov/15/the-gun-numbers-just-3-of-american-adults-own-a-collective-133m-firearms
6. https://www.npr.org/2017/10/10/556578593/the-nra-wasnt-always-against-gun-restrictions
7. https://www.theguardian.com/us-news/2015/dec/14/inside-the-nra-the-officials-keeping-gun-control-laws-off-the-us-agenda
8. http://www.businessinsider.com/gun-industry-funds-nra-2013-1
9. https://www.alternet.org/tea-party-and-right/how-gun-industry-made-fortune-stoking-fears-obama-would-take-peoples-guns-ammo
10. https://www.thenation.com/article/does-nra-represent-gun-manufacturers-or-gun-owners/
11. https://www.npr.org/2011/01/27/133247508/the-history-and-growing-influence-of-the-nra
12. https://www.nytimes.com/2016/01/07/business/after-mass-shootings-some-on-wall-st-cash-in-on-gun-shares.html?_r=0
13. http://fortune.com/2017/09/11/trump-gun-sales-decline/
14. https://www.democracynow.org/2017/10/3
after_las_vegas_massacre_republicans_in
15. https://www.theatlantic.com/business/archive/2017/10/gun-sales-mass-shooting/541809/
16. https://www.alternet.org/tea-party-and-right/how-gun-industry-made-fortune-stoking-fears-obama-would-take-peoples-guns-ammo
17. https://www.theguardian.com/commentisfree/2014/apr/28/nra-war-on-america-wayne-lapierre-indianapolis

18. http://beta.latimes.com/science/sciencenow/la-sci-sn-united-states-mass-shooting-20150824-story.html
19. http://news.gallup.com/poll/20098/gun-ownership-use-america.aspx
20. http://www.pewresearch.org/fact-tank/2017/02/21/5-facts-about-crime-in-the-u-s/
21. http://www.hoplofobia.info/wp-content/uploads/2014/05/Tragic-but-not-Random-The-Social-Contagion-of-nonfatal-gunshot-injuries.pdf
22. http://fortune.com/2017/09/11/trump-gun-sales-decline/
23. https://www.usatoday.com/story/news/politics/2017/02/28/trump-sign-bill-blocking-obama-gun-rule/98484106/
24. http://money.cnn.com/2017/09/27/news/companies/gun-exports-trump/index.html
25. https://www.nytimes.com/2016/01/07/business/after-mass-shootings-some-on-wall-st-cash-in-on-gun-shares.html?_r=0
26. http://www.cnn.com/2017/12/13/opinions/gun-sales-mass-shootings-opinion-levine-mcknight/index.html
27. https://www.businessinsider.com.au/nra-lobbying-money-national-rifle-association-washington-2012-12
28.https://www.opensecrets.org/outsidespending/recips.php?cmte=National+Rifle+Assn&cycle=2016
29. http://www.mcclatchydc.com/news/nation-world/national/article195231139.html
30. https://www.huffingtonpost.com/2011/01/18/poll-americans-gun-owners-stronger-laws_n_810069.html
31. https://www.theatlantic.com/international/archive/2017/10/australia-g
32. https://papers.ssrn.com/sol3/papers.cfm?abstract_id=1631130
33. https://www.nytimes.com/2015/12/05/world/australia/australia-gun-ban-shooting.html
34. https://www.theatlantic.com/international/archive/2017/10/australia-gun-control/541710/
35. https://www.huffingtonpost.com/2011/01/18/poll-americans-gun-owners-stronger-laws_n_810069.html
36. http://www.motherjones.com/politics/2012/12/mass-shootings-mother-jones-full-data/
37. https://www.alecexposed.org/wiki Guns,_Prisons,_Crime,_and_Immigration
38. https://www.thenation.com/article/how-alec-thwarts-honest-debate-about-gun-violence/
39. https://everytownresearch.org/fact-sheet-stand-your-ground/
40. Cheng, C., & Hoekstra, M. (2012). Does strengthening self-defense law deter crime or escalate violence? Evidence from Castle Doctrine. Cambridge, MA: National Bureau of Economic Research.

41. Humphreys, D., Gasparrini, A., Wiebe, D. (2016). Evaluating the Impact of Florida's "Stand Your Ground" Self-defense Law on Homicide and Suicide by Firearm: An Interrupted Time Series Study. JAMA Internal Medicine. Published online November 14, 2016.
42. Fisher, M., & Eggen, D. (2012, April 7). 'Stand your ground' laws coincide with jump in justifiable-homicide cases. Washington Post. Retrieved from http://wapo.st/2fZbeSM..
43. http://beta.latimes.com/business/hiltzik/la-fi-hiltzik-gun-research-funding-20160614-snap-story.html
44. https://www.washingtonpost.com/opinions/we-wont-know-the-cause-of-gun-violence-until-we-look-for-it/2012/07/27/gJQAPfenEX_story.html?utm_term=.33ba3c8ec784
45. https://www.scientificamerican.com/article/gun-science-proves-arming-untrained-citizens-bad-idea/
46. Garen J. Wintemute, Guns, Fear, the Constitution, and the Public's Health, 358 New England J. Med. 1421-1424 (Apr. 2008).
47. Linda L. Dahlberg et al., Guns in the Home and Risk of a Violent Death in the Home: Findings from a National Study, 160 Am. J. Epidemiology 929, 935 (2004)
48. Charles C. Branas, et al, Investigating the Link Between Gun Possession and Gun Assault, 99 Am. J. Pub. Health 2034 (Nov. 2009), at http://www.ncbi.nlm.nih.gov/pmc/articles/PMC2759797/pdf/2034.pdf.
49. http://lawcenter.giffords.org/dangers-of-gun-use-for-self-defense-statistics/
50. http://ajph.aphapublications.org/doi/abs/10.2105/AJPH.2013.301409
51. https://www.scientificamerican.com/article/gun-science-proves-arming-untrained-citizens-bad-idea/
52. J.C. Campbell and others, "Risk factors for femicide within physically abusive intimate relationships: results from a multi-site case control study," American Journal of Public Health 93 (7) (2003): 1089–1097.
53. Center for American Progress analysis of the National Crime Victimization Survey. While guns were used for self-defense in 85,000 crimes per year from 2010 to 2015, roughly 162,000 guns are stolen each year.
54. Federal Bureau of Investigation, "A Study of Active Shooter Incidents in the United States Between 2000 and 2013," available at https://www.fbi.gov/news/stories/fbi-releases-study-on-active-shooter-incidents (last accessed October 2017).
55. Everytown For Gun Safety, "Analysis of Mass Shootings" (2017), available at https://everytownresearch.org/reports/mass-shootings-analysis/
56. http://www.apa.org/news/press/releases/2016/08/media-contagion-effect.pdf

57. http://journals.sagepub.com/doi/abs/10.1177/0002764217739660
58. https://www.psychologytoday.com/blog/saving-normal/201405/the-mind-the-mass-murderer
59. https://www.cnn.com/2018/02/21/politics/trump-listening-sessions-parkland-students/index.html
60. Chelsea Parsons and Eugenio Weigend, "America Under Fire" (Washington: Center for American Progress, 2016), available at https://www.americanprogress.org/issues/guns-crime/reports/2016/10/11/145830/america-under-fire/.
61. https://www.scientificamerican.com/article/gun-science-proves-arming-untrained-citizens-bad-idea/
62. John Donohue, Abhay Aneja and Kyle Weber, "Right-to-Carry Laws and Violent Crime: A Comprehensive Assessment Using Panel Data and a State-Level Synthetic Controls Analysis." Working Paper 23510 (The National Bureau of Economic Research 2017), available at http://www.nber.org/papers/w23510.
63. Daniel Webster, Cassandra Kercher Crifasi, and Jon S. Vernick, "Effects of the Repeal of Missouri's Handgun Purchaser Licensing Law on Homicides," Journal of Urban Health: Bulletin of the New York Academy of Medicine 91 (3) (2014): 293–302; Daniel Webster, Cassandra Kercher Crifasi, and Jon S. Vernick, "Erratum to: Effects of the Repeal of Missouri's Handgun Purchaser Licensing Law on Homicides," Journal of Urban Health: Bulletin of the New York Academy of Medicine 91 (3) (2014): 598–601.
64. https://www.ncbi.nlm.nih.gov/pmc/articles/PMC4211925/
65.https://ps.psychiatryonline.org/doi/abs/10.1176/ps.2008.59.2.184?url_ver=Z39.88-2003&rfr_id=ori%3Arid%3Acrossref.org&rfr_dat=cr_pub%3Dpubmed&
66. https://www.ncbi.nlm.nih.gov/pmc/articles/PMC1140960/
67. http://behavioralscientist.org/myth-mental-illness-causes-mass-shootings/
68.http://onlinelibrary.wiley.com/doi/10.1002/bsl.2172/abstract;jsessionid=09D1D7D0EA798D7D280D3209C92A3FCC.f03t01
69. http://www.latimes.com/science/sciencenow/la-sci-sn-angry-impulsive-gun-access-20150408-story.html?lang=en&utm_campaign=SendToFriend&uid=0&utm_content=article&utm_source=email&part=sendtofriend&utm_medium=article&position=0&china_variant=False
70. https://news.nd.edu/news/holding-a-gun-makes-you-think-others-are-too-new-research-shows/
71. https://www.ncbi.nlm.nih.gov/pmc/articles/PMC3815007/
72.http://archive.sph.harvard.edu/cas/Documents/Gunthreats2/gunspdf.pdf

73. http://www.psypost.org/2017/11/significant-number-mass-shooters-may-aggrieved-narcissists-50227
74. https://www.mediamatters.org/blog/2015/06/18/nra-board-member-blames-murdered-reverend-for-d/204057
75. https://www.huffingtonpost.com/entry/florida-high-school-students-stage-walkout-to-protest-gun-violence_us_5a87067be4b004fc3191a117
76. https://papers.ssrn.com/sol3/papers.cfm?abstract_id=2776657
77. https://www.gunowners.org/sk0802htm.htm
78. See David Hemenway, Policy and Perspective: Survey Research and Self-Defense Gun Use: An Explanation of Extreme Overestimates, 87 J. Crim. L. & Criminology 1430, 1432 (1997)
79. http://beta.latimes.com/opinion/opinion-la/la-ol-guns-self-defense-charleston-20150619-story.html
80. David Hemenway, Deborah Azrael & Matthew Miller, Gun Use in the United States: Results from Two National Surveys, 6 Inj. Prevention 263, 263 (2000).
81. http://beta.latimes.com/opinion/opinion-la/la-ol-guns-self-defense-charleston-20150619-story.html
82. https://www.scientificamerican.com/article/gun-science-proves-arming-untrained-citizens-bad-idea/
83. https://www.rollingstone.com/politics/news/the-trump-russia-nra-connection-heres-what-you-need-to-know-w515615
84. https://www.vox.com/world/2017/6/29/15892508/nra-ad-dana-loesch-yikes
85. https://www.huffingtonpost.com/entry/nra-dana-loesch-media-shooting_us_5a8ed8d2e4b077f5bfec1a85
86. http://www.psypost.org/2017/11/significant-number-mass-shooters-may-aggrieved-narcissists-50227
87. https://www.youtube.com/watch?v=PrnIVVWtAag
88. https://www.wired.com/2016/06/ar-15-can-human-body/
89. https://www.theatlantic.com/politics/archive/2018/02/what-i-saw-treating-the-victims-from-parkland-should-change-the-debate-on-guns/553937/
90. https://www.nbcnews.com/news/us-news/family-ar-15-inventor-speaks-out-n593356
91. https://theconversation.com/if-lawful-firearm-owners-cause-most-gun-deaths-what-can-we-do-48567
92. https://www.brisbanetimes.com.au/national/queensland/more-people-will-die-police-union-berates-gun-law-overhaul-20120827-24vdz.html#ixzz3kOV8zPFU
93.https://www.gunpolicy.org/firearms/region/united-states#total_number_of_gun_deaths

References

Chapter Thirteen

1. 'Toxic groundwater – Bhopal's Second Disaster' First Published in 'Pesticides News' 87, March 2010.
2. Varma, D.R, 'Epidemiological and experimental studies on the effects of methyl isocyanate on the course of pregnancy.' Environmental Health Perspectives 1987
3. http://www.bhopal.org/what-happened/health-issues/a-dark-backdrop/
4. The Lancet, Volume 360, Number 9336 14 September 2002)
5. http://news.bhopal.net/2004/08/27/us-government-shields-anderson/
6. http://news.bhopal.net/2004/08/27/us-government-shields-anderson/
7. http://www.independent.co.uk/news/obituaries/warren-anderson-chief-executive-of-union-carbide-at-the-time-of-the-catastrophic-gas-leak-at-their-9832080.html
8. http://www.bhopal.org/what-happened/bhopals-second-disaster/
9. http://www.bhopal.org/wp-content/uploads/2010/08/a5c85e881e.gif
10. http://www.bhopal.org/what-happened/health-issues/a-child-is-born/
11. http://www.thehindu.com/news/international/world/former-union-carbide-chief-warren-anderson-wanted-in-bhopal-gas-tragedy-dies/article6551966.ece
12. https://www.motherjones.com/environment/2014/06/photos-bhopal-india-union-carbide-sanjay-verma-pesticides-explosion/
13.https://www.bhopal.org/second-poisoning/union-carbides-chemical-trail/
14.https://www.bhopal.org/second-poisoning/bhopal-second-poisoning/
15.https://www.theguardian.com/environment/2009/dec/04/bhopal-25-years-indra-sinha

Chapter Fourteen

1. https://www.knoxnews.com/story/news/crime/2018/03/28/tva-coal-ash-spill-cleanup-roane-county-lawsuits-dead-dying-workers/458342002/
2. https://www.knoxnews.com/story/news/crime/2018/09/20/kingston-coal-ash-spill-cleanup-worker-dies-lawsuit-hearing-nears/1344072002/
3. https://www.knoxnews.com/story/news/crime/2018/08/20/2008-coal-ash-spill-kingston-perjury-claim-jacobs-engineering/977361002/
4. https://www.knoxnews.com/story/news/crime/2018/10/08/sickened-kingston-coal-ash-spill-workers-want-faulty-tests-barred/1502268002/
5. https://www.knoxnews.com/story/news/crime/2018/06/29/kingston-coal-ash-epa-workers-lawsuit/723222002/
6. http://www.nashvillepublicradio.org/post/workers-who-cleaned-kingston-coal-ash-spill-say-they-were-misled-about-danger#stream/0

7. https://www.knoxnews.com/story/news/crime/2018/11/07/verdict-reached-favor-sickened-workers-coal-ash-cleanup-lawsuit/1917514002/

Chapter Fifteen

1.http://web.worldbank.org/WBSITE/EXTERNAL/NEWS/0,,contentMDK:22068931~pagePK:64257043~piPK:437376~theSitePK:4607,00.html
2. http://www.un.org/apps/news/story.asp?NewsID=30070
3. http://www.washingtonsblog.com/2013/05/more-americans-committing-suicide-than-during-the-great-depression.html
4. http://www.dailymail.co.uk/news/article-2424285/UK-suicide-rate-rose-300-extra-men-killed-financial-crisis.html
5.http://bjp.rcpsych.org/content/early/2014/05/23/bjp.bp.114.144766.abstract
6. http://money.cnn.com/2013/09/18/news/economy/financial-crisis-suicide/
7. http://www.nytimes.com/2012/04/15/world/europe/increasingly-in-europe-suicides-by-economic-crisis.html?pagewanted=all
8.http://www.alternet.org/story/123563/the_financial_crisis_is_driving_hordes_of_americans_to_suicide?page=0%2C1&paging=off¤t_page=1#bookmark
9.http://www.nytimes.com/2011/01/26/business/economy/26inquiry.html?_r=0
10. http://nymag.com/daily/intelligencer/2014/02/i-crashed-a-wall-street-secret-society.html
11. http://www.cbsnews.com/news/enron-traders-caught-on-tape/
12. https://www.telegraph.co.uk/news/2016/05/25/financial-crisis-caused-500000-extra-cancer-death-according-to-l/
13. http://www.psychologicalscience.org/index.php/news/releases/social-class-as-culture.html
14. http://www.foxnews.com/opinion/2011/05/19/powerful-people-think-rules-dont-apply/
15. http://www.huffingtonpost.com/2012/11/25/deficit-reduction-council-fiscal-cliff_n_2185585.html
16. Boddy C. The Corporate Psychopaths Theory of the Global Financial Crisis. Journal of Business Ethics (2011) 102:255–259 DOI 10.1007/s10551-011-0810-4
17. https://www.cbsnews.com/news/enron-traders-caught-on-tape/

Chapter Sixteen

1.http://community.seattletimes.nwsource.com/archive/?date=19910609&slug=1288069

References

2. https://www.washingtonpost.com/archive/politics/1980/02/25/firms-exporting-products-banned-as-risks-in-us/821dfa34-13e1-4f7d-9c0f-2bb10ab1cbf5/?utm_term=.5ac670c78035
3. http://www.motherjones.com/politics/1979/11/karl-marx-and-pajama-game
4.http://community.seattletimes.nwsource.com/archive/?date=19910609&slug=1288069
5.http://digitalcommons.law.umaryland.edu/cgi/viewcontent.cgi?article=1155&context=mjil
6. http://www.motherjones.com/politics/1979/11/corporate-crime-century/
7.http://digitalcommons.law.umaryland.edu/cgi/viewcontent.cgi?article=1155&context=mjil
8. https://www.nytimes.com/1972/03/09/archives/mercury-poisoning-in-iraq-is-said-to-kill-100-to-400.html
9. https://www.nytimes.com/1982/08/22/weekinreview/products-unsafe-at-home-are-still-unloaded-abroad.html
10. http://www.nytimes.com/1987/12/06/magazine/the-sad-legacy-of-the-dalkon-shield.html
11. http://www.motherjones.com/politics/1979/11/charge-gynocide/2/
12.http://digitalcommons.law.umaryland.edu/cgi/viewcontent.cgi?article=1155&context=mjil
13. http://www.motherjones.com/politics/1979/11/corporate-crime-century/
14.https://www.aljazeera.com/programmes/specialseries/2016/11/circle-poison-pesticides-developing-world-161115084547144.html
15. http://articles.latimes.com/2012/dec/17/opinion/la-ed-bangladesh-20121217
16. http://www.washingtonpost.com/wp-dyn/content/article/2007/08/31/AR2007083101877.html
17.http://inthesetimes.com/working/entry/18066/out_of_sight_erik_loomis
18. https://www.bloomberg.com/news/articles/2014-11-13/bangladesh-leather-industrys-workers-exposed-to-toxic-chemicals
19. http://world.time.com/2013/09/03/hell-for-leather-bangladeshs-toxic-tanneries-ravage-lives-and-environment/
20. https://www.reuters.com/article/us-bangladesh-tanneries/toxic-tanneries-drive-bangladesh-leather-exports-report-idUSBRE89816C20121009
21. https://www.seattletimes.com/business/report-examines-grim-b
22. https://www.pbs.org/newshour/show/bangladeshs-leather-industry-exposes-workers-and-children-to-toxic-hazards
23. http://anthrojournal.com/issue/october-2011/article/the-tanning-

industry-of-medieval-britain

Chapter Seventeen

1. https://www.theguardian.com/business/2015/sep/26/volkswagen-scandal-emissions-tests-john-german-research
2. https://www.reuters.com/article/us-volkswagen-emissions/ex-volkswagen-ceo-winterkorn-charged-in-u-s-over-diesel-scandal-idUSKBN1I42I3
3. https://www.reuters.com/article/us-volkswagen-emissions-stadler/head-of-vws-audi-arrested-in-germany-over-diesel-scandal-idUSKBN1JE0R3
4. http://fortune.com/2018/02/06/volkswagen-vw-emissions-scandal-penalties/
5. http://fortune.com/inside-volkswagen-emissions-scandal/
6. https://www.nature.com/articles/nature22086
7. https://www.theguardian.com/environment/2017/may/15/diesel-emissions-test-scandal-causes-38000-early-deaths-year-study
8. https://www.nature.com/articles/s41598-017-01135-2
9. https://doctorsagainstdiesel.uk

Chapter Eighteen

1. https://www.theatlantic.com/business/archive/2016/04/evil-corporation-trope/479295/
2. https://www.cbsnews.com/news/60-minutes-massey-coal-don-blankenship-king-of-coal/
3. https://www.cbsnews.com/news/60-minutes-massey-coal-don-blankenship-king-of-coal/
4. http://beta.latimes.com/business/hiltzik/la-fi-hiltzik-blankenship-20170523-story.html
5. http://www.motherjones.com/politics/2017/11/disgraced-coal-baron-don-blankenship-is-running-for-senate-in-west-virginia/
6. https://www.huffingtonpost.com/entry/umwa-don-blankenship-tv-advertisement_us_59b6c255e4b036fd85cce450
7. https://www.psychologytoday.com/blog/what-mentally-strong-people-dont-do/201602/5-traits-actual-psychopaths
8. http://www.telegraph.co.uk/books/non-fiction/spot-psychopath/
9. http://www.telegraph.co.uk/news/2016/09/13/1-in-5-ceos-are-psychopaths-australian-study-finds/
10. https://hbr.org/2004/10/executive-psychopaths
11. https://www.theguardian.com/commentisfree/2014/jun/03/how-i-discovered-i-have-the-brain-of-a-psychopath

12. https://www.psychologytoday.com/blog/our-humanity-naturally/201103/why-corporations-are-psychotic
13. http://www.nytimes.com/1994/03/20/magazine/how-do-tobacco-executives-live-with-themselves.html?pagewanted=all
14. https://www.theatlantic.com/magazine/archive/2017/07/power-causes-brain-damage/528711/
15. https://www.oveo.org/fichiers/power-changes-how-the-brain-responds-to-others.pdf
16. https://insight.kellogg.northwestern.edu/article/losing_touch
17. http://onlinelibrary.wiley.com/doi/10.1111/jofi.12432/full
18. https://academic.oup.com/brain/article/132/5/1396/354862
19. Babiak P, Neumann CS, Hare RD. Corporate psychopathy: Talking the walk. Behav Sci Law. 2010 Mar-Apr;28(2):174-93.
20. https://www.wsj.com/articles/california-becomes-first-state-to-mandate-female-board-directors-1538341932
21. https://www.ncbi.nlm.nih.gov/pmc/articles/PMC3379858/
22. https://www.huffingtonpost.com/2012/10/09/mitt-romney-bain-tobacco_n_1949812.html
23.https://www.washingtonpost.com/opinions/2019/01/02/mitt-romneys-op-ed-crystallizes-all-reasons-old-gop-establishment-has-been-pushed-aside/?utm_term=.2414b1af7af4

Chapter Nineteen

1. http://foreignpolicy.com/2016/03/15/these-25-companies-are-more-powerful-than-many-countries-multinational-corporate-wealth-power/
2.http://law2.wlu.edu/deptimages/Powell%20Archives/PowellMemorandumPrinted.pdf
3. https://www.theatlantic.com/business/archive/2015/04/how-corporate-lobbyists-conquered-american-democracy/390822/
4. https://newint.org/blog/2016/09/16/corporations-running-the-world-used-to-be-science-fiction
5. https://www.thehindubusinessline.com/economy/top-10-global-firms-richer-than-180-poor-countries-combined/article9482603.ece
6. http://foreignpolicy.com/2016/03/15/these-25-companies-are-more-powerful-than-many-countries-multinational-corporate-wealth-power/
7. https://www.forbes.com/sites/frederickallen/2012/07/23/super-rich-hide-21-trillion-offshore-study-says/#54ee76d76ba6
8. https://www.forbes.com/sites/mikecollins/2015/07/14/the-big-bank-bailout/#7e08b4712d83
9. https://www.forbes.com/sites/mikecollins/2015/07/14/the-big-bank-bailout/#7e08b4712d83
10. https://www.thehindubusinessline.com/economy/top-10-global-

firms-richer-than-180-poor-countries-combined/article9482603.ece
11. https://www.morriscreative.com/6-corporations-control-90-of-the-media-in-america/
12. http://www.truth-out.org/speakout/item/29112-control-the-corporations-before-they-completely-control-us
13. https://imgur.com/pnMMj
14. https://www.cnbc.com/2015/04/15/5-biggest-banks-now-own-almost-half-the-industry.html
15. https://www.theguardian.com/commentisfree/2014/dec/08/taming-corporate-power-key-political-issue-alternative
16. https://www.foei.org
17. https://www.citizen.org/our-work/globalization-and-trade/nafta-wto-other-trade-pacts/trans-pacific-partnership
18. https://www.citizen.org/our-work/globalization-and-trade/more-power-corporations-attack-nations
19. https://www.motherjones.com/food/2012/10/fda-barely-inspects-imported-seafood/
20. http://www.presidency.ucsb.edu/ws/index.php?pid=41535
21.https://www.forbes.com/sites/teresaghilarducci/2018/09/28/who-benefits-from-the-tax-cut-10-months-later/#6b3ab3e426bb
22.https://www.epi.org/publication/ceo-compensation-surged-in-2017/
23.https://pdfs.semanticscholar.org/a998/60707178f6d241137a32720abe9a8aa24b77.pdf
24.https://www.fastcompany.com/90223130/how-elizabeth-warrens-accountable-capitalism-act-works
25.https://www.brookings.edu/blog/up-front/2018/08/20/sen-warrens-accountable-capitalism-act-rightfully-challenges-a-central-tenet-of-corporate-governance-theory/
26. http://www.let.rug.nl/usa/presidents/grover-cleveland/state-of-the-union-1888.php
27.https://technical.ly/delaware/2014/09/23/why-delaware-incorporation/
28.https://www.alternet.org/2019/01/its-time-to-bring-back-the-corporate-death-penalty/
29. https://papers.ssrn.com/sol3/papers.cfm?abstract_id=1572085

Made in the USA
Monee, IL
26 August 2019